Nanoceramics in Clinical Use

From Materials to Applications
2nd Edition

RSC Nanoscience & Nanotechnology

Editor-in-Chief:
Professor Paul O'Brien FRS, *University of Manchester, UK*

Series Editors:
Professor Ralph Nuzzo, *University of Illinois at Urbana-Champaign, USA*
Professor Joao Rocha, *University of Aveiro, Portugal*
Professor Xiaogang Liu, *National University of Singapore, Singapore*

Honorary Series Editor:
Sir Harry Kroto FRS, *University of Sussex, UK*

Titles in the Series:
1: Nanotubes and Nanowires
2: Fullerenes: Principles and Applications
3: Nanocharacterisation
4: Atom Resolved Surface Reactions: Nanocatalysis
5: Biomimetic Nanoceramics in Clinical Use: From Materials to Applications
6: Nanofluidics: Nanoscience and Nanotechnology
7: Bionanodesign: Following Nature's Touch
8: Nano-Society: Pushing the Boundaries of Technology
9: Polymer-based Nanostructures: Medical Applications
10: Metallic and Molecular Interactions in Nanometer Layers, Pores and Particles: New Findings at the Yoctolitre Level
11: Nanocasting: A Versatile Strategy for Creating Nanostructured Porous Materials
12: Titanate and Titania Nanotubes: Synthesis, Properties and Applications
13: Raman Spectroscopy, Fullerenes and Nanotechnology
14: Nanotechnologies in Food
15: Unravelling Single Cell Genomics: Micro and Nanotools
16: Polymer Nanocomposites by Emulsion and Suspension
17: Phage Nanobiotechnology
18: Nanotubes and Nanowires, 2nd Edition
19: Nanostructured Catalysts: Transition Metal Oxides
20: Fullerenes: Principles and Applications, 2nd Edition
21: Biological Interactions with Surface Charge Biomaterials
22: Nanoporous Gold: From an Ancient Technology to a High-Tech Material
23: Nanoparticles in Anti-Microbial Materials: Use and Characterisation
24: Manipulation of Nanoscale Materials: An Introduction to Nanoarchitectonics
25: Towards Efficient Designing of Safe Nanomaterials: Innovative Merge of Computational Approaches and Experimental Techniques
26: Polymer–Graphene Nanocomposites

How to obtain future titles on publication:
A standing order plan is available for this series. A standing order will bring
delivery of each new volume immediately on publication.

For further information please contact:
Book Sales Department, Royal Society of Chemistry, Thomas Graham
House, Science Park, Milton Road, Cambridge, CB4 0WF, UK
Telephone: +44 (0)1223 420066, Fax: +44 (0)1223 420247
Email: booksales@rsc.org
Visit our website at www.rsc.org/books

Nanoceramics in Clinical Use
From Materials to Applications
2nd Edition

María Vallet-Regí
Universidad Complutense de Madrid, Spain
Email: vallet@farm.ucm.es

Daniel Arcos Navarrete
Universidad Complutense de Madrid, Spain
Email: arcosd@farm.ucm.es

THE QUEEN'S AWARDS
FOR ENTERPRISE:
INTERNATIONAL TRADE
2013

RSC Nanoscience & Nanotechnology No. 39

Print ISBN: 978-1-78262-104-1
PDF eISBN: 978-1-78262-255-0
ISSN: 1757-7136

A catalogue record for this book is available from the British Library

Published by The Royal Society of Chemistry,
Thomas Graham House, Science Park, Milton Road,
Cambridge CB4 0WF, UK

Registered Charity Number 207890

For further information see our web site at www.rsc.org

Printed and bound by CPI Group (UK) Ltd, Croydon, CR0 4YY

Preface to the 1st Edition

The research on nanoceramics for biomedical applications responds to the challenge of developing fully biocompatible implants, which exhibit biological responses at the nanometric scale in the same way that biogenic materials do. Any current man-made implant is not fully biocompatible and will always set off a foreign body reaction involving inflammatory response, fibrous encapsulation, *etc.* For this reason, great efforts have been carried out in developing new synthetic strategies that allow tailoring implants surfaces at the nanometric scale. The final aim is always to optimize the interaction at the tissue/implant interface at the nanoscale level, thus improving the life's quality of the patients with enhanced results and shorter rehabilitation periods.

The four chapters that constitute this book can be read as a whole or independently of each other. In fact, the author's purpose has been to write a book useful for students of Biomaterials (by developing some basic concepts of biomimetic nanoceramics), but also as a reference book for those specialists interested in specific topics of this field. At the beginning of each chapter, the introduction provides insight on the corresponding developed topic. In some cases, the different introductions deal with some common topics. However, even at the risk of being reiterative, we have decided to include some fundamental concepts in two or more chapters, thus allowing the comprehension of each one independently.

Chapter 1 deals with the description of biological hard tissues in vertebrates, from the point of view of mineralization processes. For this aim, the concepts of hard tissues mineralization are applied to explain how Nature works. This chapter finally provides an overview about the artificial alternatives suitable to be used for mimicking Nature.

In Chapter 2 we introduce general considerations of solids reactivity, which allows tailoring strategies aimed at obtaining apatites in the laboratory. These

RSC Nanoscience & Nanotechnology No. 39
Nanoceramics in Clinical Use: From Materials to Applications, 2nd Edition
By María Vallet-Regí and Daniel Arcos Navarrete
© M. Vallet-Regi and D. Arcos Navarrete 2016
Published by the Royal Society of Chemistry, www.rsc.org

strategies must be modified and adapted in such a way that artificial carbonated calcium deficient nanoapatites can be obtained resembling as much as possible the biological apatites. For this purpose, a review on the synthesis methods applied for apatite obtention are collected in the bibliography.

In Chapter 3 we have focused on the specific topic of hard tissue-related biomimetism. To reach this goal, we have dealt with nanoceramics obtained as a consequence of biomimetic processes. The reader will find information about the main topics related with the most important bioactive materials and the biomimetic apatites growth onto them. Concepts and valuable information about the most widely used biomimetic solutions and biomimetism evaluation methods are also included.

Finally, Chapter 4 reviews the current and potential clinical applications of apatite-like biomimetic nanoceramics, intended as biomaterials for hard tissue repair, therapy and diagnosis.

The authors wish to thank the Royal Society of Chemistry for the opportunity provided to write this book, as well as their comprehensive technical support. Likewise, we want to express our greatest thanks to Dr Fernando Conde, Pilar Cabañas and José Manuel Moreno for their assistance during the elaboration of this manuscript. We are also thankful to Dr M. Colilla, Dr M. Manzano, Dr B. Gónzalez and Dr A. J. Salinas for their valuable suggestions and scientific discussions. Finally, we would like to express our deepest gratitude to all our co-workers and colleagues that have contributed over the years with their effort and thinking to these studies.

María Vallet-Regí
Daniel Arcos

Preface to the 2nd Edition

Nanoceramics in Clinical Use: from Materials to Applications arises as a revision of our previous book *Biomimetic Nanoceramics in Clinical Use*. This first work tackled the biomedical applications of those nanoceramics that mimic the mineral component of the bones (mainly nanoapatites) and also those bioceramics that reproduce the bone mineralization processes after being implanted *in vivo*, such as some calcium phosphates and bioactive glasses. Over recent years, new nanoceramics have been added to the arsenal of biomaterials available for applications in bone tissue regeneration. Moreover, the incorporation of supramolecular chemistry to the low-temperature synthesis of bioceramics has expanded their concept beyond "materials of ceramic origin of which a single unit is sized (in at least one dimension) between 1 and 100 nm". The addition of structure-directing agents allows for the design of patterns, porosities and textures at the nanometrical scale in the macro objects, thus significantly changing their *in vivo* behaviour. Of course, these nanostructures can also be tailored over nanoparticles, which represent a revolution in the therapeutic field of nanomedicine as well as new diagnosis agents.

In view of this new scenario, we have felt impelled to deal with all these topics beyond the role that some nanoceramics play in biomineralization skeletal processes. For these reason the title has been changed and expanded to *Nanoceramics in Clinical use: from Materials to Applications* and incorporates three new chapters to the already existing, but revised, ones of the previous volume. Therefore this work is addressed to the same readership but in a more ambitious way, adapting its contents to the requirements of the under- and post-graduate subject "nanoparticles in biomedicine" implanted in different universities all over the world.

RSC Nanoscience & Nanotechnology No. 39
Nanoceramics in Clinical Use: From Materials to Applications, 2nd Edition
By María Vallet-Regí and Daniel Arcos Navarrete
© M. Vallet-Regi and D. Arcos Navarrete 2016
Published by the Royal Society of Chemistry, www.rsc.org

The authors thank the Royal Society of Chemistry for the proposal to carry out this project, as well as José Manuel Moreno, Pilar Cabañas and Fernando Conde for their assistance during the elaboration of this book. Finally, we would like to express our deepest gratitude to all our coworkers and colleagues who have contributed to these studies over the years with their effort and thinking.

María Vallet-Regí
Daniel Arcos

Contents

RSC Nanoscience & Nanotechnology No. 39
Nanoceramics in Clinical Use: From Materials to Applications, 2nd Edition
By María Vallet-Regí and Daniel Arcos Navarrete
© M. Vallet-Regi and D. Arcos Navarrete 2016
Published by the Royal Society of Chemistry, www.rsc.org

CHAPTER 1

Biological Apatites in Bone and Teeth

1.1 Hard-Tissue Biomineralization: How Nature Works

Most biominerals are inorganic/organic composite materials.[1] This is also the case for the bones and teeth of all vertebrates, which are formed by the combination of an inorganic calcium phosphate phase and an organic matrix[2] (Figure 1.1). The inorganic component is a nanocrystalline solid with apatite structure and the chemical composition of a carbonated, basic calcium phosphate, hence it can be termed a carbonate-hydroxy-apatite. It comprises 65% of the total bone mass, with the remaining mass formed by organic matter and water.[3] The benefits that the inorganic component brings to this combination are toughness and the ability to withstand pressure.

In contrast, the organic matrix formed by collagen fibres, glycoproteins and mucopolysaccharides provides elasticity and resistance to stress, bending and fracture. Such symbiosis of two very different compounds, with markedly different properties, confers to the final product (i.e. the biomineral) some properties that would be unattainable for each of its individual components per se. This is a fine example in nature of the advantages that a composite material can exhibit, reaching new properties with added value. In fact due to this evidence, a large portion of the field of modern materials science is currently focused on the development of composite materials.

RSC Nanoscience & Nanotechnology No. 39
Nanoceramics in Clinical Use: From Materials to Applications, 2nd Edition
By María Vallet-Regí and Daniel Arcos Navarrete
© M. Vallet-Regi and D. Arcos Navarrete 2016
Published by the Royal Society of Chemistry, www.rsc.org

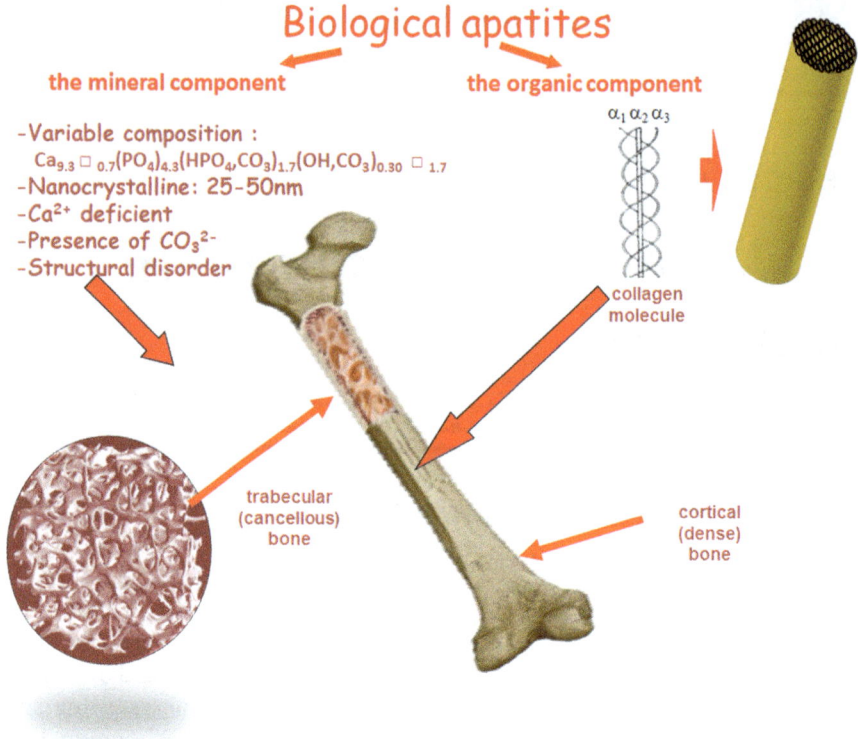

Biological apatites

the mineral component   the organic component

$\alpha_1\,\alpha_2\,\alpha_3$

-Variable composition :
  $Ca_{9.3\,\square\,0.7}(PO_4)_{4.3}(HPO_4,CO_3)_{1.7}(OH,CO_3)_{0.30}\,\square\,1.7$
-Nanocrystalline: 25-50nm
-Ca^{2+} deficient
-Presence of CO_3^{2-}
-Structural disorder

collagen
molecule

trabecular
(cancellous)
bone

cortical
(dense)
bone

Figure 1.1   Inorganic–organic composite nature of trabecular and cortical bone.

1.1.1   Bone Formation

The bone exhibits some physical and mechanical properties which are rather unusual. It is able to bear heavy loads, to withstand large forces and to flex without fracture within certain limits. Besides this, the bone also acts as an ion buffer both for cations and anions. From the material point of view, bone could be simplified as a three-phase material formed by *organic fibers*, an *inorganic nanocrystalline phase* and a *bone matrix*. Its unique physical and mechanical properties are the direct consequence of intrinsic atomic and molecular interactions within this very particular natural composite material.

Bone is not uniformly dense. It has a hierarchical structure. Due to its true organic–inorganic composite nature, it is able to adopt different structural arrangements with singular architectures, determined by the properties required from it depending on its specific location in the skeleton. Generally speaking, most bones exhibit a relatively dense outer layer, known as *cortical* or *compact* bone, which surrounds a less dense and porous layer, termed *trabecular* or *spongy* bone, which is in turn filled with a jelly tissue: the *bone marrow*.[4] This complex tissue is the body deposit of non-differentiated trunk cells, the precursors of most repairing and regenerating cells produced after formation of the embryonic subject.[5,6] The bone fulfils critical functions in terms of *structural* material and ion *reservoir*. Both functions strongly depend on the size, shape, chemical composition and crystalline structure of the mineral phase, and also on the mineral distribution within the organic matrix.

The main constituents of bone are: *water*; a mineral phase, *calcium phosphate* in the form of carbonated apatite with low crystallinity and nanometric dimensions, which accounts for roughly two-thirds of the bone's dry weight; and an organic fraction, formed of *several proteins*, among which type I collagen is the main component, which represents approximately the remaining one-third of bone dry weight. The other intervening proteins, such as proteoglycans and glycoproteins, total more than 200 different proteins, known as non-collagen proteins; however, their total contribution to the organic constituent falls below 10% of the organic fraction. These bone constituents are hierarchically arranged, with at least five levels of organization. At the molecular level, the polarized triple helix of tropocollagen molecules are grouped in microfibers, with small cavities between their edges, where small apatite crystals – ~5 nm × 30 nm sized – nucleate and grow. These microfibers unite to form larger fibers which constitute the microscopic units of bone tissue. These fibers are arranged according to different structural distributions to form the full bone.[7]

It was traditionally believed that the inorganic phase was mainly amorphous calcium phosphate which, in the aging process, evolved towards nanocrystalline hydroxyapatite. However, results of solid-state ^{31}P nuclear magnetic resonance spectroscopy evidenced that the amorphous phase is never present in large amounts during the bone development process.[6] Besides, this technique did detect acid phosphate groups. Phosphate functions correspond to proteins with *O*-phosphoserine and *O*-phosphothreonine groups, which are probably used to link the inorganic mineral component and the organic matrix. Phosphoproteins are arranged in the collagen fibers so that Ca^{2+} can be bonded at regular intervals, in agreement with the inorganic crystal structure, hence providing a repeating condition which leads to an ordered sequence of the same unit, *i.e.* the crystallinity of the inorganic phase. The cells responsible for most of the assembling process are termed *osteoblasts*. When the main assembling process is completed, the osteoblasts keep differentiating in order to form *osteocytes*, which are responsible for the bone maintenance process. The controlled nucleation and growth of the mineral take place at the microscopic voids formed in the collagen matrix. The type I collagen molecules, segregated by the osteoblasts, are grouped in microfibers with a specific tertiary structure, exhibiting a periodicity of 67 nm and 40 nm cavities or orifices between the edges of the molecules.[7] These orifices constitute microscopic environments with free Ca^{2+} and PO_4^{3-} ions, as well as groups of side chains eligible for bonding, with a molecular periodicity that allows the nucleation of the mineral phase in a heterogeneous fashion. Ca^{2+} ions deposited and stored in the skeleton are constantly renewed with dissolved calcium ions. The bone growth process can only be produced under a relative excess of Ca^{2+} and its corresponding anions, such as phosphates and carbonates, at the bone matrix. This situation is achieved due to the action of efficient ATP-powered ionic pumps, such as Ca^{2+} ATPases for active transportation of calcium.[8–10] In terms of physiology, carbonate and phosphate are present in the forms of HCO_3^-, HPO_4^{2-} and $H_2PO_4^-$ anions. When incorporated to the bone, the released protons can move throughout the bone tissue and leave the nucleation and mineralization area. The nucleation of thin, platelet-shaped apatite crystals takes place at the bone within discrete spaces inside the collagen fibers, hence restricting a potential primary growth of these mineral crystals, and imposing their discrete and discontinuous quality (Figure 1.2).

Calcium phosphate nanocrystals in bone, formed as mentioned at the spaces left between the collagen fibers exhibit the particular feature of being monodispersed and nanometer-sized platelets of *carbonate-hydroxyl-apatite*. There is no other mineral phase present, and the crystallographic axis *c* of these crystals is arranged in parallel to the collagen fibers and to the largest dimension of the platelet. In the mineral world, the thermodynamically stable form of calcium phosphate under standard conditions is the hydroxyapatite.[11] Generally speaking, this phase grows in needle-like forms, with the

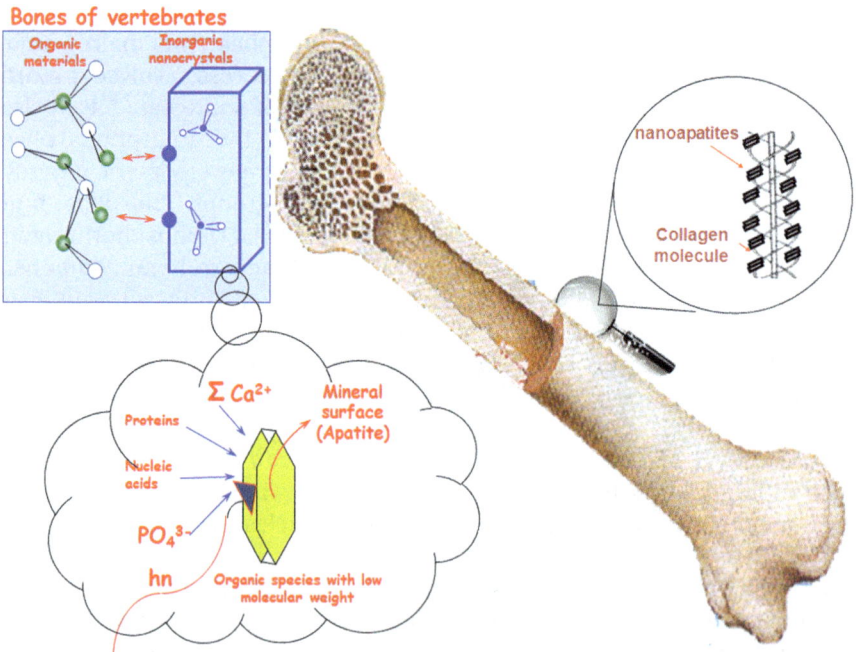

Figure 1.2   Interaction between biological nanoapatites and organic fraction of bone at the molecular scale. At the bottom of the scheme: formation of nanoapatite crystallites with the factors and biological moieties present in the process. A magnified scheme of the apatite crystallites location into collagen fibers is also displayed.

c axis parallel to the needle axis. Figure 1.3 shows the crystalline structure of hydroxyapatite, $Ca_{10}(PO_4)_6(OH)_2$, which belongs to the hexagonal system, space group $P6_3/m$ and lattice parameters $a = 9.423$ Å and $c = 6.875$ Å.

Besides the main ions Ca^{2+}, PO_4^{3-} and OH^-, the composition of biological apatites always includes CO_3^{2-} at ~4.5%, and also a series of minority ions, usually including Mg^{2+}, Na^+, K^+, Cl^- and F^-.[12] These substitutions modify the lattice parameters of the structure as a consequence of the different size of the substituting ions, as depicted in Figure 1.3. This is an important difference between minerals grown in an inorganic or biological environment.

The continuous formation of bone tissue is performed at a peripheral region, formed by an external crust and an internal layer with connective tissue and *osteoblast* cells. These osteoblasts are phosphate-rich and exude a jelly-like substance, the osteoid. Due to the gradual deposit of inorganic material, this osteoid becomes stiffer and the osteoblasts are finally confined and transformed in bone cells, the *osteocytes*. The bone transformation mechanism, and the ability to avoid an excessive bone growth, are both catered for by certain degradation processes which are performed simultaneously to the bone formation. The osteoclasts, which are giant multinucleated cells, are able to catabolyze the bone purportedly using citrates as chelating agent. The control of the osteoclast activity is verified through the action of the parathyroid hormone, a driver for demineralization, and its antagonist, tireocalcitonin.

The collagen distribution with the orifices previously described is *necessary* for the controlled nucleation and growth of the mineral, but it might not *suffice*. There are conceptual postulations of various additional organic components, such as the phosphoproteins, as an integral part of the nucleation core and hence directly involved in the nucleation mechanism. Several immuno-cyto-chemical studies of bone, using techniques such as optical microscopy and high-resolution electron microscopy have clearly shown that the phosphoproteins are

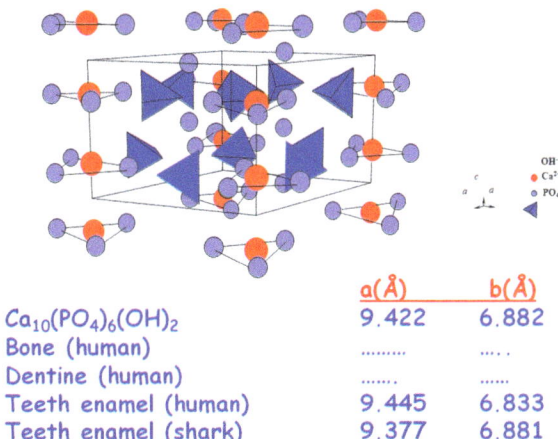

	a(Å)	b(Å)
$Ca_{10}(PO_4)_6(OH)_2$	9.422	6.882
Bone (human)	....	.....
Dentine (human)	....	......
Teeth enamel (human)	9.445	6.833
Teeth enamel (shark)	9.377	6.881

Figure 1.3 Crystalline structure and unit cell parameters for different biological hydroxyapatites.

restricted, or at least largely concentrated at the initial mineralization location, intimately related to the collagen fibers. It seems that the phosphoproteins are enzymatically phosphored prior to the mineralization.[13]

The crystallization of the complex and hardly soluble apatite structures evolves favourably through the kinetically controlled formation of metastable intermediate products. Under *in vitro* conditions, amorphous calcium phosphate is transformed into octacalcium phosphate which, in turn, evolves to carbonate hydroxyapatite; at lower pH values, the intermediate phase seems to be dehydrated dicalcium phosphate.[14,15]

The mechanisms of bone formation are highly regulated processes,[7] which seem to verify the following statements:

- Mineralization is restricted to those specific locations where crystals are constrained in size by a compartmental strategy.
- The mineral formed exhibits specific chemical composition, crystalline structure, crystallographic orientation and shape. The chemical phase obtained is controlled during the stages of bone formation. In vertebrates, said chemical phase is a hydroxyl-carbonate-apatite, even though the thermodynamically stable form of calcium phosphate in the world of minerals, under standard conditions, is hydroxyapatite.
- Since the mineral deposits onto a biodegradable organic support, complex macroscopic forms are generated with pores and cavities. The assembling and remodelling of the structure are achieved by cell activity, which builds or erodes the structure layer by layer.

Without a careful integration of the whole process, bone formation would be an impossible task. The slightest planning mistake by the body, for instance in its genetic coding or cell messengers, is enough to provoke building errors that would weaken the osseous structure.

The hard tissues in vertebrates are bones and teeth. The differences between them reside in the amounts and types of organic phases present, the water content, the size and shape of the inorganic phase nanocrystals and the concentration of minor elements present in the inorganic phase, such as CO_3^{2-}, Mg^{2+}, Na^+, *etc.*[12] The definitive set of teeth in higher order vertebrates has an outer shell of dental enamel that, in an adult subject, *does not contain* any living cells.[16] Up to 90% of said enamel can be inorganic material, mainly *carbonate-hydroxyl-apatite*. Enamel is the material that undergoes more changes during the tooth development process. At the initial stage, it is deposited with a mineral content of only 10–20%, with the remaining 80–90% proteins and special matrix fluids. In subsequent development stages, the organic components of the enamel are almost fully replaced by inorganic material. The special features of dental enamel when compared with bone material are its much larger crystal domains, with prismatic shapes and strongly oriented, made of *carbonate-hydroxyl-apatite* (Figure 1.4). There is no biological material that could be compared to enamel in terms of hardness and long life. However, it cannot be regenerated.

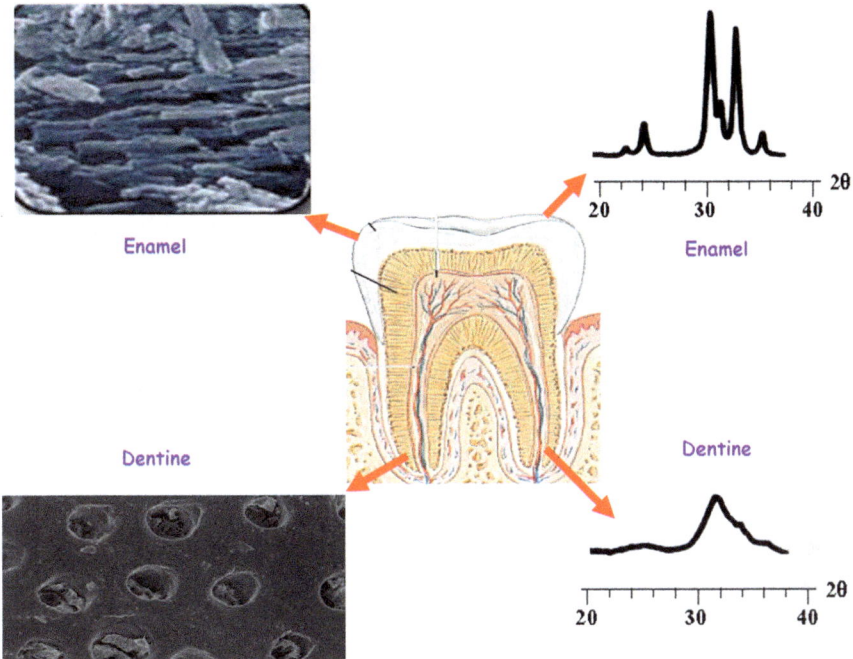

Figure 1.4 Different apatite crystallinity degrees in teeth. Enamel (top) is formed by well-crystallized apatite, whereas dentine (bottom) contains nano-crystalline apatite within a channelled protein structure.

The bones, the body-supporting scaffold, can exhibit different types of integration between organic and inorganic materials, leading to significant variations in their mechanic properties. The ratio of both components reflects the compromise between toughness (high inorganic content) and resiliency or fracture strength (low inorganic content). All attempts to synthesize bone replacement materials for clinical applications featuring physiological tolerance, biocompatibility and long-term stability have, up to now, had only relative success; the superiority and complexity of the natural structure shows where, for instance, the human femur can withstand loads of up to 1650 kg.[17]

The bones of vertebrates, as opposed to the shells of molluscs, can be considered as *living biominerals*, since there are cells inside them under permanent activity. Bone also constitutes a storage and hauling mechanism for two essential elements, phosphorus and calcium, which are mainly stored in the bones. Most of what has been described up to this point regarding the nature of bone tissue, could be summed up stating that *the bone is a highly structured porous matrix, made of nanocrystalline and non-stoichiometric apatite, calcium deficient and carbonated, intertwined with collagen fibers and blood vessels.*

Bone functions are controlled by a series of hormones and bone growth factors. Figure 1.5 attempts to depict these phenomena in a projection from our *macro* scale point of view, to the 'invisible' *nano* scale.

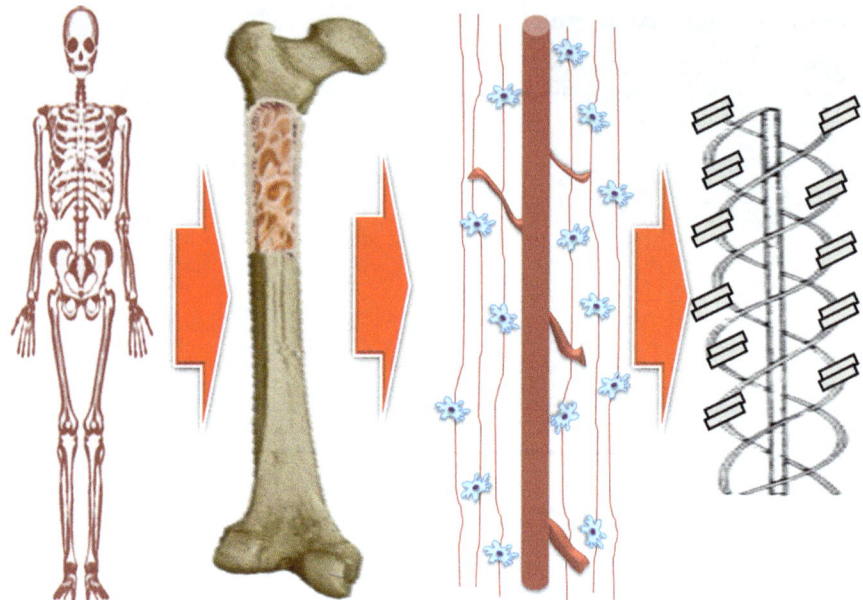

Figure 1.5   Hierarchical organization of bone tissue.

Bone's rigidity, resistance and toughness are directly related to its mineral content.[18] Although resistance and rigidity increase linearly with the mineral content, toughness does not exhibit the same trend, hence there is an optimum mineral concentration which leads to a maximum in bone toughness. This tendency is clearly the reason why the bone exhibits a restricted amount of mineral within the organic matrix. But there are other issues affecting the mechanical properties of bone, derived from the microstructural arrangement of its components. In this sense, the three main components of bone exhibit radically different properties. From this point of view, the biomineral is clearly a *composite*.[19] The organic scaffold exhibits a fibrous structure with three levels: the individual triple helix molecules, the small fibrils and its fiber-forming aggregates. These fibers can be packed in many different ways; they host the platelet-shaped hydroxyl-carbonate-apatite crystals. In this sense, the bone could be described as a composite reinforced with platelets, but the order–disorder balance determines the microstructure and, as a consequence, the mechanical properties of each bone. In fact, bones from different parts of the body show different arrangements, depending on their specific purpose.

Bone crystals are extremely small, with an average length of 50 nm (in the 20–150 nm range), 25 nm in average width (10–80 nm range) and thickness of just 2–5 nm. As a remarkable consequence, a large part of each crystal is surface; hence their ability to interact with the environment is outstanding.

Apatite phase contains 4–8% in weight of carbonate, properly described as *dahllite*. Mineral composition varies with age and it is always calcium

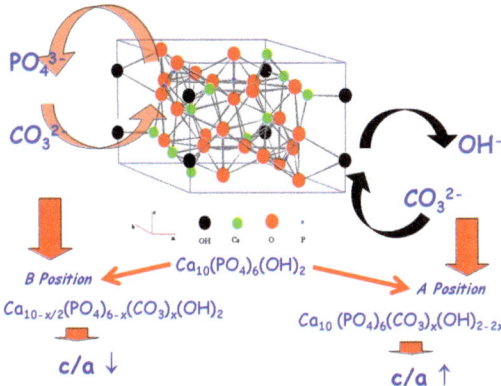

Figure 1.6   Crystalline structure and likely ionic substitutions in carbonate apatites.

deficient, with phosphate and carbonate ions in the crystal lattice. The formula $Ca_{8.3}(PO_4)_{4.3}(CO)_{3x}(HPO_4)_y(OH)_{0.3}$ represents the average composition of bone, where y decreases and x increases with age, while the sum $x + y$ remains constant and equal to 1.7.[12] Mineral crystals grow under a specific orientation, with the c axes of the crystals approximately parallel to the long axes of the collagen fibers where they are deposited. Electron microscopy techniques enabled this information to be obtained.[20]

The bones are characterized by their composition, crystalline structure, morphology, particle size and orientation. The apatite structure hosts carbonate in two positions: the OH^- sub-lattice producing so-called type A carbonate apatites or the $[PO_4]^{3-}$ sub-lattice (type B apatites) (Figure 1.6).

The small apatite crystal size is a very important factor related to the solubility of biological apatites when compared with mineral apatites. Small dimensions and low crystallinity are two distinct features of biological apatites which, combined with their non-stoichiometric composition, inner crystalline disorder and presence of carbonate ions in the crystal lattice, explain their special behaviour.

The structure of apatite allows for wide compositional variations, with the ability to accept many different ions in its three sub-lattices (Figure 1.7).

Biological apatites are calcium-deficient; hence their Ca/P ratio is always <1.67, which corresponds to a stoichiometric apatite. No biological hydroxyapatite shows a stoichiometric Ca/P ratio, but they all move towards this value as the organism ages, which is linked to an increase in crystallinity. These trends have a remarkable physiological meaning, since the younger, less crystalline tissue can develop and grow faster, while storing other elements that the body needs during its growth; this is due to the highly non-stoichiometric quality of HA, which caters for the substitutional inclusion of different amounts of several ions, such as Na^+, K^+, Mg^{2+}, Sr^{2+}, Cl^-, F^-, HPO_4^{2-}, *etc.*[21] (Figure 1.8).

Two frequent substitutions are the inclusion of sodium and magnesium ions in calcium lattice positions. When a magnesium ion replaces a calcium

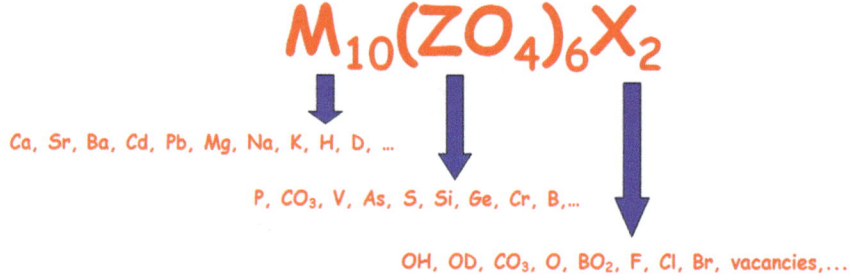

$$M_{10}(ZO_4)_6X_2$$

Ca, Sr, Ba, Cd, Pb, Mg, Na, K, H, D, ...

P, CO$_3$, V, As, S, Si, Ge, Cr, B,...

OH, OD, CO$_3$, O, BO$_2$, F, Cl, Br, vacancies,...

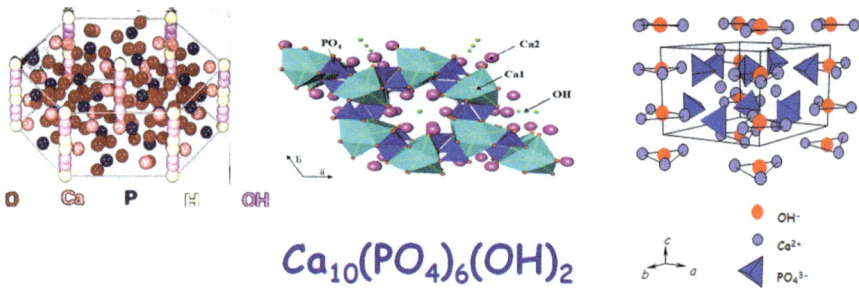

O   Ca   P   H   OH

$$Ca_{10}(PO_4)_6(OH)_2$$

OH⁻
Ca²⁺
PO₄³⁻

Figure 1.7 Compositional possibilities that can fit into the apatite-like structure, which provide high compositional variations corresponding to its non-stoichiometric character. Bottom: three different schemes and projections of the hydroxyapatite unit cell.

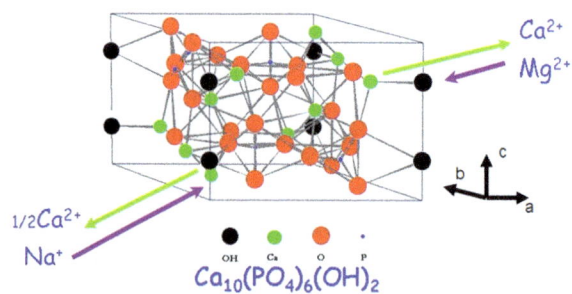

Ca²⁺
Mg²⁺

1/2Ca²⁺
Na⁺

OH   Ca   O   P

$$Ca_{10}(PO_4)_6(OH)_2$$

Figure 1.8 Likely substitutions in the cationic sub-lattice for biological apatites.

ion, the balance of charge and position is unaffected. However, if a sodium ion replaces a calcium ion, this balance is lost and the electrical neutrality of the lattice can only be restored through the creation of vacancies, therefore increasing the internal disorder.

The more crystalline the hydroxyapatite, the more difficult interchanges and growth become. In this sense, it is worth stressing that the bone is probably a very important detoxicating system for heavy metals due to the ease of their substitution in apatites; heavy metals, in the form of insoluble phosphates, can be retained in the hard tissues without important alterations of their structural properties.

However, the ability to exchange ions in this structure is not a coincidence. Nature designed it, and the materials scientist can use it as a blueprint to design and characterize new and better calcium phosphates for certain specific applications. It is known that the bone regeneration rate depends on several factors such as porosity, composition, solubility and presence of certain elements released during the resorption of the ceramic component that facilitate the bone regeneration performed by the osteoblasts. Thus, for instance, small amounts of strontium, zinc or silicates stimulate the action of these osteoblasts and, in consequence, new bone formation. Carbonate and strontium favour the dissolution and therefore the resorption of the implant.[12] Silicates increase mechanical strength, a very important factor in particular for porous ceramics, and also accelerate the bioactivity of apatite.[22] Therefore, the current trend is to obtain calcium phosphate bioceramics partially substituted by these elements. In fact, bone and enamel are some of the most complex biomineralized structures. Attempts to synthesize bone in the laboratory are devoted to obtaining biocompatible prosthetic implants, with the ability to leverage natural bone regeneration when inserted into the human body. Its formation might imply certain temporary structural changes of its components, which demand in turn the presence, at trace levels, of additional ions and molecules in order to enable the mineralization process. This is the case, for instance, with bone growth processes, where the localized concentration of silicon-rich materials coincides precisely with areas of active bone growth. The reason is yet unknown, although the evidence is clear; the possible explanation of this phenomenon would also justify the great activity observed in certain silicon-substituted apatite phases and in some glasses obtained by the sol–gel method, regarding cell proliferation and new bone growth.

1.1.2 A Discussion on Biomineralization

Biomineralization is the controlled formation of inorganic minerals in a living body; the minerals might be crystalline or amorphous, and their shape, symmetry and ultrastructure can reach high levels of complexity. Bioinorganic solids have been *replicated* with high precision throughout the evolution process, *i.e.* they have been reproduced identically to the primitive original. As a consequence, they have been systematically studied in the fields of biology and palaeontology. However, the chemical and biochemical processes of biomineralization were not studied until quite recently. Such studies are currently providing new concepts in materials science and engineering.[17]

Biomineralization studies the mineral formation processes in living entities. It encompasses the whole animal kingdom, from single-cell species to humans. Biogenic minerals are produced in large scale at the biosphere, their impact in the chemistry of oceans is remarkable and they are an important component in sea sediments and in many sedimentary rocks.

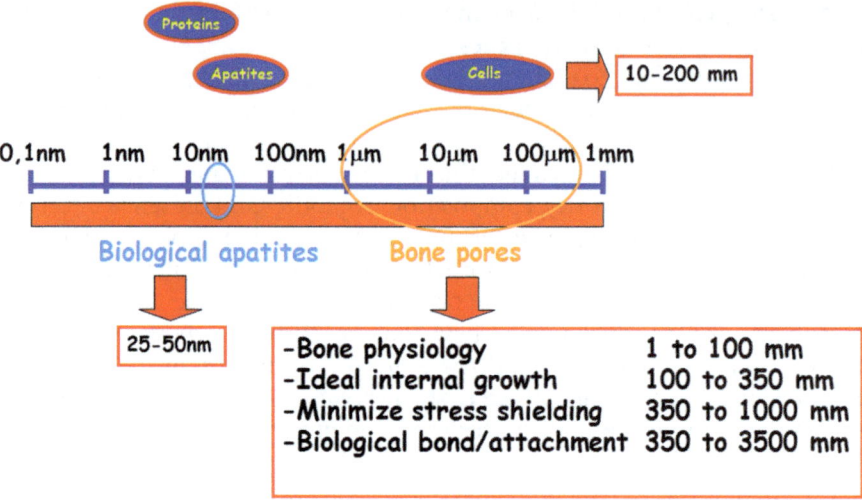

Figure 1.9   Scheme of the different scales for the most important hard-tissue-related biological moieties.

It is important to distinguish between mineralization processes under strict biological – *genetic* – control, and those induced by a given biological activity that triggers a fortuitous precipitation. In the first case, these are crystal-chemical processes aimed at fulfilling specific biological functions, such as *structural support* (bones and shells), *mechanical rigidity* (teeth), *iron storage* (ferritin) and *magnetic* and *gravitational navigation*, while in the second case there are minerals produced with heterogeneous shapes and dimensions, which may play different roles in the *increase of cell density* or as means of *protection* against predators.[23]

At the nanometer scale, *biomineralization* implies the molecular building of specific and self-assembled supramolecular organic systems (micelles, vesicles, *etc.*) which act as an environment, previously arranged, to control the formation of finely divided inorganic materials, of ~1–100 nm in size (Figure 1.9). The production of consolidated biominerals, such as *bones* and *teeth*, also requires the presence of previously arranged organic structures, at a higher length scale (micrometers).

The production of discrete or expanded architectures in biomineralization frequently includes a hierarchical process: the building of organic assemblies made of molecules confers structure to the synthesis of arranged biominerals, which act in turn as preassembled units in the generation of higher order complex microstructures. Although different in complexity, bone formation in vertebrates (support function) and shell formation in molluscs (protection function) bear in common the crystallization of inorganic phases within an organic matrix, which can be considered as a bonding agent arranging the crystals in certain positions in the case of bones, and as a bonding and grouping agent in shells (Figure 1.10).

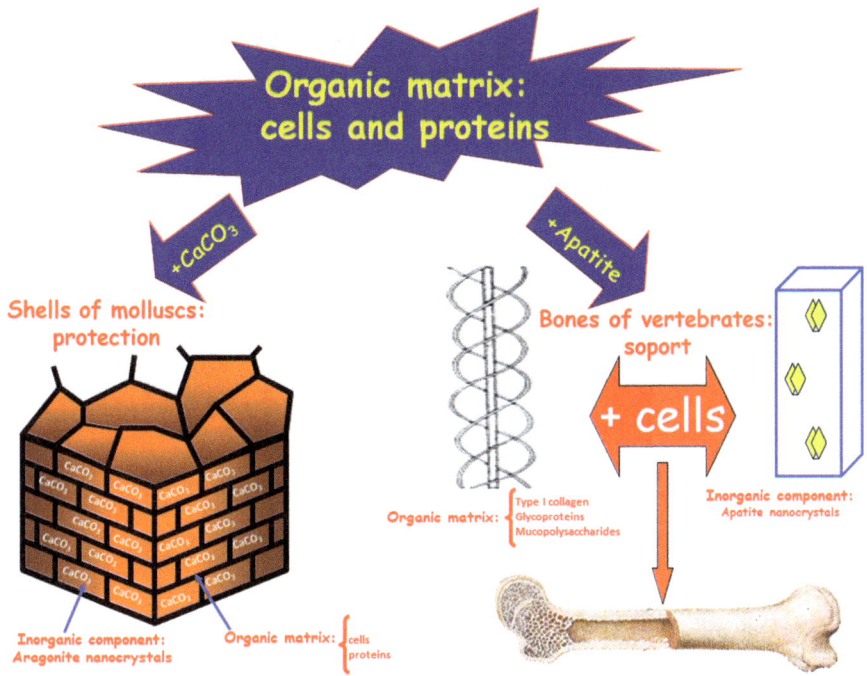

Figure 1.10 The structure–function relationship in different biominerals.

Our knowledge of the most primitive forms of life is greatly based upon biominerals, more precisely in fossils, which accumulated in large amounts. Several mountain chains, islands and coral reefs are formed by biogenic materials, such as limestone. This vast bioinorganic production over hundreds of millions of years has critically determined the development conditions of life.[24] For instance, CO_2 is combined in carbonate form, decreasing initially the greenhouse effect on the earth's crust. Leaving aside the shells, teeth and bones, there are many other systems which can be classified as biominerals: aragonite pellets generated by molluscs, the outer shells and spears of diatomea, radiolarian and certain plants, crystals with calcium, barium and iron content in the gravity and magnetic field sensors formed by certain species, and the stones formed in the kidney and urinary system, although the latter are pathological biominerals. The protein ferritin, responsible for iron storage, can also be considered a biomineral, taking into account its structure and inorganic content.

Bones, horns and teeth perform very different biological functions and their external shapes are highly dissimilar. But all of them are formed by many calcium phosphate crystals, small and isolated, with non-stoichiometric *carbonate-hydroxyl-apatite* composition and structure, grouped together by an organic component. Nucleation and growth of the mineral crystals is regulated by the organic component, the *matrix*, segregated in turn by the cells located near the growing crystals (Figure 1.11).

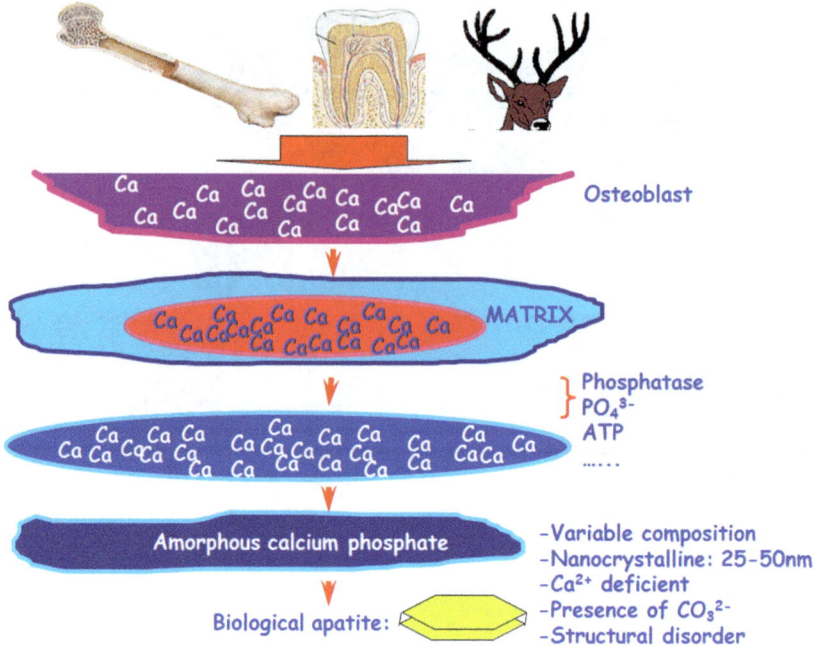

Figure 1.11   Stages of calcium phosphate maturation during the formation of different mineralized structures.

This matrix defines the space where the mineralization will take place. The main components of the organic matrix are *cellulose* in plants, *pectin* in diatomea, *chitin* and *proteins* in molluscs and arthropods and *collagen* and *proteoglycans* in vertebrates.

1.1.3   Biomineralization Processes

Different levels of biomineralization can be distinguished according to the type and complexity of control mechanisms. The most primitive form corresponds to biologically *induced* biomineralization, which is mainly present in bacteria and algae.[23] In these cases, biominerals are formed by spontaneous crystallization, due to supersaturation provoked by ion pumps, and then polycrystalline aggregates are formed in the extracellular space. Gases generated by the biological processes, by bacteria for instance, can (and often do) react with metal ions from the environment to form biomineral deposits.

More complex mechanisms involve processes with higher biological control. The obtained, well-defined bioinorganic products are formed by *inorganic* and *organic* components. The organic phase is usually made of fibrous proteins, lipids or polysaccharides, and its properties will affect the resulting morphology and the structural integrity of the composite.

Whatever the case, the formation of an inorganic solid from an aqueous solution is achieved with the combination of three main physicochemical stages: *supersaturation*, *nucleation* and *crystal growth*.

Nucleation and *crystal growth* are processes that take place in a supersaturated medium and must be properly controlled in any mineralization process. A living body is able to mineralize provided that there are well-regulated and active transport mechanisms available. Some examples of transport mechanisms are ion flows through membranes, formation or dissociation of ion complexes, enzyme-catalyzed gas exchanges (CO_2, O_2 or H_2S), local changes of redox potential or pH and variations in the medium's ionic strength. All these factors allow the creation and maintenance of a supersaturated solution in a biological environment.[23]

Nucleation is related to the kinetics of surface reactions such as *cluster* formation, growth of anisotropic crystals and phase transformations. In the biological world, however, there are certain surface structures that specifically avoid an unwanted nucleation, such as those exhibited by some kinds of fish in polar waters to avoid ice formation in body fluids.

The *growth* of a crystal or amorphous solid from a phase nucleus can be directly produced by the surrounding solution or by a continuous contribution of the required ions or molecules. Besides, diffusion can be drastically altered by any significant change in viscosity of medium.

The controlled *growth* of biominerals can be also produced by a sequence of stages, through phase transformations or by intermediate precursors that lead to the solid-state phase.

Biomineralization processes can be classified in two large groups: the first includes those phenomena where some kind of control seems to exist over the mineralization process, while the second encompasses those where control seems to be non-existent. According to Mann,[23] biomineralization processes can be described as *biologically induced* when said biomineralization is due to the withdrawal of ion or residual matter from cells, and is verified in an open environment, *i.e.* not in a region purposely restricted. It is produced as a consequence of a slight chemical or physical disturbance in the system. The crystals formed usually give rise to aggregates of different sizes, with similar morphology to mineral inorganic crystals. Besides, the kind of mineral obtained depends on both the environmental conditions of the living organism and on the biological processes involved in the formation, since the same organism is able to produce different minerals in different environments. This is particularly the case in single-cell species, although some higher order species also verify this behaviour.

However, there are situations where a specific mechanism is acting, which are then described as *biologically controlled*. An essential element of this process is space localization, whether in a membrane closed compartment, or confined by cell walls, or by a previously formed organic matrix. The biologically controlled process of biomaterials formation can be considered as the opposite to a biologically induced process. It is much more complex and implies a strict chemical and structural control.

Most biominerals formed under *controlled* conditions precipitate from solutions which are in turn *controlled* in terms of composition by the cells in charge; hence the contents of trace elements and stable isotopes in many mineralized areas are not balanced with the concentrations present in the initial medium.

Nucleation in controlled biomineralization requires low supersaturation combined with active interfaces. Supersaturation is regulated by ion transport and processes involving reaction inhibitors and/or accelerators. The active interfaces are generated by organic substrates in the mineralization area. Molecules present in the solution can directly inhibit the formation of nuclei from a specific mineral phase, hence allowing the growth of another phase.

Crystal *growth* depends on the supply of material to the newly formed interface. Low supersaturation conditions will favour the decrease in number of nuclei and will also restrict secondary nucleation, limiting somehow the disorder in the crystal phase. Under these low supersaturation conditions, growth rate is determined by the rate of ion bonding at the surface. In this scenario, foreign ions and large or small biomolecules can be incorporated to the surface, modifying the crystal growth and altering its morphology.

The final stage in the formation of a biomineral is its *growth interruption*. This effect may be triggered by a lack of ion supply at the mineralization site, or because the crystal comes in contact with another crystal, or else because the mineral comes into contact with the previously formed organic phase.

Whatever the cause, biomineralization processes are extremely complex, and not yet well known. One of the prevailing issues not yet fully elucidated is the mechanism at the molecular level that controls the crystal formation process. If we consider the features of many organism-grown minerals, it seems that such control can be exerted at various levels. The lowest level of control would be exemplified by the less specific mineralization phenomena, such as in many bacteria. These processes are considered more of an *induction* than a *control* of crystallization. The opposite case would be the most sophisticated composites of crystals and organic matter, where apparently there is a total control on crystal *orientation* during its *nucleation*, and on its *size* and *shape* during the *growth stage*. This would be the case in the bones of vertebrates. In between these two examples we can find plenty of intermediate situations where some, but not all, crystal parameters are controlled.[23]

A basic strategy performed by many organisms to control mineralization is to seal a given space in order to regulate the composition of the culture medium. This is usually done by forming barriers made of lipid bilayers or macromolecular groups. Subsequently, the sealed space can be divided into smaller spaces where individual crystals can be grown, adopting the shape of the said compartment. An additional strategy is also to introduce specific acidic glycoproteins in the sealed solution, which interact with the growing crystals and regulate their growth patterns. There are many other routes to exert control, such as introducing ions at very precise intervals, eliminating certain trace elements, introducing specific enzymes, *etc.* All these phenomena are due to the activity of specialized cells that regulate each process throughout its duration.

The stereochemical and structural relationship between macromolecules from the organic matrix and from the crystalline phase is a very important aspect in the complex phenomenon of biomineralization. These macromolecules are able to control crystal formation processes. It is already known that there is a wide range of biomineralization processes in nature, and that it is not possible to know *a priori* the specific mechanism of each one. It seems, however, that there are certain basic common rules regarding the control of crystal formation and the *interactions* involved. The term *interaction* refers here to the structure and stereochemistry of the phases involved, *i.e. nano-crystals* and *macromolecules*.

As already mentioned, the inorganic and organic components are forced to interact in order to produce a biomineral. They are not two independent elements; the specific extent and method of this interaction can be extremely varied, and the same variability applies to the biomineral's functionality.

1.1.4 Biominerals

The biominerals, natural composite materials, are the result of millions of years of evolution. The mineral phases present in living species can be also obtained in the laboratory or by geochemical routes. The synthesis conditions, however, are very different because the enforcement of those conditions in the biological environment is not as strict. It is worth noting that biogenic minerals usually differ from their inorganic counterparts in two very specific parameters: *morphology* and *order* within the biological system. It is quite likely that some general mechanisms exist that govern the formation of these minerals, and if our knowledge of these potential general principles improved, new options in material synthesis or modifications of already existing materials could be used in a wide range of applications in materials science.

The biominerals or organic/inorganic composites used in biology exhibit some unique properties which are not just interesting *per se*: the study of the formation processes of these minerals can lead to reconsider the world of industrial *composites*, to review their synthesis methods and to try and improve their properties. For instance, comparative studies with biominerals have provided new thinking on improvements of the physical–chemical properties in cements. In fact, the most noticeable property of minerals in biology is to provide *physical rigidity* to their host. But biological minerals are not just building materials, as we could consider their role in shells, bones and teeth; they also fulfil many other purposes, for instance, in sensing devices. The biomineralization process is responsible for bone formation, growth of teeth, shells, eggshells, pearls, coral and many other materials which form parts of living species. *Biomineralization* is hence responsible for the controlled formation of minerals in living organisms. These biominerals can be either crystalline or amorphous, and they belong in the *bioinorganic* family of solids.

Bioinorganic solids are usually (a) remarkably non-stoichiometric, that is, with frequent variations in their composition, allowing including impurities as *interstitial* and/or *substitutional* defects; and (b) they can be present in amorphous and/or crystalline form, and in some situations several polymorphs of the same crystalline solid can coexist. Besides, the inorganic component is just a part of the resulting biomineral, which actually is a composite material, or more precisely a *nanocomposite*, formed by an organic matrix which restricts the growth of the inorganic component at perfectly defined and delimited areas in space, determining a strict shape and size control.

The organic component might be a vesicle, perhaps a protein matrix; whatever the case, biominerals are formed by very different chemical systems, since they require the combined participation of *mineral components* and *organic molecules*. Vesicles give rise to *three-dimensional* structures, and are able to fill cavities, while the organic molecules can form *linear* or *layered* structures, and also can interact with the inorganic matrix, generating the voids to be filled with minerals.

Almost half of known biominerals include the element *calcium* among their constituents. This is the reason why the term *calcification* is often used to describe the processes where an inorganic material is produced by a living organism. But this generalization is not always true, since there are many biominerals without any calcium content. Therefore, the term biomineralization is not only much more generic but also more adequate, encompassing all inorganic phases regardless of their composition; the outcome is the biomineral, that is, a mineral inside a living organism, which is a truly *composite material*.

Biomineralization processes give rise to many inorganic phases; the four most abundant are *calcite*, *aragonite*, *apatite* and *opal*.

In load-bearing biominerals, such as bones, some stress induced changes may appear and induce in turn certain consequences on their properties, in the crystal growth for instance. The *growth* of biominerals is related to one of the great unsolved issues in biology: *the morphology of its nano- or microcrystals*. The skeletons of many species exhibit peculiarities that are clearly a product of their morphogenesis, with direct effects on it, since the gametes of biological systems never or hardly ever produce a biomineral precipitate.

Another question to be considered is the relevance of biominerals from a *chemical* point of view. Many of these minerals act as deposits that enable the regulation of the presence of cations and free anions in cell systems. Concentrations of iron, calcium and phosphates, in particular, are strictly controlled. A biomineral is the best possible *regulator* of homeostasis. It is important to recall that exocytosis of mineral deposits is a very simple function for cells, enabling them to eliminate the excess of certain elements. In fact, some authors believe that calcium metabolism is mainly due to the need to reject or eliminate calcium excess, leading to the development and temporary storage of this element in different biominerals. However, some evidences deny the validity of this point of view: many living species build their skeletons with elements that do not have to be eliminated, such as silicon.[25]

Mineral deposits such as iron and manganese oxides are used as *energy sources* by organisms moving from oxic to anoxic areas. Therefore, biominerals are also used by some living species as an *energy source* to carry out certain biological processes. This fact has been verified in marine bacteria.[26]

Although silicon in silicate form is the second most abundant element in the earth's crust, it plays a minor part in the biosphere. It may be due in part to the low solubility of silicic acid, H_4SiO_4, and of amorphous silica, $SiO_n(OH)_{4-2n}$. In an aqueous medium, at pH 1–9, its solubility is ~100–140 ppm. In the presence of cations such as calcium, aluminium or iron, the solubility decreases markedly, and solubility in sea water is just 5 ppm. In the biosphere, amorphous silicon is dissolved and then easily reabsorbed into the organism; it will then polymerize or connect with other solid structures.

Amorphous silicon biomineral is mainly present in single-cell organisms, in silicon sponges and in many plants, where is located in fitolith form in the cell membranes of grain plants or types of grass, with a clear deterrent purpose. The fragile tips of stings in some plants, such as nettles, are also made of amorphous silicon.

There is a wide range of biological systems with biomineral content, from humans to single-cell species. Modern molecular biology indicates that single-cell systems may be the best object for research in order to improve our knowledge of biological structure.

1.1.5 Inorganic Components: Composition and Most-Frequent Structures

At present, a wide range of known inorganic solids are included among the so-called biominerals. The main metal ions deposited in single-cell or multiple-cell species are the divalent alkaline earth cations Mg, Ca, Sr and Ba, the transition metal Fe and the semimetal Si. They usually form solid phases with anions such as carbonate, oxalate, sulphate, phosphate and oxides/hydroxides. The metals Mn, Au, Ag, Pt, Cu, Zn, Cd and Pb are less frequent and generally deposited in bacteria, in sulfide form. More than 60% of known minerals contain hydroxyl groups and/or water bonds, and are easily dissolved, releasing ions. The crystal lattice of the mineral group including metal phosphates is particularly prone to the inclusion of several additional ions, such as fluorides, carbonates, hydroxyls and magnesium. In some cases, this ability allows for the modification of the material's crystal structure and hence of its properties.

The field of biominerals encompasses a wide range of inorganic salts with many different functions, which are present in several species in nature. For instance, calcium in carbonate or phosphate form is important for nearly all species, while calcium sulphate compounds are essential for very few species. All along the evolution of species, there has been a constant development of the *control of selective precipitation*, that is, of *nucleation* and *growth processes*, as well as the *shape of the precipitates* and their exact *location within a living body*.[27]

The minerals in structures aimed at providing support or external protection can be crystalline or amorphous. The generation of amorphous materials in any kind of biological system is undoubtedly a favourable process from an energetic perspective, and is present in several examples such as carbonates and biological phosphates. This *amorphous phase* usually leads to a series of transformations, either as consequence of *recrystallization* processes – which give rise to a crystalline phase, likely to transform itself into other phases due to *in situ* structural modifications – or due to redissolution of the amorphous phase, enabling the *nucleation* of a new phase. If the minerals are crystalline, the biological control can be exerted over several parameters: *chemical composition, polymorph formation* and *crystal size and shape*. Each one of these parameters is in turn closely related to the organic matrix controlling *concentration of elements*, crystal *nucleation* and *growth*. If the mineral is amorphous, the chemical composition allows for almost infinite variations, although a certain concentration of the essential elements remains crucial. A typical amorphous biomineral is hydrated silica, $SiO_n(OH)_{4-2n}$, where n can be any value in the range 0–2. Several forms of hydrated silica can be found in living organisms, both in the *sea world* – such as sponges, diatomea, protozoa and single cell algae – and in the *vegetable kingdom*, present in *amorphous* form. The actions performed by these species to mineralize silicic acids are extremely complex. It seems that this process first involves the transportation of silicic acid towards the inside of the cell, and then to the deposition locations where the monomer will be polymerized to silica. For any silicon structure to be generated, the preliminary essential requirement is the availability of silicic acid, which must also be transported in adequate concentrations. If this stage is verified, the nucleation and polymerization processes may begin, which will eventually lead to the development of strict and specific morphological features, both at the microscopic and macroscopic scales. Little is known about the early stages, previous to deposition. Several mechanisms have been suggested to try to explain biosilication, but none of them is yet conclusive.

Some biominerals perform a very specific function within the biological world: they work as sensors, both for positioning and attitude or orientation. The inorganic minerals generated by some species to carry out this task are *calcite, aragonite, barite* and *magnetite*.

1.1.6 Organic Components: Vesicles and Polymer Matrices

The most common cell organ is the *vesicle*.[27] It is an aqueous compartment surrounded by a lipid membrane, impervious to all ions and most organic molecules. The ions required to form the biomineral are accumulated in the vesicle by a pumping action. These ions are, among others, Ca^{2+}, H^+, SO_4^{2-}, HPO_4^{2-} and HCO_3^-. Understanding biomineral formation depends upon the knowledge of cell vesicles and ion pumps.

Proteins or *polysaccharides* are able to build another kind of *receptacle, mould* or *sealed container*, more or less impervious to ions and molecules, depending on the particular system. This receptacle might be inside the cell itself, as in the case of ferritin, or outside the cell, such as bone collagen. The exact shape of the protein receptacle for ferritin is fixed, and the open spaces in collagen where apatite grows always exhibit the same shape. In contrast, the available space in a typical vesicle is not controlled by the organic structure, since vesicles do not have internal crosslinks in their membranes. In fact, vesicle space is very different from cytoplasmic space, which usually includes crossed fiber structures. As a consequence, when the *mould* is made of protein or polysaccharides, precipitation must be controlled through the regulation of cytoplasmic or extracellular homeostasis. Extracellular fluids have a sustained chemical composition due to the actions of control organs such as the kidney, which actually works as a macro-pump.

Most of the controlled mineralization processes performed by organisms exhibit associated macromolecules. These macromolecules perform important tasks in tissue formation and modification of the biomechanical properties of the final product. Although there are thousands of different associated macromolecules, Williams[27] stated that they all can be classified into two types: structural *macromolecules* and *acid macromolecules*. The main structural macromolecules are *collagen*, *α-* and *β-quitine* and *quitine–protein complexes*. The main acid macromolecules are not very well defined in some organisms, but we may include in this group *glycoproteins*, *proteoglycans*, *Gla-rich proteins* and *acid polysaccharides*. Little is known about the secondary conformation of acid macromolecules, apart from the fact that all acid glycoproteins with high contents of glutamic and aspartic acids partially adopt *in vitro* the β-layer conformation, in the presence of calcium. Although the composition of these macromolecules shows little variation between species, the opposite can be said of structural macromolecules. They vary from one tissue to another, and there are even some hard mineralized pieces that do not *seem* to have any kind of acid macromolecule at all. This lack of presence in some tissues allows the inference that their purpose might be to modify the mechanical properties of the final product, not to regulate biomineralization. The main means of control over biominerals are the independent areas in the cytoplasmatic space or in the extracellular zones in multiple cell species, where the organic structures develop well defined volumes and external shapes.

There are different physical and chemical controls in the development of a mineral phase. Physical controls are determined by the physics of our world and by biological source fields, in the same way that biological chemistry is restricted by the properties of chemical elements in the periodic table.

Mineral and vesicle grow together under the influence of many macroscopic fields. It should be taken into account that the functional values often depend on the interactions with these fields, due to the density, magnetic properties, ion mobility in the crystal lattice, elastic constants and other material properties. These properties do not fall under a strict biological control. Microscopic shape is restricted by the rules of symmetry in crystalline materials, but not in amorphous ones. Any crystal-based biomineral exhibits many restrictions in shape, and the organism adapts itself to them.[23]

1.2 Alternative Ways to Obtain Nanosized Calcium-Deficient Carbonate-Hydroxy-Apatites

Hydroxyapatite, $Ca_{10}(PO_4)_6(OH)_2$, is the most widely used synthetic calcium phosphate for the implant fabrication because is the most similar material, from the structural and chemical point of view, to the mineral component of bones.[28] Hydroxyapatite, with hexagonal symmetry S.G. $P6_3/m$ and lattice parameters $a = 0.95$ nm and $c = 0.68$ nm, exhibits excellent properties as biomaterial, such as biocompatibility, bioactivity and osteoconductivity. When apatites are synthesized with the aim of mimicking biological ones, the main characteristics required are small particle size, calcium deficiency and the presence of $[CO_3]^{2-}$ ions in the crystalline network. Two different strategies can be applied with this purpose.

The first strategy is based in the use of chemical synthesis methods to obtain solids with small particle sizes. There are plenty of options among these wet-route processes, which shall be generally termed as the *synthetic route*.[29]

The other strategy implies the collaboration of physiological body fluids.[30] In fact, certain ceramic materials react chemically with the surrounding medium when inserted into the vertebrate organism, yielding biological-like apatites through a *biomimetic process*.

1.2.1 The Synthetic Route

Synthetic strategies used to obtain sub-micrometric particles include the aerosol synthesis technique,[31] methods based on precipitation of aqueous solutions,[32,33] applications of the sol–gel method, or some of its modifications such as the liquid mix technique, which is based on the Pechini patent.[34,35] In these methods, variation of the synthesis parameters yields materials with different properties. Quantum/classical molecular mechanics simulations have been used to understand the mechanisms of calcium and phosphate association in aqueous solution.[36]

Conversely, it is difficult to synthesize in the laboratory calcium apatites with carbonate content analogous to those in bone. Indeed, it is difficult to avoid completely the presence of some carbonate ions in the apatite network, but the amount of these ions is always inferior to values in natural bone (4–8% wt) and/or they are located in different lattice positions.[12,37] It must be taken into account that biological apatites are always of type B, but if the synthesis of the ceramic material takes place at high temperatures, type A apatites are obtained. Synthesis at low temperatures obtains apatites with carbonate ions in phosphate positions but in lower amounts than in the mineral component of bones,[38,39] although new efforts to prepare B-type carbonated nanoapatites with higher substitution and in an easier manner are still under development.[40,41]

There is currently a large body of knowledge about the role of carbonates in apatites intended for bone grafting. However, research into the development of new carbonated apatites is an ongoing research field. Very recently, the presence of carbonates in nanocrystalline apatites have been demonstrated to enhance the proliferation and differentiation towards osteoblast pathways of human bone marrow cells.[42]

1.2.2 The Biomimetic Process

As in any other chemical reaction, the product obtained when a substance reacts with its environment might be an unexpected or unfavourable result, such as corrosion of an exposed metal, but it could also lead to a positive reaction product that chemically transforms the starting substance into the desired final outcome. This is the case of *bioactive ceramics*, which chemically react with body fluids towards the production of newly formed bone. When dealing with the repair of a section of the skeleton, there are two different basic options to consider: *replacing* the damaged part, or *substituting* it and regenerating the bone tissue. This is the role played by bioactive ceramics.[43]

Calcium phosphates, glasses and glass ceramics, the three families of ceramic materials from which several bioactive products have been obtained, have given rise to the starting materials used to obtain mixtures of two or more components, in order to improve the bioactive response in a shorter period of time.

These types of ceramics are also studied to define shaping methods enabling implant pieces in the required shapes and sizes to be obtained, with a given porosity, according to the specific role of each ceramic implant. Hence, if the main requirement is to verify in the shortest possible time a chemical reaction leading to the formation of nanoapatites as precursors of newly formed bone, it will be necessary to design highly porous pieces, which must also include a certain degree of macropores to ensure bone oxygenation and angiogenesis.

However, these requirements are often discarded when designing the ceramic piece. As a result, the chemical reaction only takes place on the external surface of the piece (if made of bioactive ceramics) or it simply does not occur if the piece is made of an inert material. In both cases, the inside of the piece remains a solid monolith able to fulfil bone replacement functions, but without the regenerative role associated with bioactive ceramics. In order to achieve a chemical reaction throughout the whole material, it is important to design pieces with a bone-like hierarchical structure of pores. In this way, the fluids will be in contact with a much larger specific surface, reaching a higher reactivity phase that allows a full reaction between the bioactive ceramic and the fluids to be achieved, to yield newly formed bone as reaction product.

From these reasons we can realize that manufacturing bioceramics following a biomimetic strategy involves different factors at different levels, *i.e.* chemical composition, microstructure and hierarchical macrostructures. Figure 1.12 summarizes these three concepts. First, the apatite intended to mimic the mineral component of the bone must be calcium-deficient, carbonated and non-stoichiometric where phosphate and hydrogenphosphate groups coexist. The microstructural characteristics of the biomimetic apatites

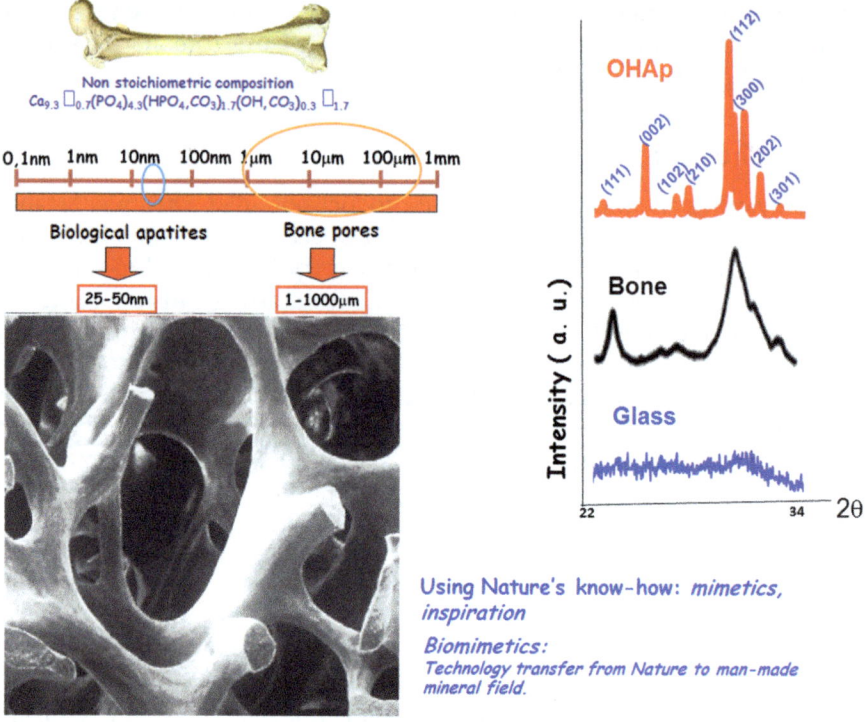

Figure 1.12 Relevant considerations when mimicking the natural synthesis of bone-like apatites: chemical composition, crystallite size and macro-porous architecture.

should comprise crystallites of 20–50 nm in size with needle-like morphology along the *c* axis. The X-ray diffraction patterns in Figure 1.12 (right) evidence that the apatites in bone exhibit a degree of crystallinity, which is an intermediate situation between the amorphous structure of a glass and the highly crystalline ordering of hydroxyapatites obtained by high temperature ceramic methods. Finally, the macrostructural characteristic of the bone involves a macroporous architecture, with pores of hundreds of micrometers in size, which allow for the ingrowth of blood vessels and the diffusion of nutrients, as can be seen in Figure 1.12 (bottom). However, this is a very simplified description of the hierarchical porosity of bones. Bones differ from each other depending on anatomical location, sex, age, *etc.*[44] Furthermore, we have already seen that for the same bone, cancellous and cortical components posses a very different porosity. Consequently, establishing how or what must be mimicked when designing biomimetic apatites is a very difficult task.

Finally, we must take into account that the goal of biomimetic apatites is commonly to restore the structure and functionality of an impaired bone, not a healthy one. In other words, the bone graft will be in contact with a tissue that, commonly, exhibits an impaired structure. Therefore a new dilemma arises: should whe mimic the properties of a healthy bone or the impaired

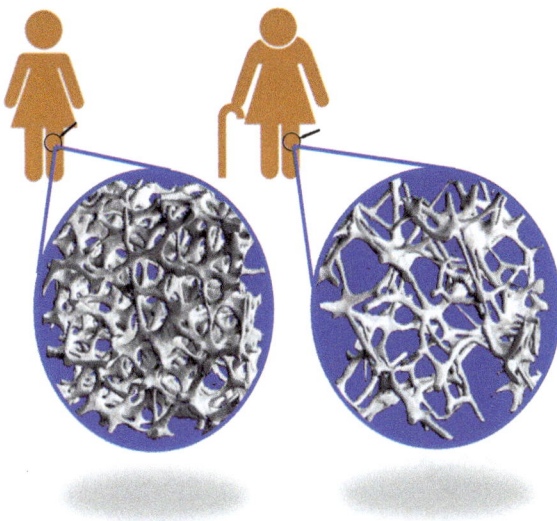

Figure 1.13 Microtomography images of healthy (left) and osteoporotic (right) bones.

characteristics of the receptor location, thus matching the implant–tissue interface? This dilemma is especially serious in the case of osteoporotic patients.[45] Bone weakening due to osteoporosis is far from satisfactory resolution. As a consequence of poor bone quality, surgical procedures performed to implant a device into weakened bone often lead to a clinical result that is worse than if such an intervention was performed on a young and strong bone. The risk of fracture increases exponentially with age, and the recovery process from a fracture is often slow, difficult and may lead to a disability or even to the death of the patient. Figure 1.13 shows micro-computed tomography images obtained of a healthy bone of a young, active woman and for a low-density bone of an older osteoporotic patient. Recent developments in biomaterials converge on the needs of biological enhancement of biomaterials fostering osteoinduction and osteogenesis to support and augment bone healing. Indeed, the reconstruction of osteoporotic bone yields significant difficulties with the solutions available today.

Another alternative is to provide a scaffolding material to the bone, aimed at enhancing its self-healing mechanisms.[46] The design and development of porous ceramics have attracted much attention in the last years. Pore distribution, as well as pore size can play a fundamental role in bone regeneration, angiogenesis and implant degradation. The incorporation of free-form preparation methods such as three-dimensional printing into the biomaterials field allow the design of hierarchical pore structures to facilitate these processes[47] (Figure 1.14). An interconnected macropore structure of 150–1000 µm allows cell colonization and enhances the diffusion rates to and from the centre of a scaffold, as well as angiogenesis and bone ingrowth. Small pores

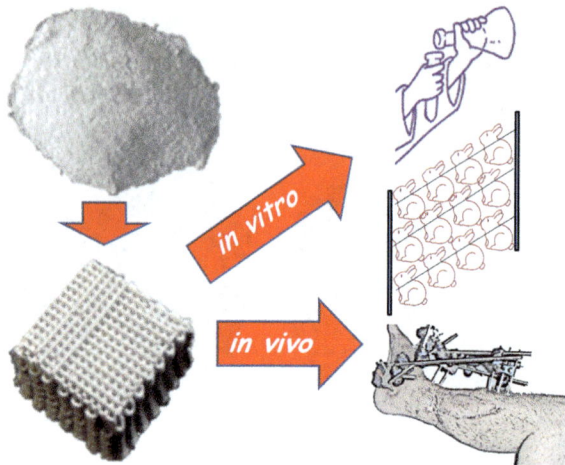

Figure 1.14   Stages of scaffolds preparation for *in situ* bone tissue regeneration. Powder synthesis, scaffolds printing, *in vitro* testing, *in vivo* studies and clinical applications.

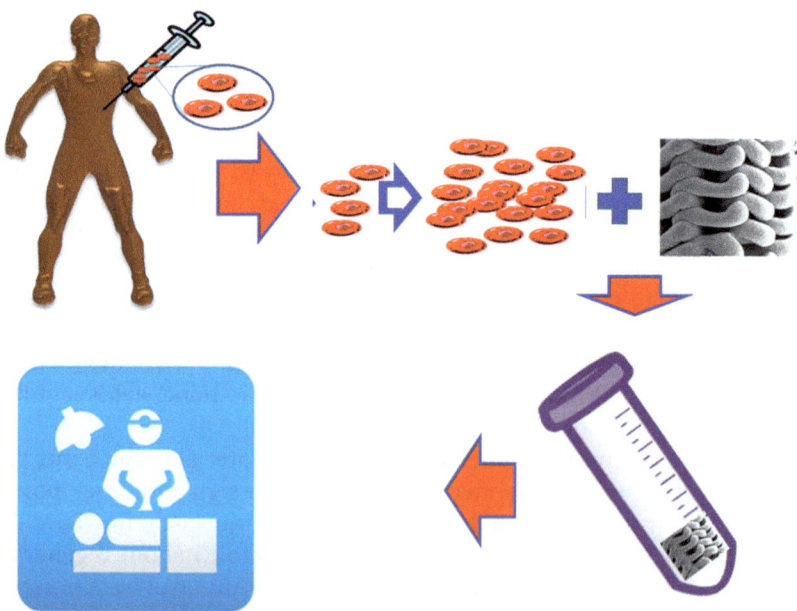

Figure 1.15   Tissue engineering stages for bone regenerative therapies. Cells are harvested from patients and seeded in the scaffolds. After *in vitro* bone tissue formation the construct is implanted.

allow phagocytic cells to adhere and resorb the scaffolds, whereas larger pores encourage the invasion of new vessels and ingrowth of bone tissue.

Finally, tissue engineering techniques provide excellent tools to mimic apatite biomineralization and bone tissue formation.[48] In this case, mesenchymal stem cells harvested from patient marrow are seeded onto bioceramic-based scaffolds. By means of adding osteoinductive agents, the osteoblast pathway is activated and osteogenesis is aroused. Finally, the construct inorganic scaffold–bone tissue is implanted to regenerate bone defects (Figure 1.15).

References

1. S. Mann, J. Webb and R. J. P. Williams, *Biomineralization. Chemical and Biochemical Perspectives*, VCH, Weinheim, Germany, 1989.
2. D. Lee and M. J. Glimcher, *J. Mol. Biol.*, 1991, **217**, 487.
3. M. Vallet-Regí and J. González-Calbet, *Prog. Solid State Chem.*, 2004, **32**, 1.
4. A. J. Friedenstein, *Int. Rev. Cytol.*, 1976, **47**, 327.
5. M. J. Glimcher, in *Disorders of Bone and Mineral Metabolism*, ed. F. L. Coe and M. J. Favus, Raven Press, New York, 1992, pp. 265–286.
6. M. J. Glimcher, in *The Chemistry and Biology of Mineralized Connective Tissues*, ed. A. Veis, Elsevier. Amsterdam, 1981, pp. 618–673.
7. L. T. Kuhn, D. J. Fink and A. H. Heuer, Biomimetic Strategies and Materials Processing, in *Biomimetic Materials Chemistry*, ed. S. Mann, Wiley-VCH, United Kingdom, 1996, pp. 41–68.
8. S. P. Bruder, A. I. Caplan, Y. Gotoh, L. C. Gerstenfeld and M. J. Glimcher, *Calcif. Tissue Int.*, 1991, **48**, 429.
9. M. D. McKee, A. Nanci, W. J. Landis, Y. Gotoh, L. C. Gertenfeld and M. J. Glimcher, *Anat. Rec.*, 1990, **228**, 77.
10. Y. Gotoh, L. C. Gerstenfeld and M. J. Glimcher, *Eur. J. Biochem.*, 1990, **228**, 77.
11. J. C. Elliott, *Structure and Chemistry of the Apatites and Other Calcium Orthophosphates*, Elsevier, London, 1994.
12. R. Z. LeGeros, in *Monographs in Oral Science, Vol. 15: Calcium Phosphates in Oral Biology and Medicine*, ed. H. M. Myers and S. Karger, Basel, 1991.
13. D. G. Pechak, M. J. Kujawa and A. I. Caplan, *Bone*, 1986, 7, 441.
14. E. D. Eanes and J. L. Meyer, *Calcif. Tissue Res.*, 1977, **23**, 259.
15. H. Nancollas, *In vitro* Studies of Calcium Phosphate Crystallization, in *Biomineralization. Chemical and Biochemical Perspectives*, ed. S. Mann, J. Webb and R. J. P. Williams, VCH, Weinheim, Germany, 1989, pp. 157–188.
16. A. Veis, Biochemical Studies of Vertebrate Tooth Mineralization, in *Biomineralization. Chemical and Biochemical Perspectives*, ed. S. Mann, J. Webb and R. J. P. Williams, VCH, Weinheim, Germany, 1989, pp. 189–222.
17. J. D. Birchall, The Importance of the Study of Biominerals to Materials Technology, in *Biomineralization. Chemical and Biochemical Perspectives*, ed. S. Mann, J. Webb and R. J. P. Willians, VCH, Weinheim, Germany, 1989, 491–508.

18. J. B. Park and R. S. Lakes, Structure-Property Relationships of Biological Materials, in *Biomaterials. An Introduction*, ed. Plenum Press, New York and London, 2nd edn, 1992, pp. 185–222.

19. J. B. Park and R. S. Lakes, Composites as Biomaterials, in *Biomaterials. An Introduction*, ed. Plenum Press, New York and London, 2nd edn, 1992, pp. 169–183.

20. J. Christofferson and W. J. Landis, *Anat. Rec.*, 1991, **230**, 435.

21. M. Vallet-Regí, *An. Quim. Int. Ed.*, 1997, **93**(suppl. 1), S6.

22. M. Vallet-Regí and D. Arcos, *J. Mater. Chem.*, 2005, **15**, 1509.

23. S. Mann, Crystallochemical Strategies in Biomineralization, in *Biomineralization. Chemical and Biochemical Perspectives*, ed. S. Mann, J. Webb and R. J. P. Williams. VCH, Weinheim, Germany, 1989, 35–62.

24. M. A. Borowitzka, Carbonate Calcification in Algae-Initiation an Control, in *Biomineralization. Chemical and Biochemical Perspectives*, ed. S. Mann, J. Webb and R. J. P. Williams, VCH, Weinheim, Germany, 1989, pp. 63–94.

25. C. C. Perry, Chemical Studies of Biogenic Silica, in *Biomineralization. Chemical and Biochemical Perspectives*, ed. S. Mann, J. Webb and R. J. P. Williams, VCH, Weinheim, Germany, 1989, pp. 223–256.

26. S. Mann and R. B. Frankel, Magnetite Biomineralization in Unicellular Microorganisms, in *Biomineralization. Chemical and Biochemical Perspectives*, ed. S. Mann, J. Webb and R. J. P. Williams, VCH, Weinheim, Germany, 1989, pp. 389–426.

27. R. J. P. Williams, The Functional Forms of Biominerals, in *Biomineralization. Chemical and Biochemical Perspectives*, ed. S. Mann, J. Webb and R. J. P. Williams, VCH, Weinheim, Germany, 1989, pp. 1–34.

28. M. Vallet-Regí, *J. Chem. Soc., Dalton Trans.*, 2001, 97.

29. M. Vallet-Regí, Preparative Strategies for controlling structure and morphology of metal oxides, in *Perspectives in Solid State Chemistry*, ed. K. J. Rao, Narosa Publishing House, India, 1995, pp. 37–65.

30. M. Vallet-Regí, C. V. Ragel and A. J. Salinas, *Eur. J. Inorg. Chem.*, 2003, **6**, 1029.

31. M. Vallet-Regí, M. T. Gutiérrez-Ríos, M. P. Alonso, M. I. de Frutos and S. Nicolopoulos, *J. Solid State Chem.*, 1994, **112**, 58.

32. M. Vallet-Regí, L. M. Rodríguez Lorenzo and A. J. Salinas, *Solid State Ionics*, 1997, **101–103**, 1279.

33. L. M. Rodríguez-Lorenzo and M. Vallet-Regí, *Chem. Mater.*, 2000, **12**(8), 2460.

34. M. P. Pechini, U. S. Patent 3,330,697, 1967.

35. J. Peña and M. Vallet-Regí, *J. Eur. Ceram. Soc.*, 2003, **23**, 1687.

36. D. Zahn and Z. Anorg, *Allg. Chem.*, 2004, **630**, 1507.

37. J. C. Elliot, G. Bond and J. C. Tombe, *J. Appl. Crystallogr.*, 1980, **13**, 618.

38. M. Okazaki, T. Matsumoto, M. Taira, J. Takakashi and R. Z. LeGeros, in *Bioceramics, vol. 11*, ed. R. Z. LeGeros and R. Z. LeGeros, World Scientific, New York, 1998, p. 85.

39. Y. Doi, T. Shibutani, Y. Moriwaki, T. Kajimoto and Y. J. Iwayama, *J. Biomed. Mater. Res.*, 1998, **39**, 603.

40. M. L. Gualtieri, M. Romagnoli, M. Hanuskova, E. Fabbri and A. F. Gualtieri, *J. Solid State Chem.*, 2014, **220**, 60.
41. J. G. Liao, Y. Q. Li, X. Z. Duan and Q. Liu, *Spectrosc. Spect. Anal.*, 2014, **34**, 3011.
42. H. Nagai, M. Kobayashi-Fujioka, K. Fujisawa, G. Ohe, N. Takamaru, K. Hara, D. Uchida, T. Tamatani, K. Ishikawa and Y. Miyamoto, *J. Mater. Sci. Mater. Med.*, 2015, **26**, 99.
43. A. J. Salinas and M. Vallet-Regí, *Z. Anorg. Allg. Chem.*, 2007, **633**, 1762.
44. F. Duboeuf, B. Burt-Pichat, D. Farlay, P. Suy, E. Truy and G. Boivin, *Bone*, 2015, **73**, 105.
45. D. Arcos, A. R. Boccaccini, M. Bohner, A. Díez-Pérez, M. Epple, E. Gómez-Barrena, A. Herrera, J. A. Planell, L. Rodríguez-Mañas and M. Vallet-Regí, *Acta Biomater.*, 2014, **10**, 1793.
46. M. J. Feito, R. M. Lozano, M. Alcaide, C. Ramírez-Santillán, D. Arcos, M. Vallet-Regí and M. T. Portolés, *J. Mater. Sci. Mater. Med.*, 2011, **22**, 405.
47. M. Manzano, D. Lozano, D. Arcos, S. Portal-Núñez, C. Lopez-Laorden, P. Esbrit and M. Vallet-Regí, *Acta Biomater.*, 2011, **7**, 3555.
48. R. Detsch, S. Alles, J. Hum, P. Westenberger, F. Sieker, D. Heusinger, C. Kasper and A. R. Boccaccini, *J. Biomed. Mater. Res. A*, 2015, **103**, 1024.

CHAPTER 2

Synthetic Nanoapatites

2.1 Introduction

2.1.1 General Remarks on the Reactivity of Solids

The most common reactions that a chemist needs to know in order to obtain a solid are those starting from two reactants in solution, leading to a new compound that is insoluble in the solvent used, usually water. There are, however, many other types of reactions that also lead to the synthesis of a solid (Figure 2.1). The main difference between classical synthesis from a solution and all the other synthesis routes depicted in the figure is the lack of a solvent, *i.e.* of an easy transport medium for the reactants, although its presence imposes a restriction on the feasible temperature range for the reaction, since it cannot exceed the boiling point of said solvent.

According to Figure 2.1, it is possible to obtain solids from reactants in solid, melted or even gaseous states, increasing remarkably the temperature range available; this fact allows us to prepare solids that would be otherwise unfeasible by a conventional method. These principles can be directly applied to the laboratory synthesis of apatites. Although there are obvious differences between the four alternative routes depicted, which can be even more complex if the reactants themselves are in dissimilar phases (liquid/solid, gas/solid, *etc.*), the common feature in all these processes is the synthesis and outcome of a new phase. This means that a new interface has appeared, with associated thermodynamical restrictions to its formation (nucleation stage), which are not present in homogeneous processes. Besides, the wider temperature range associated with solvent-free synthesis, while clearly advantageous, does impose remarkable restrictions from a kinetic point of view on solid–solid synthesis reactions. This process is determined by the low mobility of the reactants.

RSC Nanoscience & Nanotechnology No. 39
Nanoceramics in Clinical Use: From Materials to Applications, 2nd Edition
By María Vallet-Regí and Daniel Arcos Navarrete
© M. Vallet-Regi and D. Arcos Navarrete 2016
Published by the Royal Society of Chemistry, www.rsc.org

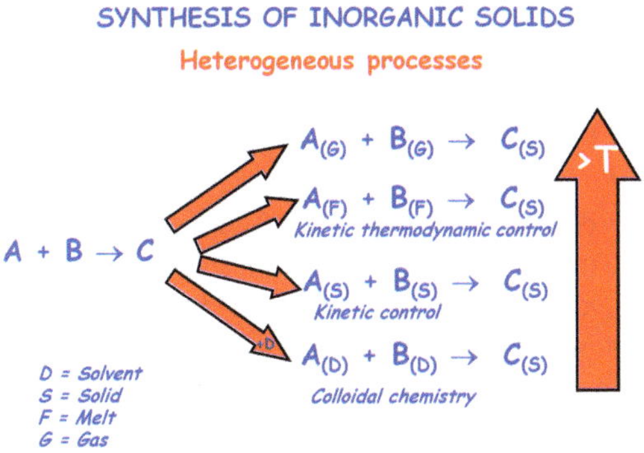

SYNTHESIS OF INORGANIC SOLIDS

Heterogeneous processes

$A_{(G)} + B_{(G)} \rightarrow C_{(S)}$

$A_{(F)} + B_{(F)} \rightarrow C_{(S)}$
Kinetic thermodynamic control

$A + B \rightarrow C$

$A_{(S)} + B_{(S)} \rightarrow C_{(S)}$
Kinetic control

$A_{(D)} + B_{(D)} \rightarrow C_{(S)}$
Colloidal chemistry

D = Solvent
S = Solid
F = Melt
G = Gas

Figure 2.1   Scheme of possible reactions that lead to solid product formation.

Solid formation reactions are usually classified into five groups:

- Solid → products
- Solid + gas → products
- Solid + solid → products
- Solid + liquid → products
- Surface reactions in solids.

The first group includes decomposition of solids and polymerization. The second group corresponds to oxidation or reduction reactions. The solid–solid reactions of the third group take place for instance in the ceramic method, the most traditional synthesis method in the world of cements and ceramic materials. The fourth group includes reactions such as intercalation and percolation, while the fifth group holds all those reactions occurring on the surface of solids.

Solid-state reactions may include one or more elementary stages such as adsorption or desorption of gas phases onto the solid surface, chemical reactions at the atomic scale, nucleation of a new phase and transportation phenomena through the solid. Besides, external factors such as temperature, surrounding environment, irradiation, *etc.*, significantly affect reactivity.

Multiple factors influence the reactivity of solids. In fact, features such as particle size, gas atmosphere and external additives, as well as dopants and impurities, play a predominant role in reactivity. For instance, reactivity increases when the particle size decreases. In this sense also, the use of solid reactants with small particle sizes leads to more homogeneous solid products. The atmosphere where the reaction takes place has clear effects on reactivity, even more so if the gas is also an exchangeable component of the solid phases. Doping with certain species also determines the reaction kinetics, and impurities lower the temperature required for a given reaction.

According to these observations, it seems clear that the *previous history* of any solid is extremely important for its future reactivity. The preparation method used may have determined a certain particle size, impurities, defects, *etc.*, which forcefully affect the subsequent reactivity of this solid. A mechanical treatment in a mortar or ball mill, for instance, greatly affects the treated solid, creating different types of defects that may determine the kinetics of the whole process.

The synthesis of tailored solids, with predetermined structure and properties, is the main and most difficult challenge in solid-state chemistry, which plays a crucial role in the fields of materials science and technology.

In the last few years, scientists working in solid-state chemistry have put special efforts into the study and development of new synthesis methods. Due to the vast number of theoretically possible solids obtainable, the synthesis tools to be used may vary with the issues to be solved in each particular case. Luckily, at present there are adequate techniques available to control both the structure and morphology of many different materials. A well-designed synthesis process does require in all cases a profound knowledge of crystallochemistry together with a good control of the particular thermodynamics, phase diagram and reaction kinetics involved; all this, added to the information available in the literature, is the first and vital step in the design and synthesis of new apatites with tailored properties.

Theoretically speaking, it is possible to design properties using the classical tools: control of structure and composition. Besides, the properties of apatites are closely related to their previous history; it is important to choose carefully the synthesis method and to perform a detailed microstructural characterization in order to correlate the influence of structure and defects on its properties.

2.1.2   Objectives and Preparation Strategies

In order to modify the properties of apatites, two strategies may be followed: (a) producing structural changes preserving the chemical composition; and (b) introducing compositional changes avoiding changes in the average structure. The latter may allow a systematic search of new compositions to obtain new and better properties, hence designing tailored apatites. A valid motto for a solid-state chemist would be "to understand all available synthesis methods to obtain a given solid, in order to always choose the optimum one". This is the strategy that has to be applied with apatites, using different synthesis methods and opening up new expectations in the field of applications.

In the words of Rao and Gopalakrishnan,[1] *it is useful to distinguish between synthesis of new solids and synthesis of solids by new methods*. To obtain a new solid, it is not always compulsory to apply a new method. It could be very useful, however, to synthesise already-known materials using different routes that allow modification of their *texture* and *microstructure*.

There are plenty of methods nowadays to obtain apatites. Once again, it is very important to establish first our objectives, before initiating a synthesis process.

2.2 Synthesis Methods

Synthesis of apatites from solid precursors implies slow solid-state reactions, and it is usually difficult to achieve a complete reaction. Long treatment periods and high temperatures are needed, in order to improve the diffusion of the atoms implied in the reaction throughout their respective solid precursors, reaching the interface where the reaction is actually happening. It is also possible to produce a solid-state transformation from a given phase to another one with equal composition, whether under high temperature or pressure, or under a combination of both.

The synthesis of solids from liquids occurs by solidification of the melted product, obtaining single crystals when the cooling rate is low enough, or noncrystalline materials, glasses, when the cooling rate is high enough to avoid the ordered arrangement of atoms, and hence crystallization. This is not the most common way to obtain apatites. There is a more adequate alternative for apatite synthesis, namely crystallization of solids from solutions. Frequently a solid is obtained from a liquid phase, where the formation of the solid product is a purely physical process and corresponds to a phase transformation. In other cases, the synthesis incorporates a liquid. These synthesis routes may be classified according to the quality of melted matter or solution of the precursor liquid phase.

Synthesis of solids from condensation of reactants in gaseous phase gives rise usually to solids in the form of *thin films* deposited onto adequate substrates.

Obviously, several techniques have been utilized for the preparation of hydroxyapatite and other calcium phosphates,[2–4] which include precipitation, hydrothermal synthesis and hydrolysis of other calcium phosphates.[5–36] Modifications of these "classical" methods (precipitation, hydrolysis or precipitation in the presence of urea, glycine, formamide, hexamethylenetetramine, *etc.*)[37–41] or alternative techniques have been employed to prepare hydroxyapatite with morphology, stoichiometry, ion substitution or degree of crystallinity as required for a specific application. Among them, sol–gel,[42–51] microwave irradiation,[52,53] freeze-drying,[54] mechanochemical methods,[55–59] emulsion processing,[60–62] spray pyrolysis,[63–65] hydrolysis of α-tricalcium phosphate (TCP),[66] ultrasonics,[67,68] *etc.*, can be outlined.

2.2.1 Synthesis of Apatites by the Ceramic Method

The most traditional method in apatite synthesis is the ceramic method, which consists of a solid–solid reaction where both reactants and products are in the solid state. The usual starting phases are oxides, carbonates or, generally speaking, salts, with very different particle sizes and irregular morphologies (Figure 2.2). When mixed and homogenized in the stoichiometric ratio, they are subsequently submitted to an adequate thermal treatment to start the reaction. In most cases, this method requires high temperatures and long heating periods.

Figure 2.2   Scheme of the ceramic method.

The study of chemical reactions between solid materials is a fundamental aspect of solid-state chemistry, allowing an understanding of the influence of structure and defects in the reactivity of solids. It is important to determine which factors rule reactivity in the solid state, in order to obtain new solids with the desired structure and properties.

Solid-state reactions differ in a fundamental aspect from those taking place at a homogeneous fluid medium: the intrinsic reactivity of liquid- or gaseous-state reactions mainly depends on the intervening chemical species and their respective concentrations, while solid-state reactions greatly depend on the crystallinity of the chemical constituents. The fact that said constituents (atoms, ions or molecules) occupy fixed positions in determined sites of a given crystal lattice brings a new dimension to the reactivity of solids, in contrast with other physical states.

In other words, the chemical reactivity of solids is often more determined by the crystalline structure and the presence of defects, than by the intrinsic chemical reactivity of its constituents. This fact is clearly evidenced in a type of solid-state reaction termed *topochemical* or *topotactical*.

Another type of solid-state reaction occurs in intercalation processes. In addition, in catalytic reactions, or in many fields where catalysis plays a fundamental role, it is worth considering that not only the crystal order of the chemical constituents is important, but also their particle size. This is somehow implicit when considering that the chemical reactivity of solids relies mainly on their crystalline structure and defects, since the surface of any crystal particle can be considered as a plane defect. The smaller the particle, the lower the number of complete unit cells forming the crystal; as a consequence, the constituents will have a short diffusion route, reaching higher levels of reactivity. Another important factor is that the specific

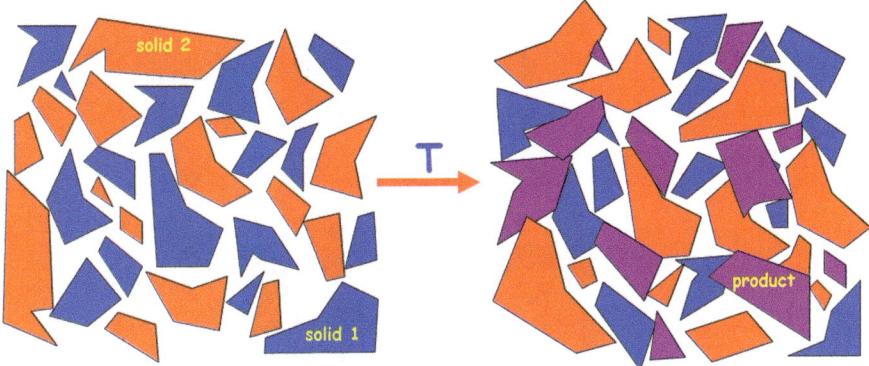

Figure 2.3   Scheme representing the heterogeneity of solid reactants. A 100% yield of reaction product is difficult to obtain due to the kinetic features of the process.

surface of small particles is significantly higher, and all these reasons combined emphasize the importance of particle size in the reactivity of solids.

Solid-state reactions begin by interphase reactions at the contact points between the reactants. The product phase represents a kinetic obstacle for the ongoing reaction, which keeps reacting due to the diffusion of the constituents, which again come into contact. These difficulties often lead to the impossibility to obtain a pure single-phase, homogeneous product by this procedure (Figure 2.3).

The ceramic method is perhaps the best example to understand the reactions between solids. As previously mentioned, in the liquid or vapor state the reacting molecules have more opportunities for contact between them and to react due to the continuous movement under conditions determined by statistical laws. To put it simply, diffusion is extremely easy in these two media.

On the contrary, in the solid state, the reactions generally take place between apparently regular crystalline structures, where the movements of the constituent species are much more restricted and depend to a complex degree on the presence of defects. Besides, said interaction can only occur at points of close contact between the reacting phases. Moreover, another difference is that in liquid-state reactions, the formed product does not affect greatly the course of the reaction.

However, in the solid state, the production of a more or less static layer of product can inhibit or at least slow down the progress of the reaction; it cannot carry on without contact, hence diffusion, albeit restricted, is the only way in which the reaction can continue in the solid state.

The study of reactions between solids could be expected to be simple, since they occur in a homogeneous state of the matter, as is the case in processes taking place with all elements in a liquid or gas phase; the truth, however, is quite different. Reaction processes between solid precursors are extremely

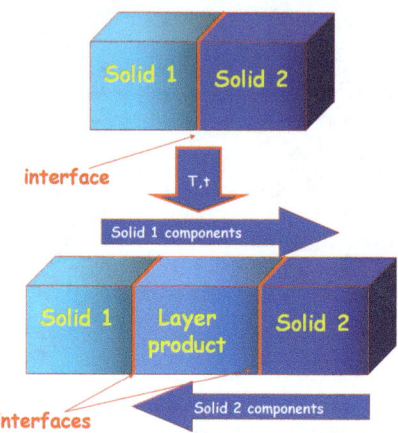

Figure 2.4   Possible formation mechanism of a solid product through the interface of the reactants.

complex, dealing with two or more reacting phases plus the reaction product(s). There are difficult theoretical and experimental issues in this study.

When studying reactions in solids and chemical reactions in general, it is important to distinguish between *thermodynamic* and *kinetic* issues; among them, it is worth considering those factors that may improve the reaction kinetics.

Since the chemical potential and the activity of a pure solid remain constant at constant temperature and pressure, ΔG is an invariant for a given reaction process. When $\Delta G > 0$, the reaction does not start spontaneously. Conversely, when $\Delta G < 0$, the reaction should be produced spontaneously. Even under these last thermodynamically favorable conditions, solid-state reactions may not be completed due to the formation of a reaction-product layer in the interphase area, which becomes larger as the reaction progresses; at last one of the reactants must trespass this layer in order to continue the reaction.

According to the two first principles of thermodynamics $\Delta G = \Delta H - T\Delta S$, where ΔH and ΔS are the enthalpy and entropy variations during the reaction. In most cases, solid-state reactions imply a regrouping of the crystalline lattice as evidenced in many examples. The degree of crystallinity does not vary much in these cases; hence ΔS always has a value close to zero. A reaction will be verified if $\Delta H < 0$, because ΔG forcefully adopts a negative value $(\Delta G < 0)$. Therefore, solid-state reactions are usually exothermic.

Generally speaking, the theoretical study of reaction *mechanisms* is carried out using a geometrical model which considers a *counterflow diffusion* of the various species forming both solids (Figure 2.4).

The production of hydroxyapatite by the ceramic method is a traditional laboratory process, which can be carried out using different precursor salts from the phosphate and carbonate families.

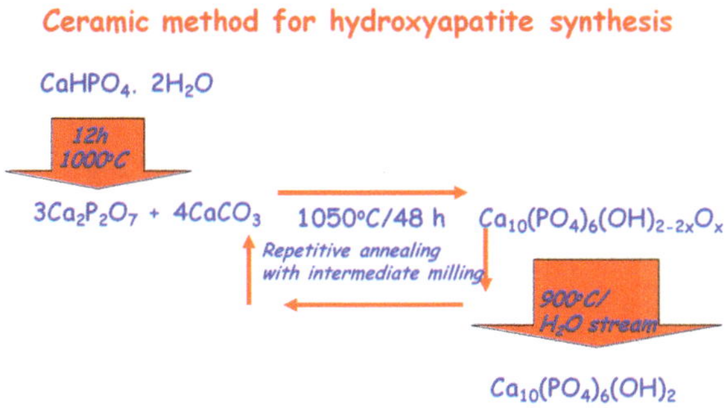

Ceramic method for hydroxyapatite synthesis

$CaHPO_4 \cdot 2H_2O$

12h
1000°C

$3Ca_2P_2O_7 + 4CaCO_3$ 1050°C/48 h $Ca_{10}(PO_4)_6(OH)_{2-2x}O_x$

Repetitive annealing
with intermediate milling

900°C/
H_2O stream

$Ca_{10}(PO_4)_6(OH)_2$

Figure 2.5 The ceramic method for hydroxyapatite synthesis.

Figure 2.5 depicts a possible route of solid-state reaction used in the laboratory synthesis of crystalline hydroxyapatite similar to mineral apatites available in nature. Using the ceramic method for this synthesis, the starting precursors exhibit a particle size of ~10 µm, which is the order of magnitude of the particle size of any chemical salt-type compound in commercial form. Ten micrometres may roughly correspond to a succession of 10 000 unit cells, which is the diffusion route to be covered by ions of each one of the reactants in order to reach the interface of the other reactant, so that the chemical reaction can take place. Therefore, the kinetic obstacles greatly restrict the ionic diffusion, rendering impossible in many cases the complete solid–solid chemical reaction. All cases require starting products in stoichiometric amounts, submitted to milling and homogenizing processes. The obtained starting mixture must then be submitted to generally very high temperatures and long treatment times. This procedure allows very crystalline apatites as opposed to biological apatites, where the particle size is in the range 25–50 nm, to be obtained (Figure 2.6). Therefore, if the aim is to obtain low-crystallinity apatites in the laboratory, with a particle size no greater than 50 nm, this is not the proper route; it is necessary to rely on wet-route methods, where the precursor salts are in solution, allowing a more homogeneous distribution of the components, almost at the atomic scale, where no kinetic impediments restrict the contact between reactants and give rise to a final product that is obtained faster and with less energy.

2.2.2 Synthesis of Apatites by Wet-Route Methods

Synthetic apatites aimed at emulating the biological scenario should exhibit small particle sizes and presence of CO_3^{2-}. In this sense, the wet route is the most adequate method of synthesis. There are several methods leading to nanometric-sized apatites.

The alternative method, to enable the reactivity of solids and alleviate the problems of their diffusion is in the wet route. Working with reactants in solution, diffusion is now a simple phenomenon that enables chemical reactions at much lower temperatures. Besides, this method based on solutions

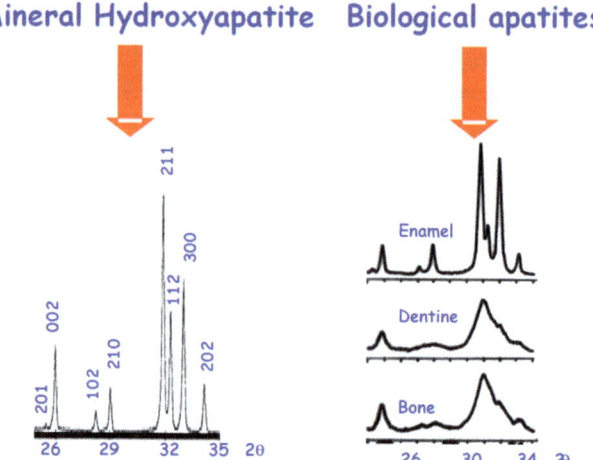

Figure 2.6 X-ray diffraction patterns evidence the different crystallinity of mineral apatites *vs.* biological ones.

not only simplifies the synthesis of the final product, it also achieves higher-quality products with a more homogeneous distribution of its components, higher reactivity, a decrease in reaction temperatures and heating periods, higher density of the final product and a smaller particle size. All this is related to the homogeneity of the starting materials and the use of lower synthesis temperatures.

Wet-route methods modify the first stages of the reaction and allow a more efficient and complete *solid–solid* reaction at the last stage, with an easier diffusion process. As a consequence, many properties of the obtained solids are significantly modified and improved with these procedures.

Several methods have been tested in an attempt to improve the homogeneity both in composition and particle size. These solution techniques can be classified into two large categories: *coprecipitation* and *sol–gel*. Both types allow solids to be obtained without previous milling and in a single calcination stage.

2.2.2.1 Sol–Gel

The chemistry of the sol–gel process is based on the hydrolysis and condensation of molecular precursors. Two different routes are described in the literature, depending on whether the precursor is formed by an aqueous solution of an inorganic salt or by a metal-organic compound. In any case, this method requires the careful study of parameters such as oxidation states, pH and concentration.

The sol–gel method is a process divided into several stages where different physical and chemical phenomena are performed, such as *hydrolysis, polymerization, drying* and *densification*. This process is known as *sol–gel* because differential viscosity increases at a given instant during the process sequence. A sudden increase in viscosity is a common feature in all *sol–gel* processes, which indicates the onset of *gel* formation. *Sol–gel* processes allow oxides to be synthesized oxides from inorganic or metal-organic precursors, the latter usually being metal alkoxides. A large part of the literature on the sol–gel process deals with synthesis from alkoxides.

The most important features of the sol–gel method are: better homogeneity, compared with the traditional ceramic method, high purity of obtained products, low processing temperatures, very uniform distribution in multicomponent systems, good control of size and morphology, the ability to obtain new crystalline or noncrystalline solids and, finally, easy production of thin films and coatings. As a consequence, the sol–gel method is widely used in ceramic technology.

The six main stages in sol–gel synthesis are depicted in Figure 2.7 and defined as follows:

- *Hydrolysis*: the hydrolysis process may start with a mixture of metal alkoxides and water in a solvent (usually alcohol) at ambient or slightly higher temperature. An acid or basic catalyst may be added to increase the reaction rate.
- *Polymerization*: neighboring molecules are condensed, water and alcohol is removed from them and the metal-oxide bonds are formed. The polymer network grows to colloidal dimensions in liquid state (*sol*).
- *Gelification*: the polymer network keeps growing until a three-dimensional network is formed through the ligand. The system becomes slightly stiff, which is a typical feature of a *gel* upon removal of the *sol* solvent. The solvent, water and alcohol molecules remain, however, inside the *gel* pores. The addition of smaller polymer units to the main network continues progressively with *gel* ageing.
- *Drying*: water and alcohol are removed at mild temperatures (<470 K), giving rise to hydroxilated metal oxides with a residual organic content. If the aim is to prepare an *aerogel* with high specific surface and low density, the solvent must be removed under supercritical conditions.
- *Dehydration*: this stage is performed between 670 and 1070 K to remove the organic residue and chemically bonded water. The result is a metal oxide in glass or microcrystalline form, with microporosity >20–30%.
- *Densification*: at temperatures >1270 K we can obtain dense materials, due to the reaction between the various components of the precursor in the previous stage.

These six stages may or may not be followed strictly in practice; this will depend on the solid to be synthesized. As Figure 2.7 illustrates, the choice will be different if the purpose is obtaining an aerosol, a glass or a crystalline material.

It is possible to prepare crystalline materials following modifications of the sol–gel route described, without the addition of metal alkoxides.

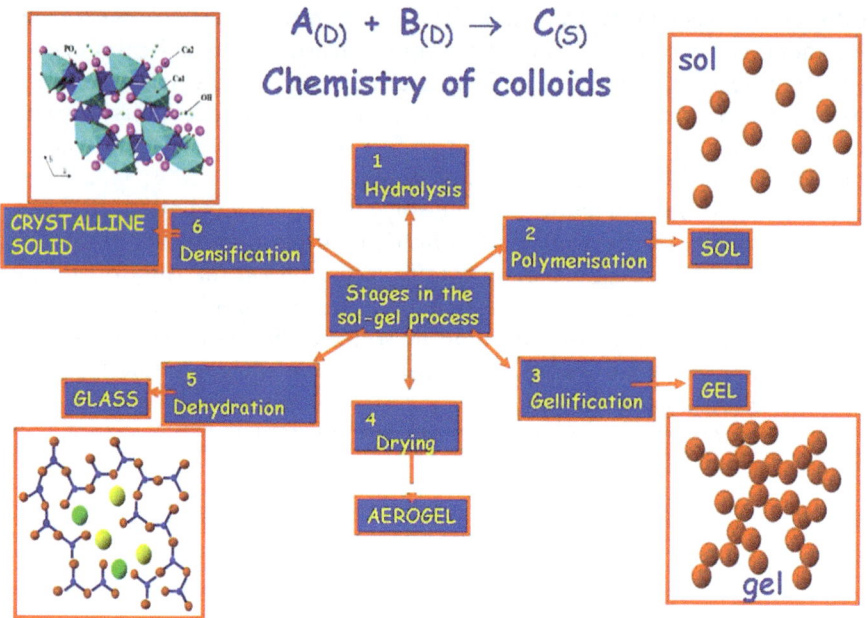

$$A_{(D)} + B_{(D)} \rightarrow C_{(S)}$$

Chemistry of colloids

Figure 2.7 The sol–gel process: stages in the production of sols, gels, aerogels, glasses or crystalline solids.

For instance, a solution of transition metal salts can be transformed to a gel adding an adequate organic reactant (*e.g.* 2-ethyl-1-hexanol). Alumina gels are prepared from ageing salts obtained by hydrolysis of aluminum butoxide followed by hydrolysis in hot water and peptization with nitric acid.

The sol–gel process for hydroxyapatite synthesis can produce a fine-grain microstructure containing a mixture of nano- to submicrometre crystals.[69] Low-temperature formation and fusion of the apatite crystals have been the main contributions of the sol–gel process in comparison with conventional methods for hydroxyapatite powder synthesis. A number of combinations of calcium and phosphorus precursors were employed for sol–gel hydroxyapatite synthesis. For instance Liu *et al.*[70] used a triethyl phosphate sol that was diluted in anhydrous ethanol and then a small amount of distilled water was added for hydrolysis. The molar ratio of water to the phosphorus precursor was kept at 3. The mixture was then sealed and stirred vigorously. After ~30 min of mixing the emulsion transformed into a clear solution, suggesting that the phosphite had been completely hydrolyzed. A stoichiometric amount of calcium nitrate was subsequently dissolved in anhydrous ethanol, and dropped into the hydrolyzed phosphorus sol. As a result of this process, a clear solution was obtained and aged at room temperature for 16 h before drying. Further drying of the viscous liquid at temperatures ~60 °C results in a white gel, which can be treated at temperatures in the range 600–1100 °C as a function of the particle size desired.

The major limitation of the sol–gel technique application is linked to the possible hydrolysis of phosphates and the high cost of the raw materials. Recently, Fathi and Hanifi[71] have developed a new sol–gel strategy to tackle this problem by using phosphoric pentoxide and calcium nitrate tetrahydrate. This sol–gel method provides a simple route for synthesis of hydroxyapatite nanopowder, where the crystalline degree and morphology of the obtained nanopowder are also dependent on the sintering temperature and time.

2.2.2.2 Solidification of Liquid Solutions

There are many variations and modifications to the sol–gel method. For instance, both simple and mixed oxides can also be synthesized by decomposition of metal salts of polybasic carboxylic acids, such as citrates. This procedure, however, can only be followed when the metal cations are soluble in organic solvents.

Solidification of this solution can be achieved by addition of a diol, for instance, which greatly increases the viscosity of the solution, due to the formation of three-dimensional ester type polymers. When the diol reacts with the citric solution, a resin is formed that avoids the partial segregation of the components, preserving the homogeneity of the solution which is now in solid form.

The organic matter is removed by calcination at temperatures >450 °C. A subsequent thermal treatment of the residue enables the solid–solid reaction in an easy and complete way, at temperatures lower than those needed for the ceramic method, as a consequence of the small particle size and the good homogeneity of all components in the matrix. This method was developed by Pechini and is known as liquid solutions solidification technique (LSST).[72] Figure 2.8 depicts the different stages of this method.

The application of the liquid mix technique is based on the Pechini patent.[72] This patent was originally developed for the preparation of multicomponent oxides, allowing the production of massive and reproducible quantities with a precise homogeneity in both composition and particle size. This method is based on the preparation of a liquid solution that retains its homogeneity in the solid state. It not only allows a precise control of the cation concentration, but also the diffusion process is enormously favored by means of the liquid solution, compared to other classical methods. Its application has now extended to the preparation of calcium phosphates. The main difficulty of this synthesis lies in the presence of PO_4^{3-} groups that cannot be complexed by citric acid, and may cause its segregation and the formation of separated phosphate phases. The success of this task would suggest the possibility of obtaining large amounts of single phases or biphasic mixtures with precise proportions of the calcium phosphates, by modifying the synthesis conditions. This method enables single-phase hydroxyapatite, β-TCP and α-TCP to be obtained; also biphasic materials whose content in β-TCP and hydroxyapatite can be precisely predicted from the Ca/P ratio in the precursor solutions[73] (Figure 2.9).

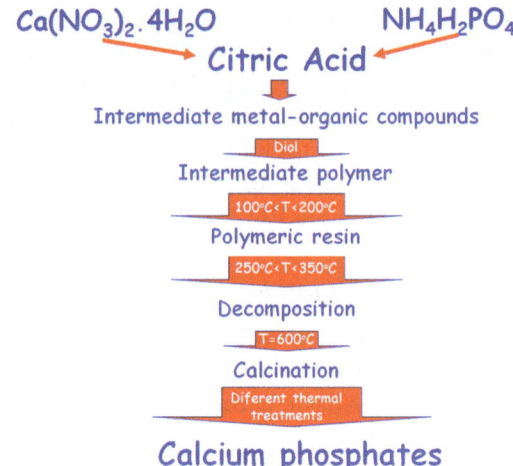

Figure 2.8   Liquid solutions solidification technique application applied to calcium phosphates preparation.

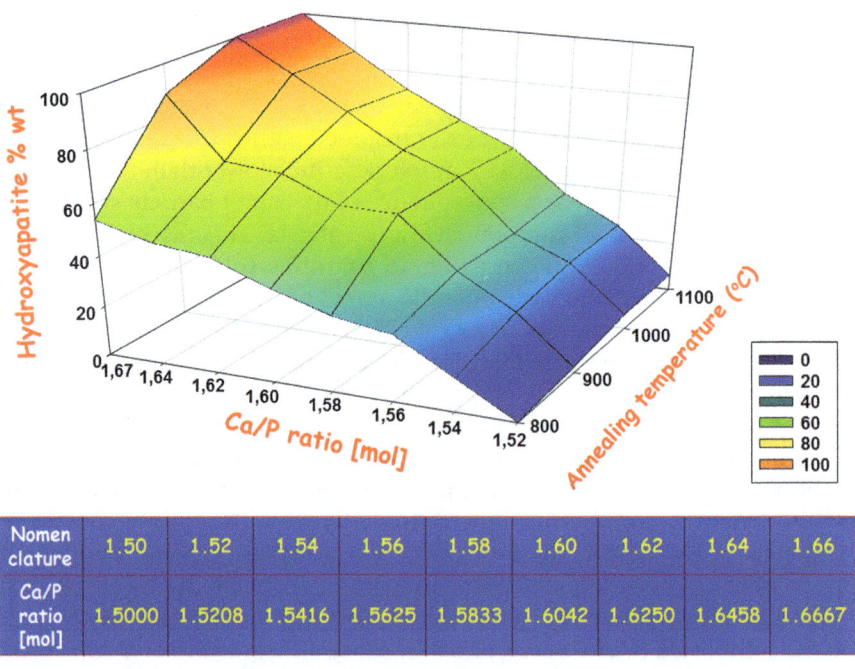

Nomen clature	1.50	1.52	1.54	1.56	1.58	1.60	1.62	1.64	1.66
Ca/P ratio [mol]	1.5000	1.5208	1.5416	1.5625	1.5833	1.6042	1.6250	1.6458	1.6667

Figure 2.9   Percentage of hydroxyapatite as a function of Ca/P ratio and annealing
temperature when the liquid mix technique is applied.

2.2.2.3   Controlled Crystallization Method

Methods based on precipitation from aqueous solutions are most suitable for
the preparation of large amounts of apatite, as needed for processing both
into ceramic bodies and in association with different matrices. The difficulty

with most of the conventional precipitation methods used is the synthesis of well-defined and reproducible orthophosphates.[8,9]

Problems can arise due to the usual lack of precise control on the factors governing the precipitation, pH, temperature, Ca/P ratio of reagents, *etc.*, which can lead to products with slight differences in stoichiometry, crystallinity, morphology, *etc.*, that could then contribute to the different "*in vivo/in vitro*" behaviors described. In this sense, it is important to develop a methodology that can produce massive and reproducible quantities of apatite, optimized for any specific application or processing requirements by controlling composition, impurities, morphology and crystal and particle size. For quantitative reactions in solutions, the reactants must be calcium and phosphate salts with ions that are unlikely to be incorporated into the apatite lattice. Since it has been claimed that NO_3^- and NH_4^+ are not incorporated into crystalline apatites, or in the case of NH_4^+ present a very limited incorporation,[74] the chosen reaction for this method was $10Ca(NO_3)_2 \cdot 4H_2O + 6(NH_4)_2HPO_4 + 8NH_4OH \rightarrow Ca_{10}(PO_4)_6(OH)_2 + 20NH_4NO_3 + 6H_2O$.

Thus, apatites with different stoichiometry and morphology can be prepared and the effects of varying synthesis conditions on stoichiometry, crystallinity, and morphology of the powder can be analyzed. The effects of varying concentration of the reagents, the temperature of the reaction, reaction time, initial pH, ageing time and the atmosphere within the reaction vessel can also be control with equipment like that represented in Figure 2.10. Temperatures in the range 25–37 °C are necessary to obtain apatites with crystal sizes in the range of adult human bone, while 90 °C is necessary to obtain apatites with crystal sizes in the range of enamel. Higher reaction times lead to apatites with higher Ca/P ratios. Ageing of the precipitated powder can lead to the incorporation of minor quantities of carbonate. It is possible to force the incorporation of carbonate ions into the apatite structure without introducing monovalent cations.[5] In short, the main results of the studied variations in the reaction conditions are that higher concentrations of reagents produce higher amounts of products with minor differences in their characteristics, allowing the production of homogeneous sets of materials.

2.2.3   Synthesis of Apatites by Aerosol Processes

Aerosol-based processes can be considered as a type of solid synthesis that involves the transformation from *gas* to *particle* or from *droplet* to *particle*. When the aerosol reaches the reaction area, different decomposition phenomena may take place, depending on the precursor features and the temperature, as shown in Figure 2.11.

At low deposition temperatures, the droplets reach the substrate in liquid form. The solvent evaporates leaving a finely divided precipitate onto the substrate (layout A). At higher temperatures, the solvent may evaporate before coming into contact with the substrate and the precipitate impacts the substrate (layout B).

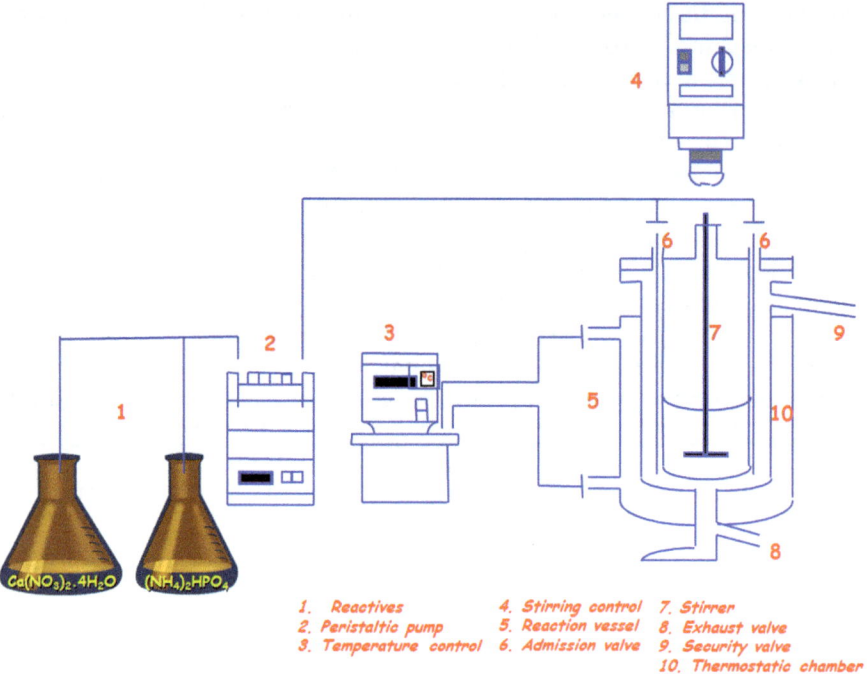

Figure 2.10   Scheme of the equipment used for the synthesis of apatites by the controlled crystallization method.

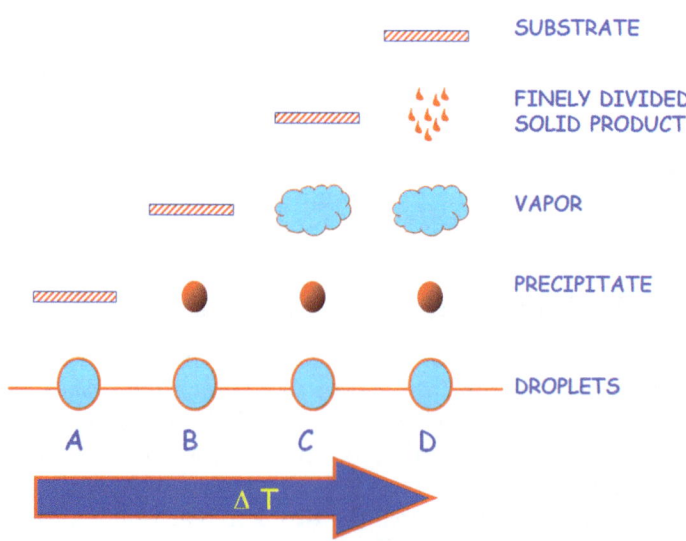

Figure 2.11   Different decomposition phenomena taking place depending on the temperature in an aerosol-assisted process.

When the deposition temperature is high enough and the precursor is volatile, there is a consecutive solvent evaporation and solute sublimation. This solute, in vapor form, diffuses to the substrate where a heterogeneous chemical reaction with its surface is performed, in the solid state. This is the so-called chemical deposition technique in vapor phase, or chemical vapor deposition (layout C).

At high temperatures, the reaction is verified before the vapors reach the substrate; hence it is a homogeneous reaction. The product of this reaction is deposited onto the substrate as a finely divided powder (layout D).

In a gas–particle transformation, gases or vapors react, forming primary particles which then start to grow by coagulation or by surface reactions. The powdered solids obtained with this process exhibit a narrow range distribution in sizes, and the method may yield spherical nonporous particles.

In the droplet-to-particle transformation processes, droplets containing the solute are suspended in a gaseous medium through a liquid atomization where the droplets react with gases or are pyrolyzed at high temperatures to form powder solids. The particle size distribution is determined by the droplet size or by the processing conditions. The most frequently used industrial methods to obtain powder solids from droplet to solid transformations are *drying* from an aerosol or *pyrolysis* from an aerosol. *Droplet freeze drying* is another technique in which powder solids are obtained by particle formation from droplets (Figure 2.12).

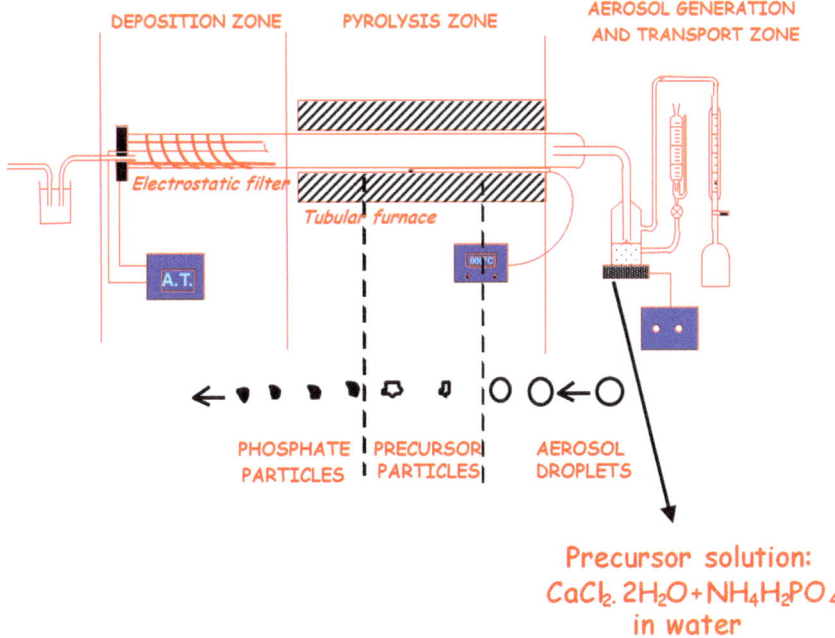

Figure 2.12 Scheme of the equipment used for the synthesis of powder solids from aerosol droplets.

There are many different methods available to prepare ceramic thin films. Each method exhibits unique features that play a crucial role in the properties of the obtained particles. Therefore, it is important to choose wisely the deposition technique and the working conditions, such as temperature, pressure, atmosphere or starting reactants.

The aerosol synthesis technique has been used to produce small particles of different materials.[75,76] Its main advantage is that this technique has the potential to create particles of unique composition, for which starting materials are mixed in a solution at atomic level (Figure 2.12). An ulterior thermal treatment can originate important modifications to morphology and texture. In consequence, hydroxyapatite preparation by this method was deemed of interest.[77] Hydroxyapatite hollow particles have been prepared by pyrolysis of an aerosol produced by ultrahigh frequency of a $CaCl_2$-$(NH_4)H_2PO_4$ solution. Hollow particles were annealed at different temperatures. Thermal treatment at 1050 °C produces the growth of nucleated crystallites in the particle surface, with remarkable morphology. The particle size range is 0.3–2.2 μm. Apatite nanocrystals grow onto this surface.

2.2.4 Other Methods Based on Precipitation from Aqueous Solutions

2.2.4.1 Calcium Phosphate Cements as Apatite Precursors

Cements based on calcium salts, phosphates or sulphates have attracted much attention in medicine and dentistry due to their excellent biocompatibility and bone-repair properties.[78–81] Moreover, they have the advantage over bioceramics in that they do not need to be delivered in prefabricated forms, because these self-setting cements can be handled by the clinician in paste form and injected into bone cavities. Depending on the formulation of the cement or the presence of additives, different properties such as setting time, porosity or mechanical behavior have been found in these materials.[82–86]

In the literature on phosphates focused on calcium phosphate cements, the technique employed for obtaining such cements is to mix the different components; one of them is responsible for curing the mixture. For instance, in the Constanz cement[87] – the first of its kind to be commercialized – the final product is a carbonate-apatite (dahllite) with low crystallinity and a carbonate content reaching 4.6%, in substitution of phosphate groups (B type carbonate-apatite) as is the case in bones. Constanz cement is obtained from a dry mixture of α-TCP, α-$Ca_3(PO_4)_2$, calcium phosphate monohydrate, $Ca(H_2PO_4)\cdot H_2O$ and calcium carbonate, $CaCO_3$. The Ca/P ratio of the first component is 1.50, and 0.5 for the second component: both values significantly lower than the Ca/P ratio of 1.67 for hydroxyapatite. A liquid component – a sodium monoacid phosphate solution – is then added to this solid mixture, which allows the formation of an easily injectable paste that will cure over time. The paste curing happens after a very reasonable period of time when considering its use in surgery. In fact, after 5 min it shows a consistency suitable for injection, and at 10 min it is solid without any exothermal response, exhibiting an initial strength of 10 MPa. 12 h later, 90% of its weight has evolved to dahllite, with compression strength of 55 MPa, and 2.1 MPa when under stress. This cement is then resorbed and gradually replaced by newly formed bone.

Calcium phosphate cements which can be resorbed and injected are being commercialized by various international corporations,[88] with slight differences in their compositions and/or preparation. Research is still under way in order to improve the deficiencies still present.

These cements are very compatible with the bone and seem to resorb slowly; during this gradual process, the newly formed bone grows and replaces the cement. However, the properties of the calcium phosphate cements are still insufficient for their reliable application. There are problems related to their mechanical toughness, the curing time, the application technique on the osseous defect and the final biological properties. New improvements in the development of these cements will soon be described, solving at least in part some of these disadvantages. For instance, the curing time will be shortened, even in contact with blood, and the toughness under compression will also improve.

Most of the injectable calcium phosphate cements used evolve to an apatitic calcium phosphate during the setting reaction. One of the main drawbacks of these apatitic cements is the slow resorption rate of the apatite. Conversely, calcium sulphate dihydrate, gypsum, has been used as bone-void filler for many years,[78,87–89] although its resorption rate is too fast to provide good support for new bone. The combination of both calcium sulphate and apatite can overcome the individual drawbacks, and studies using this biphasic material have been performed.[90–92]

Despite the advantages, all these implants can act as foreign bodies and become potential sources of infections. Then, the *in vivo* use of these materials requires a preventive therapy and this may be achieved by introducing a drug into them, which can be locally released *in situ* after implantation. In fact, different studies using bioceramics and self-setting materials containing active drugs have been performed.[93–97]

In this sense, the addition of an antibiotic to calcium sulphate-based cements has also been studied, in order to determine if the presence of the drug affects the physical–chemical behavior of the cements and to study the release kinetics of the drug from the cement. Two system types were chosen: gypsum and apatite/gypsum. The antibiotic chosen for this study was cefalexin in crystalline form, *i.e.* cefalexin monohydrate.

The presence of cefalexin into the cements does not alter the physical–chemical behavior of the cements nor does it introduce structural changes. The release of the drug is different depending on the composition of the cement. For gypsum cements, the cefalexin is quickly released, helped by a dissolution process of the matrix, whereas the drug release is more controlled by the hydroxyapatite presence in hydroxyapatite/gypsum samples.

Apatite-containing cements not only show a different drug-release process, the paste viscosity is lower and a faster formation *in vitro* of an apatite-type layer on their surface is observed.[98]

2.2.4.2   *Biphasic Mixtures of Calcium Phosphates as Apatite Precursors*

Several attempts have been made to synthesize the mineral component of bones starting from biphasic mixtures of calcium phosphates.[99] Hence bone-replacing materials based on mixtures of hydroxyapatite and β-TCP have been prepared; under physiological conditions, such mixtures evolve to carbonate hydroxyapatite. The chemical reactions are based in equilibrium conditions between the more stable phase, hydroxyapatite, and the phase prone to resorption, β-TCP. As a consequence, the mixture is gradually dissolved in the human body, acting as a stem for newly formed bone and releasing Ca^{2+} and PO_4^{3-} to the local environment. This material can be injected, used as a coating or any other form suitable for application as bulk bone replacement – forming of bulk pieces or filling of bone defects.[100] At present, a wide range of biphasic mixtures are under preparation, using various calcium phosphates, bioactive glasses, calcium sulphates, *etc.*[92,101,102]

Currently, there is an increasing interest on the preparation of mixtures of two or more calcium phosphates. These materials are commonly prepared with hydroxyapatite and a more resorbable material, such as TCP (α or β) or calcium carbonate in different proportions depending on the characteristics required for a specific application. Some examples of commercial products based on these mixtures include TRIOSITE™, MBCP™ and EUROCER®. The synthesis routes commonly employed in the preparation of these mixtures include the blending of different calcium phosphates[103,104] and precipitation.[105-107] Other techniques also employed are solid state,[108] treatment of natural bone,[109] spray pyrolysis,[110] microwave[111] and combustion.[112] Some authors have defended the superior properties of the biphasic materials "directly" prepared over those obtained by mixing two single phases.[113]

The promising results obtained with *cements* and *biphasic mixtures* seem to indicate that it is easier to obtain precursors of synthetic apatites that, when in contact with the biological environment, can evolve towards similar compositions to that of the biological apatite, than to obtain apatites in the laboratory with similar compositional and structural characteristics to those of the biological material, and in adequate quantities, *i.e.* large, industry-scale amounts with precise composition and easily repeatable batch after batch, for its use in the production of ceramic biomaterials.

Bioceramics aimed at the replacement or filling of bones could be obtained by synthesis of apatite precursors through different calcium phosphate mixtures, using a wet route. If the information gathered from the calcium cements is put to use, it would be necessary to eliminate the solution added to cure the mixture and search for compositions and ratios that allow to obtain precursors that, when in contact with the body fluids, evolve chemically towards the formation of carbonate hydroxyapatite crystals, with small particle size and low crystallinity, calcium-deficient and with a carbonate content of ~4.5% w/w, located in the PO_4^{3-} sublattice.

2.2.5 Apatites in the Absence of Gravity

Particular attention must be paid at this point to the assays performed in the absence of gravity. Suvurova and Buffat[114] have compared the results obtained when calcium phosphate specimens, in particular hydroxyapatite and triclinic octacalcium phosphate (OCP) are prepared from aqueous solutions under different conditions of precipitation. When supersaturated solutions of calcium phosphates are prepared by diffusion-controlled mixing in outer space (EURECA 1992–1993 flight) several differences are observed in crystal size, morphology and structural features with respect to those prepared on earth. It is worth stressing that space-grown OCP crystals possess a maximum growth rate in the [001] direction and a minimum rate in the [100][115] one. Space-grown and terrestrial hydroxyapatite crystals differ from each other in size: the former are ≥1–1.5 orders of magnitude bigger in length. Diffusion-controlled mixing in space seems to provide a lower supersaturation in the crystallization system compared to that on earth, promoting crystal growth in the competition between nucleation and growth. These authors conclude that similar processes may most probably arise in the human body (under definite internal conditions) during space flying when quite large hydroxyapatite crystals start to grow instead of the small natural ones. In addition, other modifications of OCP crystals with huge sizes appear. These elements may disturb the Ca dynamic equilibrium in the body, which might lead to demineralization of the bone tissue.

2.2.6 Carbonate Apatites

Biological apatites (mineral component of the bones) are difficult to synthesize in the laboratory with carbonate content equivalent to that in bone. Although the carbonate inclusion in itself is very simple (in fact, when producing stoichiometric apatites in the laboratory, a strict control of synthesis conditions is needed to avoid carbonate inclusion), the carbonate content is always different from the fraction of carbonates in the natural bone (4–8% wt)[6] and/or are located in different lattice positions.[116] At this point, it should be mentioned that this carbonate content can be slightly different when samples from other vertebrates are analyzed.[117] The carbonate easily enters into the apatite structure, but the problem lies in the amount that should be introduced, taking into account the carbonate content of biological apatites. When the aim is to obtain carbonate hydroxyapatite and the reaction takes place at high temperatures, the carbonates enter and occupy lattice positions in the OH⁻ sublattice (A-type apatites). In contrast, the carbonates in biological apatites always occupy positions in the PO_4^{3-} sublattice (that is, they are B-type apatites).[118] In order to solve this problem, low-temperature synthesis routes have to be followed, allowing obtaining carbonate hydroxyapatites with carbonates in phosphate positions.[6] But the amount entered remains to be solved, and it is usually lower than the carbonate content of the mineral component of the bones.

These calcium-deficient and carbonated apatites have been obtained in the laboratory by various techniques; nowadays, it is known that apatites with low crystallinity, calcium deficiency and carbonate content can be obtained, but with carbonate contents usually unequal to those of the natural bones.[5,119,120] Therefore, the main problem remains in the control of carbonate content and lattice positioning.

2.2.7 Silica as a Component in Apatite Precursor Ceramic Materials

One way to enhance the bioactive behavior of hydroxyapatite is to obtain substituted apatites, which resemble the chemical composition and structure of the mineral phase in bones.[121,122] These ionic substitutions can modify the surface structure and electrical charge of hydroxyapatite, with potential influences on the material in biological environments. In this sense, an interesting way to improve the bioactivity of hydroxyapatite is the addition of silicon to the apatite structure, taking into account the influence of this element on the bioactivity of bioactive glasses and glass-ceramics.[123,124] In addition, several studies have proposed the remarkable importance of silicon on bone formation and growth[125,126] at *in vitro* and *in vivo* conditions.

Several methods for the synthesis of silicon-substituted hydroxyapatites have been described. Ruys[127] suggested the use of a sol–gel procedure; however, these materials, besides the hydroxyapatite phase, include other crystalline phases depending on the substitution degree of silicon. Tanizawa and Suzuki[128] tried hydrothermal methods, obtaining materials with a Ca/(P + Si) ratio higher than that of pure calcium hydroxyapatite. Boyer *et al.*[129] conducted studies on the synthesis of silicon-substituted hydroxyapatites by solid-state reaction, but in these cases the incorporation of a secondary ion, such as La^{3+} or SO_4^{2-} was needed. In these examples, no bioactivity studies were performed on the silicon-containing apatites.

Gibson *et al.*[130] synthesized silicon-containing hydroxyapatite by using a wet method, and its *in vitro* bioactivity studies gave good results. These authors studied the effects of low substitution levels on the biocompatibility and *in vitro* bioactivity, determining the ability to form the apatite-like layer by soaking the materials in a simulated body fluid (SBF).[131] Marques *et al.*[132] synthesized, by wet method, hydroxyapatite with silicon content up to 0.15 wt%, obtaining stable materials at 1300 °C and noting that the unit cell volume and the *a* parameter length of the hydroxyapatite decreased as the silicon content increased.

Hence, the role of silicon substituting part of the phosphorus atoms present in the hydroxyapatite lattice seems to be an important factor influencing the bioactive behavior of the material. However, it is not clearly known whether the silicon present in the material substitutes completely the phosphorus in the hydroxyapatite structure, or whether the replacement is partial, or even if in any of the described syntheses the silicon species remain as an independent phase. In all the cited syntheses, the final product contains silicon, but its chemical nature is not revealed.

A similar work focused on the synthesis and bioactivity study of hydroxyapatites containing orthosilicate anions that isomorphically replace phosphate groups, aimed at improving the bioactivity of the resulting materials as compared with that of pure calcium hydroxyapatite.[133] To accomplish this purpose, two synthesis procedures were used, starting from two different calcium and phosphorus precursors and the same silicon reagent in both cases. To assess the proposed substitution, surface chemical and structural characterization of the silicon-substituted hydroxyapatites was performed by means of X-ray diffraction (XRD) and X-ray photoelectron spectroscopy (XPS). The *in vitro* bioactivity of the thus-obtained materials was determined by soaking the materials in SBF and monitoring the changes of pH and chemical composition of the solution, whereas the modification at the surface was followed by means of XPS, XRD and scanning electron microscopy (SEM). Silicon-containing hydroxyapatites were synthesized by the controlled crystallization method. Chemical analysis, N_2 adsorption, Hg porosimetry, XRD, SEM, energy-dispersive X-ray spectroscopy and XPS were used to characterize the hydroxyapatite and to monitor the development of a calcium phosphate layer onto the substrate surface immersed in a simulated body fluid, that is, *in vitro* bioactivity tests. The influence of the silicon content and the nature of the starting calcium and phosphorus sources on the *in vitro* bioactivity of the resulting materials were studied. A sample of silicocarnotite, whose structure is related to that of hydroxyapatite and contains isolated SiO_4^{4-} anions that isomorphically substitute some PO_4^{3-} anions, was prepared and used as reference material for XPS studies. An increase of the unit cell parameters with the Si content was observed, which indicated that SiO_4^{4-} units are present in lattice positions, replacing some PO_4^{3-} groups. By using XPS it was possible to assess the presence of monomeric SiO_4^{4-} units in the surface of apatite samples containing 0.8 wt% of silicon, regardless of the nature of the starting raw materials, either $Ca(NO_3)_2/(NH_4)_2HPO_4/Si-(OCOCH_3)_4$ or $Ca(OH)_2/H_3PO_4/Si(OCOCH_3)_4$. However, an increase of the silicon content up to 1.6 wt% leads to the polymerization of the silicate species at the surface. This technique shows silicon enrichment at the surface of the three samples. The *in vitro* bioactivity assays showed that the formation of an apatite-like layer onto the surface of silicon containing substrates is strongly enhanced compared to pure silicon-free hydroxyapatite (Figure 2.13). The samples containing monomeric silicate species showed higher *in vitro* bioactivity than that of silicon-rich sample containing polymeric silicate species. The use of calcium and phosphate salts as precursors lead to materials with higher bioactivity.[133]

Finally, the results revealed that controlled crystallization is a good procedure to prepare silicon-substituted hydroxyapatites that can be used as a potential material for prosthetic applications.

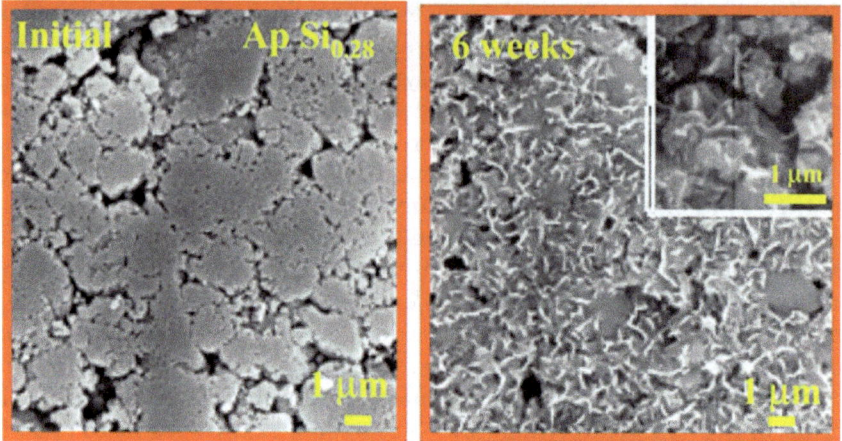

Figure 2.13 Scanning electron micrographs of silicon-substituted apatites before and after soaking for 6 weeks in simulated body fluid.

The presence of silicon (Si) in hydroxyapatite has shown an important role on the formation of bone.[122] To study the role of Si, Si-substituted hydroxyapatite (SiHA) has been synthesized by several methods,[127–130,132,133] but its structural characteristics and microstructure are still not fully understood. Most of the structural studies performed (mainly by XRD) had not demonstrated the Si incorporation into the apatite structure. In fact, the very similar scattering factor makes very difficult to determine if Si has replaced some P in the same crystallographic position. The absence of secondary phases and the different bioactive behavior were the best evidences for the Si incorporation. No positive evidence or quantitative study of P substitution by Si has yet been performed. Meanwhile, the hydroxyl groups sited at the *4e* position are one of the most important sites for the hydroxyapatite reactivity. The movement of H along the *c* axis contributes to the hydroxyapatite reactivity. However, XRD is not the optimum tool for the study of light atoms such as H; neutron diffraction is an excellent alternative to solve this problem. The Fermi length for Si and P are different enough to be discriminated, whereas neutrons are very sensitive to the H presence. Consequently XRD and neutron diffraction have combined to answer one of the most recent subjects in the dentistry and orthopedic surgery fields.

In order to explain the higher bioactivity of SiHA, synthetic ceramic hydroxyapatite and SiHA have been structurally studied by neutron scattering. The Rietveld refinements show that the final compounds are oxyhydroxyapatites, when obtained by solid-state synthesis under air atmosphere. By using neutron diffraction, the substitution of P by Si into the apatite structure has been corroborated in these compounds. Moreover, these studies also allow us to explain the superior bioactive behavior of SiHA, in terms of higher thermal displacement parameters of the H located at the *4e* site.[134]

Structure refinements by the Rietveld method indicate that the thermal treatment produces partial decomposition of the OH groups, leading to

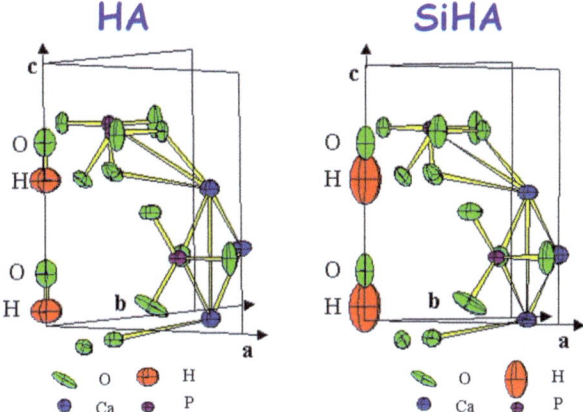

Figure 2.14 Thermal ellipsoids of hydroxyapatite (left) and silicon-substituted hydroxyapatite (Si-HA; right). The thermal ellipsoids for H atoms are more than twofold larger for Si-HA than for hydroxyapatite.

oxy-hydroxyapatites in both samples and the higher reactivity of the SiHA can be explained in terms of an increase of the thermal ellipsoid dimension parallel to the *c* axis for H atoms (Figure 2.14).

2.2.8 Apatite Coatings

The application of synthesis methods to obtain apatite coatings is a subject of great interest in the field of biomaterials, in relation to the fabrication of load-bearing implants made of metal alloys.[135] When the said metal implants are coated with a ceramic material such as apatite, the performance of the implant improves due to the barrier effect of the apatite coating against metal ion diffusion from the implant towards the body, while enabling a better adhesion to the bone tissue. There are plenty of methods in use nowadays to fabricate this type of coating. Some of the techniques used will be briefly described below, although the production of biomimetic coatings is described with in detail in Chapter 4.

2.2.8.1 *Production of Thin Films by Vapor-Phase Methods*

Some of the vapor-phase methods in use are chemical vapor deposition, chemical transportation, substrate reaction, pulverization pyrolysis, vacuum evaporation, sputtering, ion plating techniques and plasma pulverization methods.

2.2.8.2 *Production of Thin Films by Liquid-Phase Methods*

Thin films can also be obtained from liquid precursors, using procedures such as sol–gel, where a gel is prepared with metal alkoxides or from organic or inorganic salts. The films are formed onto substrates by drying and heating

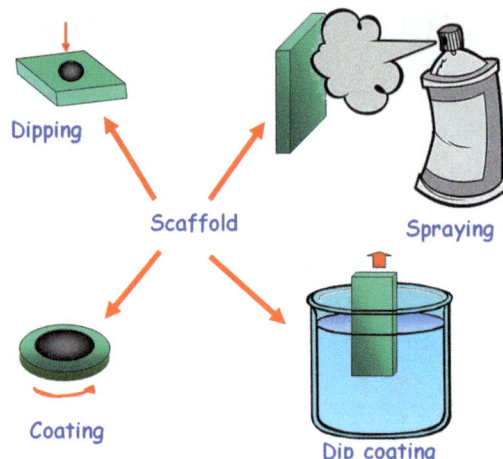

Figure 2.15 Coating techniques.

a *sol* previously used to coat the substrates. Different coating techniques are available (Figure 2.15). This is a very popular coating process due to its ability to grow films onto substrates of very different shapes and sizes.

Besides sol–gel, other procedures in this category include *liquid-phase epitaxy* and *melt epitaxy*.

2.2.8.3   Production of Thin Films by Solid-Phase Methods

Solids can also be used as precursors for thin films. For instance, the method of *thermal decomposition of a coating*, where the thin film is obtained by high-temperature decomposition of a metal organic compound dissolved in an organic solvent which covers the substrate, or the *precipitation* method, where aqueous solutions of different salts react and the less soluble compounds precipitate. Ceramic powder materials can be obtained after washing, drying and calcining the precipitates. Using different salts and controlling parameters such as temperature, it is possible to control the particle size of the obtained solids, which can be synthesized at high temperatures to obtain polycrystalline films.

This method has been successfully applied to improve the biological osteoblastic response,[136] and silicon-substituted apatite coatings have been synthesized.[137,138] Basically, this last process is carried out by coating Ti substrates or, in the case of silicon-substituted calcium phosphates, on quartz substrates. The procedure can be briefly described as follows. Ammonium phosphate solution is titrated into an aqueous solution of calcium nitrate adding ammonium hydroxide to keep the pH at 10.5, according to the following reaction:

$$10Ca(NO_3)_2 + 6NH_4H_2PO_4 + 14NH_4OH \rightarrow$$
$$Ca_{10}(PO_4)_6(OH)_2 + 20NH_4NO_3 + 12H_2O$$

The resulting calcium phosphate solution is then aged at room temperature for 1 day and then concentrated. The corresponding substrate (Ti alloy, quartz, *etc.*) is dipped into the solution. After drying, the samples can be sintered at temperatures of 700–1100 °C, depending on the final microstructure.

2.2.9 Precursors to Obtain Apatites

The different synthesis routes applied to obtain apatites require the use of precursors with certain features. Several potential precursors will be described and classified below:

2.2.9.1 Inorganic Salts

Inorganic salts are used as molecular precursors in wet-route chemical processes, such as *sol–gel*, *colloidal* or *hydrothermal* synthesis. They are ionic compounds, and some examples of these precursor salts are listed in Table 2.1.

2.2.9.2 Coordination Compounds with Organic Ligands

Coordination compounds with organic ligands are covalent or ionic coordination compounds in which the metal site is bonded to the ligand by an oxygen, sulphur, phosphorus or nitrogen atom. These compounds are used as precursors both in wet-route processes and in vapor-phase reactions. Table 2.2 shows some examples of coordination compounds with organic ligands.

2.2.9.3 Organometallic Compounds

Organometallic compounds are covalent or coordination compounds in which the ligand is bonded to the metal site by a carbon atom. As in the case of coordination compounds with organic ligands, organometallic compounds are used as precursors in wet-route processes and in vapor-phase reactions. The most commonly used organometallic compounds are listed in Table 2.3.

Table 2.1 Examples of precursor salts.

Inorganic salts	Examples
Metal halides	$MgCl_2$, LiF, KCl, $SiCl_4$, $TiCl_4$, $CuCl_2$, KBr, $ZrOCl_2$
Metal carbonates	$MgCO_3$, $CaCO_3$, Na_2CO_3, $SrCO_3$
Metal sulphates	$MgSO_4$, $BaSO_4$, K_2SO_4, $PbSO_4$
Metal nitrates	$LiNO_3$, KNO_3, $Fe(NO_3)_2$
Metal hydroxides	$Ca(OH)_2$, $Mg(OH)_2$, $Al(OH)_3$, $Fe(OH)_3$, $Zr(OH)_4$
Salts with mixed ligands	$(CH_3)SnNO_3$, $(C_2H_5)_3SiCl$, $(CH_3)_2Si(OH)_2$

Table 2.2   Examples of coordination compound precursors.

Coordination compound	General formula	Selected examples
Metal alkoxides	$-M(-OR)$, R Is an alkyl	$Al(OC_3H_7)_3$, $Si(OCH_3)_4$, $Ti(OCH_3H_7)_4$, $Zr(OC_4H_9)_4$
Metal carboxylates	$-M(-OC(O)R)_x$, R is an alkyl	$Al(OC(O)CH_3)_3$, $Pb(OC(O)CH_3)_2$ acetates, $Pb(OC(O)CH_2CH_3)_4$ propionate, $Al(OC(O)C_6H_5)_3$ benzoate
Metal ketones	$-M(-OCRCH(R')CO)_x$, R is an alkyl or aryl	$Ca(OC(CH_3)CH(CH_3)CO)_2$ pentanedionate, $Al(OC(C(CH_3)_3)CH(C(CH_3)_3)CO)_2$ heptanedionate
Metal amines		$(CH_3)_2AlNH_2$, $(C_2H_5)_2AlN(CH_3)_2$, $(CH_3)BeN(CH_3)_2$, $(iC_3H_7)_3GeNH_2$, $(C_3H_7)_3PbN(C_2H_5)_2$
Metal thiolates	$-M(-SR)_x$, R is an alkyl or aryl	$(CH_3)_2Ge(SC_2H_5)$, $Hg(C_4H_3S)_2$, $(SCH_3)Ti(C_5H_5)_2$, $(CH_3)Zn(SC_6H_5)$
Metal azides	$-MN_3$	$(CH_3)_3SnN_3$, CH_3HgN_3
Metal isothiocyanates	$-M(-NCS)_x$	$(C_2H_5)_3Sn(NCS)$
Coordination compounds with mixed functional groups		$(C_4H_9)Sn(OC(O)CH_3)_3$, $(C_5H_5)_2TiCl_2$, $(C_5H_5)Ti(OC(O)CH_3)_3$

Table 2.3   Examples of organometallic precursors.

Organometallic compounds	Selected examples
Metal alkyls	$As(CH_3)_3$, $Ca(CH_3)_2$, $Sn(CH_3)_4$ methyl
Metal aryls	$Ca(C_6H_5)_2$ phenyl
Metal alkenyls	$Al(CH=CH_2)_3$ vinyl, $Ca(CH=CHCH_3)_2$ propenyl
Metal alkynyls	$Al(CCH)_3$, $Ca(CCH)_2$ acetylene
Metal carbonyls	$Co_2(CO)_8$, $Mn_2(CO)_{12}$, $W(CO)_6$ carbonyl
Mixed organic ligands	$Ca(CCC_6H_5)_2$ phenylacetylene, $(C_5H_5)_3U(CCH)$ cyclopentadienyl/ethynyl

2.2.9.4   *Polymer Precursors*

In some processes such as *metal organic decomposition* and *sol–gel* processes, the polymer precursors can be used as starting material to obtain glasses or ceramics. These polymers are often referred to as preceramic polymers. Table 2.4 shows some examples.

2.2.10   Additional Synthesis Methods

Different synthesis methods play a substantial role in the design of apatites that resemble their biological counterparts. Procedures related to so-called *soft chemistry* are increasingly used and studied, since these methods allow the synthesis of new products, many of them metastable and hence impossible to prepare by conventional routes such as the

Table 2.4 Examples of polymer precursors.

Polymer	Formula	
Polycarbosilanes	$-[(RR')Si-CH_2-]_x$	SiC precursor in metal organic decomposition and sol–gel processes, where R is an active functional group such as a olefin, acetylene or H
Polysilazanes	$-[(RR')Si-NR-]_x$, R is an organic radical or H	Si_3N_4 or silicon carbonitride precursor, in a similar fashion to polycarbosilane
Polysiloxanes	1. $-[Si(RR')O-]_x$, linear R is an alkyl or aryl 2. Silsesquioxanes: ladder structure 3. $-[Si(CH_3)_2OSi(CH_3)_2 (C_6H_5)-]_m$ 4. Random- or block-copolymers	Used in sol–gel processes and *in situ* multiphase systems, as SiO_2 or silicon oxocarbide precursors
Polysilanes	$-[Si(RR')-]_n$, R is an alkyl or aryl	SiC precursors, for photoresistant and photoinitiator materials
Borazines	$-[BRNR'-]_n$, cyclic units or in repeated chains	BN precursors in chemical vapor deposition/metalorganic chemical vapor deposition and sol–gel processes
Carboranes	B and C cage structures	B_4C precursors in metalorganic chemical vapor deposition and metal organic decomposition processes
Polyphosphazenes	$-[N=P(R_2)-]_n$, R is an organic, organic metal or inorganic unit	Common substituents are: alkoxides, aryloxi, arylamides, carboxylates or halides
Polytinoxanes	$-[Sn(R)_2-O-R'-O-Sn(R)_2-O]_n$ chains, R is an organic group. Stair and drum structures are also possible	
Polygermanes	$-[Ge(RR')-]_n$	Can be used in microlithographic applications as polysilanes

ceramic method. This "soft chemistry" applies to simple reactions that take place at relatively low temperatures, such as *intercalation*, *ionic exchange*, *hydrolysis*, *dehydration* and *reduction*. The advantage of using soft methods is that it is possible to better control the structure, stoichiometry and phase purity.

It is worth recalling that, opposed to this clear trend of avoidance of "extreme conditions" of synthesis, there is a method that actually opts for them. This is the *mechanochemical* method, where an intense and prolonged milling process is applied to generate locally high pressures and temperatures that lead to chemical transformations in the starting products, often obtaining metastable phases related to those special conditions of temperature and pressure.[139]

2.2.11   Sintered Apatites

In *sintering* processes, the particles are agglomerated. Sintering could be defined as a process where a compact solid changes its morphology and the size of its *grains* and *pores* through an atomic transport mechanism, so that the final result is a denser ceramic (Figure 2.16).

This transformation takes place when the powder material is subjected to a given pressure or to high temperature, or when placed under both effects simultaneously.

A sintering process may produce two different results: (1) a chemical transformation; or (2) a simple geometrical rearrangement of the texture, defined here as the size and shape of the grains and pores in the solid. In this last case, the sintering process merely yields a product with identical chemical composition and crystalline structure to the initial material.

However, sintering is more often used to improve the ceramic properties of a material, and to achieve higher values of packing and density. This occurs when a certain pressure is applied, without modifying the crystal structure, in order to modify some of its mechanical properties. The most common process includes a combination of pressure and temperature.

Regardless of the presence or not of structure variations, the sintering phenomenon implies a series of changes in the properties of materials, which can be summarized in the following points:

- the agglomerate is contracted,
- the pores change their shape and can even disappear,
- the grain size increases, and
- the density increases.

Geometrical models help to understand the mechanisms involved in a sintering process with powder solids. A commonly used model considers equally sized spherical particles stacked together forming a compact packing and a three-stage sintering process.

Due to the heating effect, the particles are joined together during the first stage, and the empty spaces between them start to disappear or to decrease;

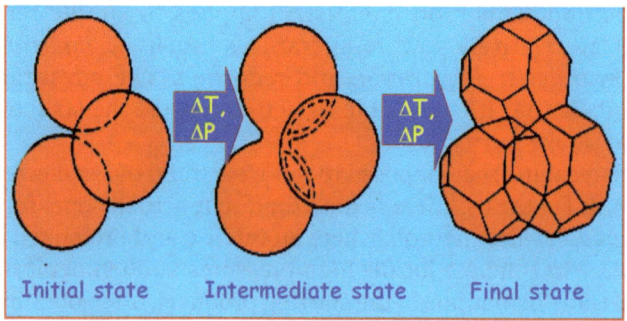

Figure 2.16   Stages of the sintering process.

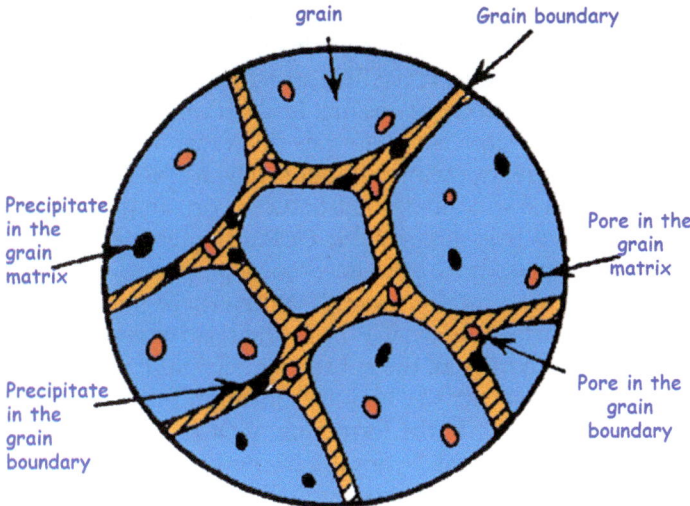

Figure 2.17 Evolution of microstructure in a solid during a sintering process.

in the second stage, *grain boundaries* start to form; and in the third stage, when the grain growth has been verified due to recrystallization, the uniformly distributed pores are placed inside the grain and not in the boundaries. In this last stage, the larger pores grow at the expense of the smaller ones, due to their different chemical potential. Parameters such as *temperature*, *grain size*, *pressure* and *atmosphere* are very important in any sintering process. Figure 2.17 depicts the evolution of microstructure in a solid during a sintering process.

These general remarks on sintering of solids are applicable to apatites, and are especially important when dealing with the fabrication of implants, since many of their features can be modified in this way, such as crystallinity, particle size and porosity.

2.2.12 Bioinspired Synthesis of Nanoapatites

Biomimetic nanoapatites feature sizes, shapes and compositions similar to those occurring in the mineral component of the bones and teeth. The interest in the preparation of biomimetic nanoapatites is based on two abutments. The first concerns the understanding of the biomineralization processes and the second deals with the preparation of new materials with biomedical applications (bone regenerative therapies and drug delivery systems), which are described further in Chapters 3 and 4.

The size and shape of apatite crystallites in bone tissue is controlled through the interaction with collagen fibrils and other noncollagenous proteins, which lead to platelet-like crystalline morphologies. Research[140,141] shows that platelet-like morphology plays an important role in bone characteristics and performance. For this reason there is a wide interest in mimicking the biosynthesis of these solids. Platelet morphology involves the partial inhibition of the crystal growth along the normal direction to the *c* axis of the hexagonal apatite crystalline structure ($P6_3/m$), breaking the crystalline symmetry so the mechanisms that lead to this situation are still under debate. Actually, the mechanism that rules the transition between the amorphous calcium phosphate (ACP) formation toward the hydroxyapatite-like phase crystallization is still unresolved. Some reports indicate that the platelet morphology arises from an unstable OCP intermediate.[142,143] Highly bioactive mesoporous glasses (MBGs) have been prepared which mimic this calcium phosphate maturation *in vitro*.[144] When these materials are soaked in SBF, after the nucleation of ACP onto their surfaces a transitory OCP phase can be observed prior to the subsequent crystallization to hydroxyapatite. Figure 2.18 represents the biomineralization mechanism trough the ACP–OCP–HA pathway. Since the formation of OCP occurs under acid environments, this phase is likely to appear favored by the environmental pH decrease during the osteoclast bone resorption in the remodelling processes. MBGs mimic this behavior because they incorporate protons onto their surfaces by calcium exchange with the SBF. In this way, the new calcium phosphate growth evolves as ACP, OCP, OCP–HA, hydroxyapatite after 1 h, 4 h, 8 h and 48 h, respectively in SBF. The synthesis and bioactive behavior of MBGs are widely discussed in Chapter 3.

In addition to the biomimetic growth of apatite onto bioactive surfaces, synthetic strategies to prepare bioinspired apatites have recently been developed.[145] One of the most interesting is the use of citrates as morphology-directing agents during nanoapatite crystallization.[146] The incorporation of citrates is somehow motivated by the recent evidence of large citrate amounts in bone (about 5.5% in weight of the total organic matter), as well as its role in the stabilization of size and shape of bone apatite.[147,148] The control exerted by citrates is explained in terms of the similar distances between the carboxylate groups in citrate and neighboring Ca^{2+} cations in the $(1\,0\,\bar{1}\,0)$ crystal faces of the apatite. The consequence of this matching between distances is that citrate strongly bonds to this face and inhibits the growth in that direction, thus resulting in the platelet morphology.

The citrate mediated preparation of bioinspired apatite consists of the controlled co-precipitation from calcium (for instance $CaCl_2$) and phosphate (for instance Na_2HPO_4) dissolutions. The calcium cations are co-dissolved with citrate, which is a strong calcium complexing agent that avoids the immediate calcium phosphate precipitation when both solutions are in contact.[149] Bioinspired citrate-functionalized apatites have been also prepared as coatings to improve the osteointegration of orthopaedic implants.[150] For this purpose Ti6Al4V alloys have been pre-coated with different layers such as silicon nitride, silicon carbide or titanium nitride. When considering nanoapatites for the manufacturing of coatings, adherence is one of the most important topics to be considered. In this sense the roughness supplied by Si_3N_4 seems to facilitate the nanoapatite adhesion, while the plasticity and compactness are function of the operating conditions during the apatite crystallization.

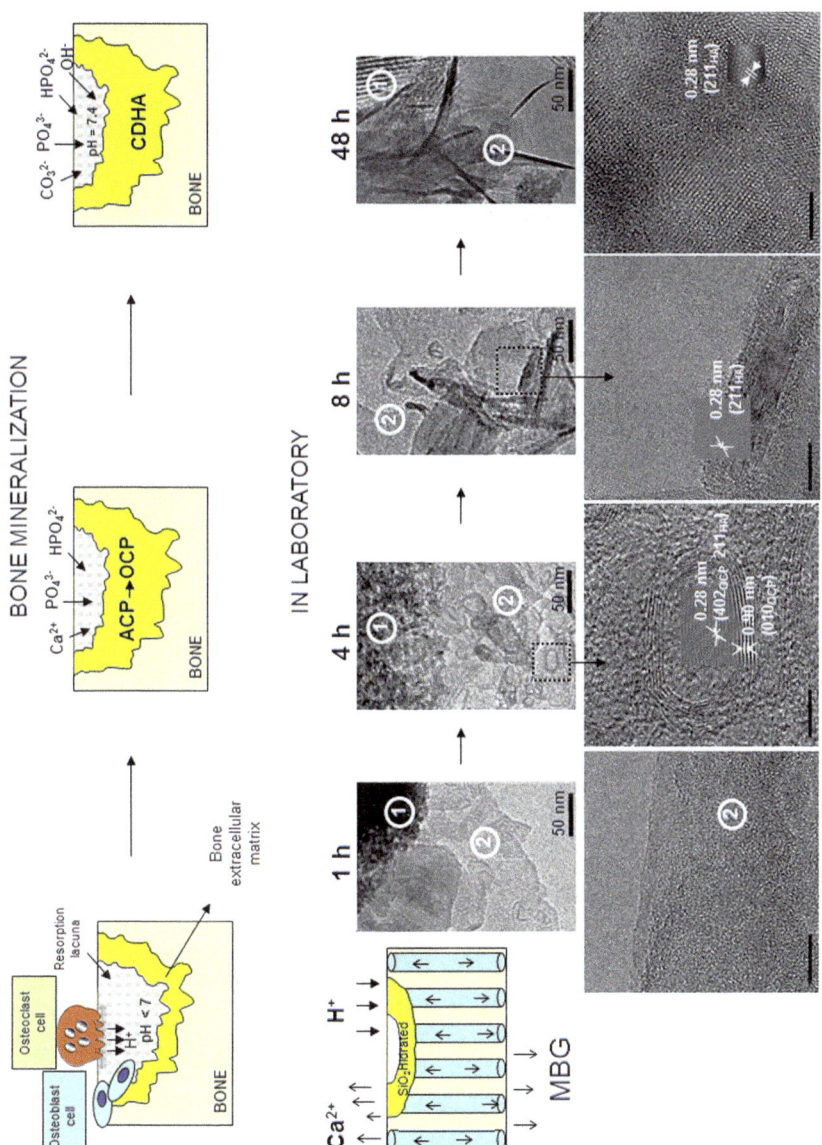

Figure 2.18 (Top) Proposed mineralization pathway during bone remodelling, considering the amorphous calcium phosphate (ACP)–octacalcium phosphate (OCP)–hydroxyapatite mineral maturation. (Bottom) Transmission electron micrograph images of a mesoporous bioactive glass (MBG) after being soaked in simulated body fluid at different times. Sites 1 and 2 correspond to MBG grains and precipitated CaP, respectively. CDHA: calcium-deficient hydroxyapatite.

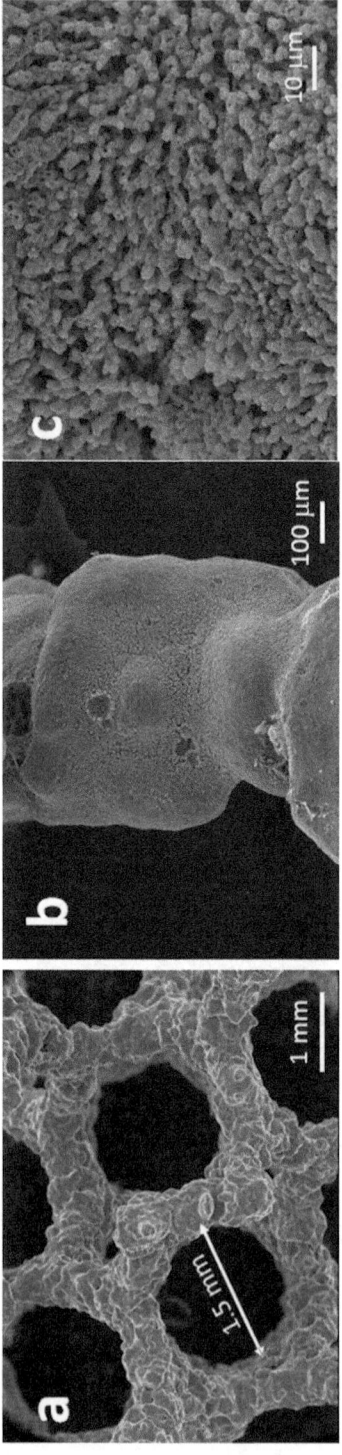

Figure 2.19 Scanning electron micrograph of a macroporous Ti6Al4V scaffold prepared by electron beam melting (a) before coating; (b) after coating with nanoapatite (low magnification); and (c) higher magnification of the apatite coating evidencing a porous microstructure.

In recent years manufacturing by rapid prototyping methods has enlarged the applications of metal alloys for orthopaedic purposes. These materials are commonly used for substitutive purposes, due to their excellent biocompatibility and mechanical properties. However, the incorporation of additive technologies, such as electron beam melting, allows the preparation of porous structures aimed at bone regenerative purposes instead of substitutive ones (Figure 2.19). In this sense, the coating of these scaffolds with osteogenic compounds is called upon to play a very important role in bone regenerative therapies. The incorporation of citrate in the biomimetic preparation of coatings not only results in apatites with similar shapes and sizes than biological ones, but also improves the adhesion of the coating. This fact has been observed with titanium scaffolds coated with biomimetic deposition by soaking the implants in SBF.[151] Further advances such as the incorporation of osteogenic agents[152] and growth factors[153] onto nanoapatites are currently under development to improve the response of these implants. This response involves not only the enhancement of the osteoblast activity but also their behavior with respect to osteoclasts[154] and immune cell responses.[155]

References

1. C. N. R. Rao and J. Gopalakrishnan, *New Directions in Solid State Chemistry*, ed. Cambridge. University Press, United Kingdom, 1997.
2. M. Vallet-Regí, *Perspectives in Solid State Chemistry*, ed. K. J. Rao, Narosa Publishing House, India, 1995, pp. 37–65.
3. M. Vallet-Regí and J. González-Calbet, *Prog. Solid State Chem.*, 2004, **32**, 1.
4. M. Vallet-Regí, *J. Chem. Soc., Dalton Trans.*, 2001, 97.
5. L. M. Rodríguez-Lorenzo and M. Vallet-Regí, *Chem. Mater.*, 2000, **12**(8), 2460.
6. R. Z. LeGeros, in *Monographs in Oral Science, Vol. 15: Calcium Phosphates in Oral Biology and Medicine*, ed. H. M. Myers, S. Karger, Basel, 1991.
7. J. C. Elliot, Recent studies of apatites and other calcium orthophosphates, in *Les Matèriaux en Phosphate de Calcium, Aspects Fondamentaux*, ed. E. Brès and P. Hardouin, Sauramps Medical, Montpellier, 1998, vol. 5, pp. 25–66.
8. W. Suchanek and M. Yoshimura, *J. Mater. Res.*, 1998, **13**(1), 94.
9. T. S. Narasaraju and D. E. Phebe, *J. Mater. Sci.*, 1996, **31**, 1.
10. K. de Groot, *Ceram. Int.*, 1993, **19**, 363.
11. M. Bohner, *J. Care Injured*, 2000, **31**, D37.
12. S. Sánchez-Salcedo, I. Izquierdo-Barba, D. Arcos and M. Vallet-Regí, *Tissue Eng.*, 2006, **12**(2), 279.
13. S. Padilla, S. Sánchez-Salcedo and M. Vallet-Regí, *J. Biomed. Mater. Res.*, 2005, **75A**, 63.
14. M. Vallet-Regí and D. Arcos, *J. Mater. Chem.*, 2005, **15**, 1509.
15. M. Vallet-Regí, J. Peña and I. Izquierdo-Barba, *Solid State Ionics*, 2004, **172**, 445.

16. D. Arcos, J. Rodríguez-Carvajal and M. Vallet-Regí, *Chem. Mater.*, 2004, **16**, 2300.

17. R. P. del Real, E. Ooms, J. G. G. Wolke, M. Vallet-Regí and J. A. Jansen, *J. Biomed. Mater. Res.*, 2003, **65A**, 30.

18. D. Arcos, R. P. del Real and M. Vallet-Regí, *J. Biomed. Mater. Res.*, 2003, **65A**, 71.

19. F. Balas, J. Pérez-Pariente and M. Vallet-Regí, *J. Biomed. Mater. Res.*, 2003, **66A**, 364.

20. S. Padilla, J. Román and M. Vallet-Regí, *J. Mater. Sci.: Mater. Med.*, 2002, **13**, 1193.

21. C. V. Ragel, M. Vallet-Regí and L. M. Rodríguez-Lorenzo, *Biomaterials*, 2002, **23**, 1865.

22. M. V. Cabañas, L. M. Rodríguez-Lorenzo and M. Vallet-Regí, *Chem. Mater.*, 2002, **14**, 3550.

23. A. Rámila, S. Padilla, B. Muñoz and M. Vallet-Regí, *Chem. Mater.*, 2002, **14**, 2439.

24. L. M. Rodríguez, M. Vallet-Regí and J. M. F. Ferreira, *J. Biomed. Mater. Res.*, 2002, **60**, 232.

25. R. P. del Real, J. G. C. Wolke, M. Vallet-Regí and J. A. Jansen, *Biomaterials*, 2002, **23**, 3673.

26. L. M. Rodríguez-Lorenzo, M. Vallet-Regí and J. M. F. Ferreira, *Biomaterials*, 2001, **22**, 583.

27. L. M. Rodríguez, M. Vallet-Regí and J. M. F. Ferreira, *Biomaterials*, 2001, **22**, 1847.

28. A. J. Salinas, M. Vallet-Regí and I. Izquierdo-Barba, *J. Sol-Gel Sci. Technol.*, 2001, **21**, 13.

29. M. Vallet-Regí, D. Arcos and J. Pérez-Pariente, *J. Biomed. Mater. Res.*, 2000, **51**, 23.

30. M. Vallet-Regí and A. Rámila, *Chem. Mater.*, 2000, **12**, 961.

31. M. Vallet-Regí, J. Pérez-Pariente, I. Izquierdo-Barba and A. J. Salinas, *Chem. Mater.*, 2000, **12**, 3770.

32. S. Sánchez-Salcedo, A. Nieto and M. Vallet-Regí, *Chem. Eng. J.*, 2008, **137**, 62.

33. M. Vallet-Regí, I. Izquierdo-Barba and A. J. Salinas, *J. Biomed. Mater. Res.*, 1999, **46**, 560.

34. M. Vallet-Regí, A. M. Romero, V. Ragel and R. Z. Legeros, *J. Biomed. Mater. Res.*, 1999, **44**, 416.

35. I. Izquierdo-Barba, A. J. Salinas and M. Vallet-Regí, *J. Biomed. Mater. Res.*, 1999, **47**, 243.

36. M. Vallet-Regí, L. M. Rodríguez Lorenzo and A. J. Salinas, *Solid State Ionics*, 1997, **101–103**, 1279.

37. A. Cuneyt Tas, *Biomaterials*, 2000, **21**, 1429.

38. M. Andrés-Vergés, C. Fernández-González, M. Martínez-Gallego, I. Solier, J. D. Cachadiña and E. Matijevic, *J. Mater. Res.*, 2000, **15**(11), 2526.

39. A. Yasukawa, T. Matsuura, M. Kakajima, K. Kandori and T. Ishikawa, *Mater. Res. Bull.*, 1999, **24**, 589.

40. J. Peña, I. Izquierdo-Barba, M. A. García and M. Vallet-Regí, *J. Eur. Ceram. Soc.*, 2006, **26**, 3631.
41. J. Peña, I. Izquierdo-Barba, A. Martínez and M. Vallet-Regí, *Solid State Sci.*, 2006, **8**, 513.
42. W. Weng and J. L. Baptista, *Biomaterials*, 1998, **19**, 125.
43. A. Jilavenkatesa and R. A. Condrate, *J. Mater. Sci.*, 1998, **33**, 4111.
44. C. S. Chai, K. A. Gross and B. Ben-Nissan, *Biomaterials*, 1998, **19**, 2291.
45. P. Layrolle, A. Ito and T. Tateishi, *J. Am. Ceram. Soc.*, 1998, **81**(6), 1421.
46. D. M. Liu, T. Trocynzki and W. J. Tseng, *Biomaterials*, 2001, **22**, 1721.
47. M. Manzano, D. Arcos, M. Rodríguez-Delgado, E. Ruíz, F. J. Gil and M. Vallet-Regí, *Chem. Mater.*, 2006, **18**, 5696.
48. M. Vallet-Regí, *Dalton Trans.*, 2006, 5211–5220.
49. M. Vallet-Regí and D. Arcos, *Curr. Nanosci.*, 2006, **2**, 179.
50. J. Peña, I. Izquierdo-Barba and M. Vallet-Regí, *Key Eng. Mater.*, 2004, **254–256**, 359.
51. M. H. Fathi and A. Hanifi, *Mater. Lett.*, 2007, **61**, 3978.
52. T. S. Kumar Sampath, I. Manjubala and J. Gunasekaran, *Biomaterials*, 2000, **21**, 1623.
53. Y. Fang, D. K. Agrawal, D. M. Roy and R. Roy, *J. Mater. Res.*, 1992, 7(2), 490.
54. K. Itatani, K. Iwafune, F. Scott Howellm and M. Aizawa, *Mater. Res. Bull.*, 2000, **35**, 575.
55. B. Yeong, J. M. Xue and J. Wang, *J. Am. Ceram. Soc.*, 2001, **84**, 465.
56. W. Kim, Q. Zang and F. Saito, *J. Mater. Sci.*, 2000, **35**, 5401.
57. B. Yeong, X. Junmin and J. Wang, *J. Am. Ceram. Soc.*, 2001, **82**, 65.
58. T. Nakano, A. Tokumura, Y. Umakoshi, S. Imazato, A. Ehara and S. Ebisu, *J. Mater. Sci.: Mater. Med.*, 2001, **12**, 703.
59. P. Shuk, W. L. Suchanek, T. Hao, E. Gulliver, R. E. Riman, M. Senna, K. S. TenHuisen and V. F. Janas, *J. Mater. Res.*, 2001, **16**, 1231.
60. G. K. Lim, J. Wang, S. C. Ng, C. H. Chew and L. M. Gan, *Biomaterials*, 1997, **18**, 1433.
61. D. Wals and S. Mann, *Chem. Mater.*, 1996, **8**, 1944.
62. T. Furuzono, D. Walsh, K. Sato, K. Sonoda and J. Tanaka, *J. Mater. Sci. Lett.*, 2001, **20**, 111.
63. S. Loher, W. J. Stark, M. Maciejewski, A. Baiker, S. E. Pratsinis, D. Reichardt, F. Maspero, F. Krumeich and D. Gunther, *Chem. Mater.*, 2005, **17**, 36.
64. M. Aizawa, T. Hanazawa, K. Itatani, F. S. Howell and A. Kishioka, *J. Mater. Sci.*, 1999, **34**, 2865.
65. D. Veilleux, N. Barthelemy, J. C. Trombe and M. Verelst, *J. Mater. Sci.*, 2001, **36**, 2245.
66. K. S. Tenhuisen and P. W. Brown, *Biomaterials*, 1998, **19**, 2209.
67. W. Kim and F. Satio, *Ultrason. Sonochem.*, 2001, **8**, 85.
68. Y. Fang, D. K. Agrawal, D. M. Roy, R. Roy and P. W. Brown, *J. Mater. Res.*, 1992, 7, 2294.
69. W. J. Weng and J. L. Baptista, *Biomaterials*, 1998, **19**, 125.
70. D. M. Liu, T. Troczynski and W. J. Tseng, *Biomaterials*, 2001, **22**, 1721.

71. M. H. Fathi and A. Hanifi, *Mater. Lett.*, 2007, **61**, 3978.
72. M. P. Pechini, US Pat., 3,330,697; 1967.
73. J. Peña and M. Vallet-Regí, *J. Eur. Ceram. Soc.*, 2003, **23**(10), 1687–1696.
74. J. C. Elliott, *Studies in Inorganic Chemistry*, Elsevier, 1994, vol. 18.
75. M. V. Cabañas, J. M. González-Calbet, M. Labeau, P. Mollard, M. Pernet and M. Vallet-Regí, *J. Solid State Chem.*, 1992, 101–265.
76. M. Vallet-Regí, V. Ragel, J. Román, J. L. Martínez, M. Labeau and J. M. González-Calbet, *J. Mater. Res.*, 1993, 8(1), 138.
77. M. Vallet-Regí, M. T. Gutierrez-Ríos, M. P. Alonso, M. I. Frutos and S. Nicolopoulos, *J. Solid State Chem.*, 1994, 8(1), 138.
78. A. S. Coetzee, *Arch. Otolaryngol.*, 1980, **106**, 405.
79. J. Lemaitre, A. Mirtchi and E. Munting, *Silicon Ind. Ceram. Sci. Technol.*, 1987, **52**, 141.
80. L. C. Chow, *J. Ceram. Soc. Jpn.*, 1991, **99**, 954.
81. T. Sugama and M. Allan, *J. Am. Ceram. Soc.*, 1992, **75**, 2076.
82. A. A. Mirtchi, J. Lemaitre and E. Munting, *Biomaterials*, 1991, **12**, 505.
83. M. Otsuka, Y. Matsuda, Y. Suwa, J. L. Fox and W. Higuchi, *J. Biomed. Mater. Res.*, 1995, **29**, 25.
84. Y. Miyamoto, K. Ishikawa, M. Takechi, T. Toh, T. Yuasa, M. Nagayama and K. Suzuki, *Biomaterials*, 1998, **19**, 707.
85. R. P. del Real, J. C. C. Wolke, M. Vallet-Regí and J. A. Jansen, *Biomaterials*, 2002, **23**, 3673.
86. M. Nilsson, E. Fernandez, S. Sarda, L. Lidgren and J. A. Planell, *J. Biomed. Mater. Res.*, 2002, **61**, 600.
87. B. R. Constanz, I. C. Ison, M. T. Fulmer, R. D. Fulmer, R. D. Poser, S. T. Smith, M. Vanwagoner, J. Ross, S. A. Goldstein, J. B. Jupiter and D. I. Rosental, *Science*, 1995, **267**, 1796.
88. S. Takagi, L. C. Chow and K. Ishikawa, *Biomaterials*, 1998, **9**, 1593.
89. W. S. Pietrzak and R. Ronk, *J. Craniofacial Surg.*, 2001, **11**, 327.
90. C. E. Rawlings III, R. H. Wilkins, J. S. Hanker, N. G. Georgiade and J. M. Harrelson, *J. Neurosurg.*, 1988, **69**, 269.
91. S. Sato, T. Koshino and T. Saito, *Biomaterials*, 1998, **19**, 1895.
92. M. V. Cabañas, L. M. Rodríguez-Lorenzo and M. Vallet-Regí, *Chem. Mater.*, 2002, **14**, 3550.
93. D. Yu, J. Wong, Y. Matsuda, J. L. Fox, W. I. Higuchi and M. Otsuka, *J. Pharm. Sci.*, 1992, **81**, 529.
94. C. Hamanishi, K. Kitamoto, S. Tanaka, M. Osuka, Y. Doi and T. Kitahashi, *J. Biomed. Mater. Res.*, 1996, **33**, 139.
95. B. Mousset, M. A. Benoit, C. Delloye, R. Bouillet and J. Guillard, *Int. Orthop.*, 1997, **21**, 403.
96. L. Meseguer-Olmo, M. J. Ros-Nicolás, M. Clavel-Sainz, V. Vicente-Ortega, M. Alcaraz-Baños, A. Lax-Pérez, D. Arcos, C. V. Ragel and M. Vallet-Regí, *J. Biomed. Mater. Res.*, 2002, **61**, 458.
97. A. Ratier, I. R. Gibson, S. M. Best, M. Freche, J. L. Lacout and F. Rodríguez, *Biomaterials*, 2001, **22**, 897.

98. J. C. Doadrio, D. Arcos, M. V. Cabañas and M. Vallet-Regí, *Biomaterials*, 2004, **25**, 2629.
99. G. Daculsi, *Biomaterials*, 1998, **19**, 1473.
100. G. Drimandi, P. Weiss, F. Millot and G. Daculsi, *J. Biomed. Mater. Res.*, 1998, **39**, 660.
101. C. V. Ragel, M. Vallet-Regí and L. M. Rodriguez-Lorenzo, *Biomaterials*, 2002, **23**, 1865.
102. A. Rámila, S. Padilla, B. Muñoz and M. Vallet-Regí, *Chem. Mater.*, 2002, **14**, 2439.
103. D. C. Tancred, B. A. O. McCormack and A. J. Carr, *Biomaterials*, 1998, **19**, 2303.
104. J. M. Bouler, M. Trecant, J. Delecrin, J. Royer, N. Passuti and G. Gaculci, *J. Biomed. Mater. Res.*, 1996, **32**, 603.
105. A. Slosarczyk and J. Piekarcyk, *Ceram. Int.*, 1999, **25**, 561.
106. N. Kivrak and A. Cuneyt Tas, *J. Am. Ceram. Soc.*, 1998, **82**, 2245.
107. O. E. Petrov, E. Dyulgerova, L. Petrov and R. Ropova, *Mater. Lett.*, 2001, **48**, 162.
108. X. Yang and Z. Wang, *J. Mater. Chem.*, 1998, **8**, 2233.
109. F. H. Lin, C. J. Liao, K. S. Chen, J. S. Sun and C. Y. Lin, *J. Biomed. Mater. Res.*, 2000, **51**, 157.
110. K. Itatani, T. Nishioka, S. Seike, F. S. Howell, A. Kishiota and M. Kinoshita, *J. Am. Ceram. Soc.*, 1994, **77**, 801.
111. I. Manjubala and M. Sivakimar, *Mater. Chem. Phys.*, 2001, **71**, 272.
112. A. Cunneyt Tas, *J. Eur. Ceram. Soc.*, 2000, **20**, 2389.
113. O. Gauthier, J. M. Bouler, E. Aguado, R. Z. LeGeros, P. Pilet and G. Daculsi, *J. Mater. Sci.: Mater. Med.*, 1999, **10**, 199.
114. E. I. Suvurova and P. A. Buffat, *Eur. Cells Mater.*, 2001, **1**, 27.
115. X. Yang and Z. Wang, *J. Mater. Chem.*, 1998, **8**, 2233.
116. J. C. Elliot, G. Bond and J. C. Tombe, *J. Appl. Crystallogr.*, 1980, **13**, 618.
117. D. Tadic and M. Epple, *Biomaterials*, 2004, **25**, 987.
118. M. Okazaki, T. Matsumoto, M. Taira, J. Takakashi and R. Z. LeGeros, *Bioceramics 11*, ed. R. Z. Legeros and J. P. LeGeros, World Scientific, New York, 1998, p. 85.
119. Y. Doi, T. Shibutani, Y. Moriwaki, T. Kajimoto and Y. Iwayama, *J. Biomed. Mater. Res.*, 1998, **39**, 603.
120. M. Vallet-Regí, A. Rámila, S. Padilla and B. Muñoz, *J. Biomed. Materials Res.*, 2003, **66**, 580.
121. R. Z. LeGeros, *Nature*, 1965, **206**, 403.
122. L. J. J. Jha, S. M. J. Best, J. C. Knowles, I. Rehman, I. D. Santos and W. Bonfield, *J. Mater. Sci.: Mater. Med.*, 1997, **8**, 185.
123. L. L. Hench, J. Wilson, L. L. Hench and J. Wilson, *An Introduction to Bioceramics*, World Scientific, Boca Raton, FL, 1992, p. 20.
124. K. Ohura, T. Nakamura, T. Yamamuro, T. Kokubo, Y. Ebisawa, Y. Kotoura and M. Oka, *J. Biomed. Mater. Res.*, 1991, **25**, 357.
125. E. M. Carlisle, *Science*, 1970, **167**, 179.

126. E. M. Carlisle, *Calcif. Tissue Int.*, 1981, **33**, 27.
127. A. J. Ruys, *J. Aust. Ceram. Soc.*, 1993, **29**, 71.
128. Y. Tanizawa and T. Suzuki, *J. Chem. Soc., Faraday Trans.*, 1995, **91**, 3499.
129. L. Boyer, J. Carpena and J. L. Lacout, *Solid State Ionics*, 1997, **95**, 121.
130. I. R. Gibson, S. M. Best and W. Bonfield, *J. Biomed. Mater. Res.*, 1999, **44**, 422.
131. T. Kokubo, H. Kushitani, S. Sakka, T. Kitsugi and T. Yamamuro, *J. Biomed. Mater. Res.*, 1990, **24**, 721.
132. P. A. A. P. Marques, M. C. F. Magalhaes, R. N. Correia and M. Vallet-Regí, *Key Eng. Mater.*, 2001, **192–195**, 247.
133. F. Balas, J. Pérez-Pariente and M. Vallet-Regí, *J. Biomed. Mater. Res.*, 2003, **66A**, 364.
134. D. Arcos, J. Rodriguez-Carvajal and M. Vallet-Regí, *Phys. Rev. B*, 2004, **350**, e607.
135. M. Vallet-Regí, *An. Quim. Int. Ed.*, 1997, **93**(1), S6.
136. S. H. Sohn, H. K. Jun, C. S. Kim, K. N. Kim, S. M. Chung, S. W. Shin, J. J. Ryu and M. K. Kim, *J. Oral Rehabil.*, 2006, **33**, 12.
137. L. Tuck, M. Sayer, M. Mackenzie, J. Hadermann, D. Dunfield, A. Pietak, J. W. Reid and A. D. Stratilatov, *J. Mater. Sci.*, 2006, **41**, 4273.
138. E. S. Thian, J. Huang, S. M. Best, Z. H. Barber and W. Bonfield, *J. Biomed. Mater. Res.*, 2006, **78A**, 121.
139. J. Peña, R. P. del Real, L. M. Rodríguez-Lorenzo and M. Vallet-Regí, *Bioceramics*, ed. H. Ohgushi, G. W. Gastings and T. Yoshihawa, World Scientific Publishing Co. Pte. Ltd, Nara, Japan, 12th edn, 1999, p. 353.
140. S. J. Eppell, W. Tong, J. L. Katz, L. Kuhn and M. J. Glimcher, *J. Orthop. Res.*, 2011, **19**, 1027.
141. P. Fratzl, H. S. Gupta, E. P. Paschalis and P. Roschger, *J. Mater. Chem.*, 2004, **14**, 2115.
142. N. J. Crane, V. Popescu, M. D. Morris, P. Steenhuis and M. A. Ignelzi Jr, *Bone*, 2006, **39**, 434.
143. S. Weiner, *Bone*, 2006, **39**, 431.
144. I. Izquierdo-Barba, D. Arcos, Y. Sakamoto, O. Terasaki, A. López-Noriega and M. Vallet-Regí, *Chem. Mater.*, 2008, **20**, 3191.
145. S. V. Dorozhkin, *Acta Biomater*, 2010, **6**, 715.
146. J. M. Delgado-López, R. Frison, A. Cervellino, J. Gómez-Morales, A. Guagliardi and N. Masciocchi, *Adv. Funct. Mater.*, 2014, **24**, 1090.
147. Y. Y. Hu, A. Rawal and K. Schmidt-Rohr, *Proc. Natl. Acad. Sci. U. S. A.*, 2010, **107**, 22425.
148. B. Xie and G. H. Nancollas, *Proc. Natl. Acad. Sci. U. S. A.*, 2010, **107**, 22369.
149. J. M. Delgado-López, M. Iafisco, I. Rodríguez, A. Tampieri, M. Prat and J. Gómez-Morales, *Acta Biomater*, 2012, **8**, 3491.
150. J. M. Delgado-López, M. Iafisco, I. Rodríguez-Ruiz and J. Gómez-Morales, *J. Inorg. Biochem.*, 2013, **127**, 261.
151. P. Yu, F. Lu, W. J. Zhu, D. Wang, X. J. Zhu, G. X. Tan, X. L. Wang, Y. Zhang, L. H. Li and C. Y. Ning, *Appl. Surf. Sci.*, 2014, **313**, 947.

152. M. Manzano, D. Lozano, D. Arcos, S. Portal-Núñez, C. Lopez Laorden, P. Esbrit and M. Vallet-Regí, *Acta Biomater.*, 2011, 7, 3555.
153. M. C. Matesanz, M. J. Feito, C. Ramírez-Santillán, R. M. Lozano, D. Arcos, M. Vallet-Regí and M. T. Portolés, *Macromol. Biosci.*, 2012, **12**, 446.
154. M. C. Matesanz, J. Linares, I. Lilue, S. Sánchez-Salcedo, M. J. Feito, D. Arcos, M. Vallet-Regí and M. T. Portolés, *J. Mater. Chem. B*, 2014, **2**, 2910.
155. M. C. Matesanz, M. J. Feito, M. Oñaderra, C. Ramírez-Santillán, C. da Casa, D. Arcos, M. Vallet-Regí, J. M. Rojo and M. T. Portolés, *J. Colloid Interface Sci.*, 2014, **416**, 59.

CHAPTER 3

Bioceramics Forming Nanoapatites

3.1 Introduction

Biomimetic materials science is an evolving field that studies how nature designs, processes and assembles/disassembles molecular building blocks to fabricate high-performance minerals, polymers and mineral–polymer composites (*e.g.*, mollusc shells, bones or teeth) and/or soft materials (*e.g.*, skin, cartilage or tendons) and then applies these designs and processes to engineer new molecules and materials with unique properties.[1] The fabrication of nanostructured materials that resemble the complex hierarchical structures of natural hard tissues present in bones and teeth is a primary objective from the point of view of biomaterials science. Chapter 1 showed that bone is an excellent example of hierarchical organization with structural and functional purposes, where the transition from the nanometric to the macroscopic scale is carefully organized.[2] However, the development of bio-materials science is still far from this objective, and perhaps a more realistic aim is to design implant surfaces at the nanometric scale to optimize the tissue–implant interface,[3] facilitating bone self-healing.

Chapter 2 describes how bone-like hydroxyapatite (HA) nanoparticles can be synthesized by a range of production methods, such as precipitation from aqueous solutions, sol–gel synthesis, aerosol-assisted methods, *etc.* In this chapter we deal with one of the most promising and developed methods: the biomimetic synthesis. In the frame of the bioceramics field, biomimetism is considered as mimicking natural manufacturing methods to generate artificial bone-like calcium phosphates (CaPs), mainly apatites, which can be used for bone- and teeth-repairing purposes. The most common process consists

RSC Nanoscience & Nanotechnology No. 39
Nanoceramics in Clinical Use: From Materials to Applications, 2nd Edition
By María Vallet-Regí and Daniel Arcos Navarrete

of the crystallization of nonstoichiometric carbonate hydroxyapatite (CHA) from simulated physiological solutions at temperatures similar to those in physiological conditions.[4] The bone-like apatite crystallization takes place through the nucleation of CaP precursors, such as amorphous calcium phosphate (ACP) or octacalcium phosphate (OCP).[5,6] These precursors subsequently maturate to calcium-deficient hydroxyapatite (CDHA) by incorporating CO_3^{2-}, OH^-, Ca^{2+}, PO_4^{3-}, *etc.* ions from the surrounding solution.

The idea of using *bioactive ceramics* as substrates for biomimetic synthesis of nanoapatites acquired great importance when in 1971 Hench *et al.*[7] discovered that the bioactive process in SiO_2-based bioceramics took place through the formation of a carbonate-containing CDHA at the implant tissue surface. Thereafter, it could be seen that the prior *in vitro* biomimetic growth of a nanocrystalline CDHA allowed the fabrication of implants with fitted-out surfaces to be colonized by bone cells.[8,9] Bone cells have been shown to proliferate and differentiate on these apatite layers, showing increased bioresponse and new bone formation.[10,11] Nowadays, among the different concepts for fabrication of highly bioresponsive nanoceramics, biomimetic methods are among the most developed strategies to produce body interactive materials, helping the body to heal, and promoting tissue regeneration. In this sense, bioceramics such as bioactive glasses, glass-ceramics and calcium-phosphate-based synthetic compounds are excellent substrates that develop CaP nanoceramics with almost identical characteristics to the biological ones, when soaked in solutions mimicking physiological conditions.

3.1.1 Biomimetic Nanoapatites and Bioactive Ceramics

The motivation to undertake the synthesis of nanostructured apatites over bioceramic surfaces arises from the understanding of the physical–chemical and biological processes that lead to the bond formation between bones and implants. When bioactive ceramics such as bioglass, apatite-wollastonite glass ceramic or HA/β-tricalcium phosphate (TCP) biphasic calcium phosphate are implanted in bone tissue, the examination of the implant site reveals the presence of a nanocrystalline calcium-deficient carbonate apatite at the bonding interface.[12] This intermediate apatite layer is similar to biological apatites in terms of calcium deficiency and carbonate substitutions, and it was believed that it would interact with osteoblasts in a similar manner to biological apatites. On the other side, when the so-called bioactive ceramics are soaked in artificial or *simulated physiological fluids*, surface analysis evidences the setting off of chemical reactions at the material surface, such as dissolution, precipitation, ionic exchange, *etc.*, together with biological material adsorption.

One of the most important works evidencing the role of the newly formed apatite layer on bioactive ceramics was carried out by Professor Kokubo's research team.[13,14] Kokubo and co-workers systematically demonstrated that the *in vivo* bioactivity of a material, measured by the rate of bone ingrowth, could be directly related to the rate at which the material forms apatite

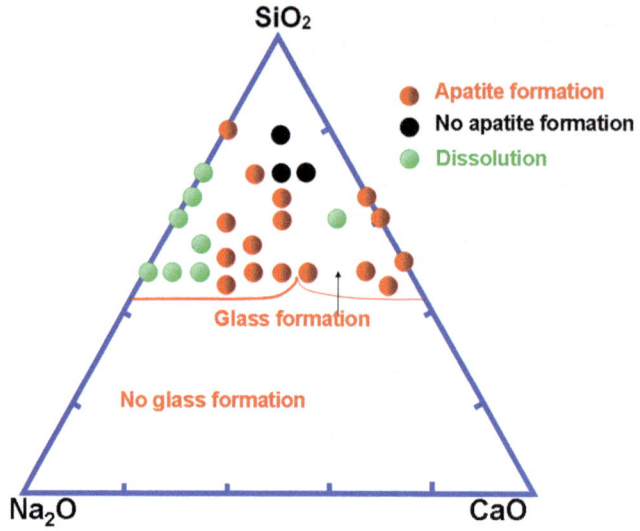

Figure 3.1 Compositional dependences of nanoapatite formation on glasses in the system Na$_2$O–CaO–SiO$_2$, after soaking in simulated body fluid.

in vitro when immersed in simulated body fluids (SBF). This work consisted of synthesizing different compositions of glass particles in the system Na$_2$O–CaO–SiO$_2$ which were packed into the bony defects of rabbit femoral condyle to evaluate their ability to induce bone ingrowth, while the same set of bioglass formulations (see Figure 3.1) were also immersed in SBF to evaluate their apatite-forming ability *in vitro*.

The results showed that the *in vivo* bioactivity was precisely reproduced by the apatite-forming ability *in vitro*. The glass formulation that induced apatite formation most efficiently and rapidly *in vitro* also stimulated the most significant bone-formation activity 3 and 6 weeks after implantation *in vivo*. Nowadays, the *in vitro* nucleation and growth of a nanocrystalline apatite onto a bioceramic is considered as a clear sign of a good *in vivo* behaviour. The implant–bone bonding ensures the materials' osteointegration, and very often also promotes the bone-tissue regeneration. In the last cases, gene activation, implant resorption and bone-ingrowth mechanisms are also involved.

3.1.2 Biomimetic Nanoapatites on Nonceramic Biomaterials; Two examples: Polyactive® and Titanium Alloys

The formation of nanocrystalline apatites at the implant surface sets off bioactive bonding and/or bone-tissue regeneration when implants are in contact with living tissues. Clear examples of the significance of the biomimetic apatite layer can be found not only in the case of ceramic compounds, but also in polymers and metals such as Polyactive® and titanium alloys. Polyactive® is a member of a series of segmented copolymers based on polyethylene oxide and polybutylene terephthalate.[15] This polymer is considered as a potential bone substitute material with bioactive properties and, consequently, with bone-bonding ability.[16,17] The capability of Polyactive® as potential bone substitute had been investigated with different animal models,[18-22] but some studies raised a concern about the clinical usage of this polymer, concerning its osteoconductive properties. The problems were tackled by performing a biomimetic growth of bone-like apatite coating, in order to stimulate or enhance the bone bonding with this polymer.[23] After being implanted in rabbit femur, abundant new bone growth with spongy appearance along the implant surface was observed after 2 weeks, and the marginal bone formation with a maximal penetration depth of ~1 mm in 4 mm diameter defects was observed after 8 weeks.

The biomimetic growth of nanoapatites has been also extended to metal alloys commonly used in orthopaedic surgery, for instance titanium alloys. The application of titanium alloys in artificial joint replacement prostheses is mainly motivated for their lower modulus, superior biocompatibility and enhanced corrosion resistance when compared to more conventional stainless steels and cobalt-based alloys. When this material is manufactured as HA-coated joint implants, the plasma spraying technique is the process commonly used in their production.[24,25] However, this technique exhibits several drawbacks related to the heterogeneity of the coating thickness, weak adherence and structural integrity as well as coating delamination, which lead to the fibrous tissue ingrowth and occasional implant loosening. Li[26] implanted multichanneled Ti6Al4V implants, in which four channels were apatite coated in an aqueous solution formulated to include HCO_3^- ions and other major inorganic ions present in the body, such as HPO_4^{2-}, Ca^{2+}, Mg^{2+}, Na^+, K^+, Cl^- and SO_4^{2-}, which could induce the formation of an apatite coating closely mimicking bone mineral, whereas the rest of the channels remained uncoated. Eight weeks after implantation into the distal femur of dogs, histological examination revealed much higher bone ingrowth through the apatite-lined channel of all implants, while the uncoated channel displayed minimal ingrowth.

3.1.3 Significance of Biomimetic Nanoapatite Growth on Bioceramic Implants

The improved clinical performance of Polyactive® and titanium alloys when coated with biomimetic nanoapatite reveals the potential of this field in orthopaedic and dental surgery. Both substrates are suitable to be coated after different chemical treatments aimed at preparing their surfaces for nanoapatite crystallization. However, in the case of bioactive ceramics, these materials not only can be coated by a newly formed CaP layer, but they strongly promote the biomimetic process and their potential for bone-tissue regeneration deserves special attention.

The pioneering works of Hench *et al.*[7] on the bioactive processes in SiO_2-based bioceramics, and the correlation established by Kokubo *et al.*[4] between

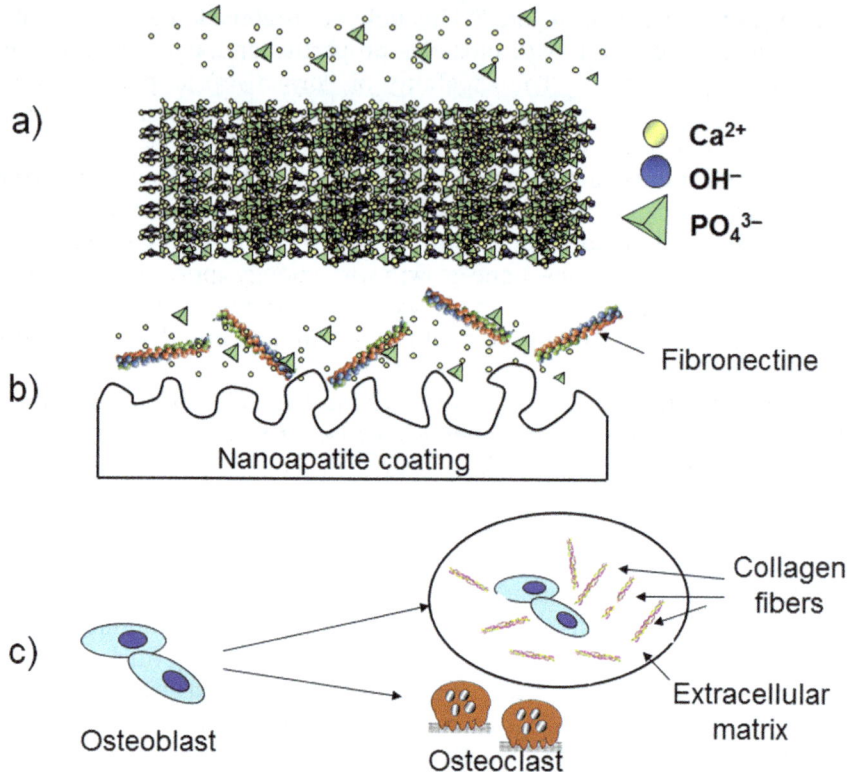

Figure 3.2 Possible biological mechanisms that rule the improved response of coated bioceramic surfaces with biomimetic nanoapatites. (a) Dissolution of the biomimetic nanoapatite leading to the saturation of surrounding fluids. (b) Increase of the surface roughness and adsorption of large amounts of cell adhesion proteins. (c) Ca^{2+} and PO_4^{3-} ions may signal cells toward the osteoblast differentiation.

the formation of biomimetic nanoapatites and their *in vivo* performance, led to the extended use of this procedure to measure the level of bioactivity of a bioceramic, by examining the *in vitro* apatite-forming ability on its surface. However, after considering the advantages of advanced nanoceramics, these nanoscaled coatings are being used to produce bioceramics with better hard- and soft-tissue attachment, higher biocompatibility and enhanced bioactivity for bone-regenerative purposes. The biological mechanisms that rule these enhanced characteristics are not fully defined. However, several performance guidelines of the biomimetic nanoapatites can be addressed[27] (Figure 3.2):

(1) *In vivo* dissolution of the biomimetic nanoapatite, leading to the saturation of surrounding fluids and thus accelerating the precipitation of truly biological apatites onto the coated implant.

(2) Adsorption of large amounts of protein from the neighbouring environment due to the surface charge of the nanoapatite, thus triggering cell differentiation.

(3) The microstructure of the substrate/apatite coating increases the surface roughness, which is beneficial for osteoinduction as compared to smooth surfaces.

(4) The apatite could be the source for Ca^{2+} and PO_4^{3-} ions which may signal cells toward the differentiation pathway and trigger bone formation.

(5) Since the biomimetic nanoapatite is similar in structure and properties to natural biological apatites, it could constitute an excellent substrate for new biological phase nucleation.

3.2 Simulated Physiological Solutions for Biomimetic Procedures

Nanocrystalline HA coatings can be easily produced on various ceramic substrates through the reaction with artificial physiological fluids. In general terms the biomimetic formation of apatite involves nucleation and growth from an ionic solution.[28,29] The composition of any solid deposited onto the surface of a bioceramic will be largely determined by the surrounding media, so choosing the correct experimental conditions and mimicking solution is mandatory. The apatite crystallization from a solution could be reached by mixing aqueous solutions containing the calcium and phosphates ions. However, this kind of process would lead to precipitates with properties very different with respect to biological apatites. There are obvious differences between the *in vivo* and *in vitro* crystallization conditions,[30] which can be summarised as follows:

(1) Depleting concentration conditions commonly occur under *in vitro* crystallization. On the contrary, the concentrations of ions and molecules are kept constant during biological mineralization.

(2) Kinetics of the precipitation reaction. Chemical crystallization is a much faster process (minutes to days), while the biological process is measured in terms of weeks and even years.

(3) The presence of inorganic, organic, biological and polymeric compounds within biological fluids, which are absent in artificial solutions. These species often act as inhibitors, seeds and templates during the growth of biological apatites.

The first and second differences can be overcome by using appropriate crystallization techniques and this topic will be discussed later. However, the third requires a more complex approximation, involving chemical and biological concepts, to fabricate appropriate crystallization solutions. Using natural fluids such as blood, saliva, *etc.* involves serious drawbacks related to the amounts available, variability and storage. Moreover, in the case of solutions able to mimic bone apatite formation, the presence of proteins and other biological entities exert a large inhibitory or delaying effect.[31–33] For this reason, inorganic ionic solutions are the most widely applied fluids for biomimetic nanoapatite purposes.

Among the different artificial biological solutions able to partially simulate the physiological conditions, the SBF developed by Kokubo *et al.* is the most widely applied solution for biomimetic purposes.[4] SBF is a metastable aqueous solution with pH of ~7.4, supersaturated with respect to the solubility product of HA. This solution only contains inorganic ions at concentrations almost equal to those of human plasma (Table 3.1). The main difference between SBF and the inorganic part of biological plasma is the bicarbonate (HCO_3^-) concentration, which is significantly lower in SBF (4.2 mM instead of 27 mM in plasma). SBF has been widely used for *in vitro* assessment of the bioactivity of artificial materials by examining their apatite-forming ability in the fluid.[34–36] SBF has been also used to prepare bioactive composites by forming bone-like apatite on various types of substrates.[37–39] In this sense, controlling the composition and structure of the apatite produced in SBF has been one of the most important aims into the frame of biomimetic synthesis, and several efforts were made in order to precipitate apatites equal (or very similar) to those occurring in bones.

Kim *et al.*[40,41] reported that the apatite produced in a conventional SBF differs from bone apatite in its composition and structure. They attributed this difference to the higher Cl^- and lower HCO_3^- concentrations of the SBF than those of blood plasma[40,41] (see Table 3.1), and they demonstrated that an apatite with a composition and structure similar to that of bone would be produced if the SBF had ion concentrations almost equal to those of human

Table 3.1 Human plasma and ion concentration of some of the most applied artificial solutions for biomimetic processes[a] (mM).

	Na^+	K^+	Ca^{2+}	Mg^{2+}	HCO_3^-	Cl^-	HPO_4^{2-}	SO_4^{2-}
Human plasma (total)	142.0	5.0	2.5	1.5	27.0	103.0	1.0	0.5
Human plasma (dissociated)	142.0	5.0	1.3	1.0	27.0	103.0	1.0	0.5
SBF	142.0	5.0	2.5	1.5	4.2	148.0	1.0	0.5
i-SBF	142.0	5.0	1.6	1.0	27.0	103.0	1.0	0.5
m-SBF	142.0	5.0	2.5	1.5	10.0	103.0	1.0	0.5
r-SBF	142.0	5.0	2.5	1.5	27.0	103.0	1.0	0.5
n-SBF	142.0	5.0	2.5	1.5	4.2	103.0	1.0	0.5
HBSS[45]	142.0	5.8	1.3	0.8	4.2	145.0	0.8	0.8
PECF[46]	145.0	5.0			30.0	118.0	1.0	
EBSS[47]	144.0	5.4	1.8	0.8	30.0	125.0	1.0	
PBS[48,49]	146.0	4.2				141.0	9.5	

[a]SBF: simulated body fluid; i: ionized; m: modified; r: revised; HBSS: Hanks' balanced salt solution; PECF: Homsy's pseudoextracellular fluid; EBSS: Earl's balanced salt solution; PBS: phosphate buffered saline.

plasma. When tailoring new SBFs, it must be taken into account that of the calcium ions in blood plasma (2.5 mM), 0.9 mM of Ca^{2+} are bound to proteins, and 0.3 mM of Ca^{2+} are bound to inorganic ions, such as carbonate and phosphate ions.[42] Keeping this in mind, Oyane *et al.*[43] prepared new SBFs denoted:

- ionized (i-)SBF, designed to have concentrations of dissociated ions equal to those of blood plasma;
- modified (m-)SBF, designed to have concentrations of ions equal to those of blood plasma, excepting HCO_3^-, the concentration of which is decreased to the level of saturation with respect to calcite ($CaCO_3$); and
- revised (r-)SBF, designed to have a concentration of ions all of which are equal to those of blood plasma, including Cl^- and HCO_3^-.

The main drawback of i-SBF and r-SBF is their instability. These two fluids are less stable than SBF and m-SBF in terms of calcium carbonate cluster formation. For these reasons, these two fluids are not suitable for long-term use in the bioactivity assessment of materials, although they can be used for biomimetic synthesis of bone-like apatite. In contrast, m-SBF is stable for a long time with respect to changes in ion concentrations and, in contact with bioceramics, m-SBF better mimics biological apatite formation compared with conventional SBF.

In 2004, Takadama *et al.* proposed a newly improved (n-)SBF in which they decreased only the Cl^- ion concentration to the level of human blood plasma, leaving the HCO_3^- ion concentration equal to that of the conventional SBF.[44] n-SBF was compared with conventional SBF in terms of stability and reproducibility of apatite formation, evidencing that SBF does not differ from n-SBF and both solutions could be indifferently used for biomimetic studies.

Further attempts to improve the biomimetic properties of SBF have been undertaken. Some efforts have been made to replace artificial buffers by simultaneously increasing of the hydrogen carbonates concentration of SBF or avoiding CO_2 losses from SBF through the permanent bubbling of CO_2. Addition of the most important organic and biological compounds such as glucose and albumin is another direction taken to improve biomimetic properties of SBFs, although the presence of proteins can seriously impede the HA crystallization.

Occasionally, condensed solutions of SBF (1.5×, 2×, 5× and even 10× concentration) are used to accelerate the precipitation.[50-54] The use of condensed solutions is controversial since it leads to changes in the chemical composition of the biomimetically grown CaP. Commonly, the crystallised apatite exhibits different microstructures and lower phosphate amounts due to a higher carbonate ions incorporation, which could affect to the osteoblast response when a biomimetic nanoapatite makes contact with them. This effect has been studied on culture-grade polystyrene.[55] For instance, biomimetic treatments consisting of 1 day in SBF followed by 14 days in more concentrated SBF (1.5×) lead to nanocrystalline CHA, *i.e.* conventional biomimetic apatite commonly observed on the surface of bioactive ceramics after a few days in SBF. Nanocrystalline OCP or even ACP, considered as HA precursors during the biomineralization process, can be obtained at the implant surface by homogeneous precipitation within highly supersaturated SBF (5×). These kinds of solutions cannot be prepared at physiological pH 7.4, and acid pH values are required to avoid immediate precipitation. Once these precursors are formed, they can be converted into biomimetic apatite by soaking the substrates in SBF depleted of HA crystal growth inhibitors, *i.e.* without Mg^{2+} and HCO_3^-. At this point, it is important to highlight that, depending on the pH at which the precursors were precipitated, the microstructure of the final HA can vary from large plate-shaped crystals (CaP precursor precipitated at pH ~6.5) to small platelet crystallites (CaP precursor precipitated at pH <6). Table 3.2 summarizes the conditions to control the biomimetic process through the combination of solutions.

Table 3.2 shows how the chemical composition and microstructure of the biomimetic CaP is determining for an appropriated osteoblastic cells response. Viability *in vitro* cell culture studies indicate that biomimetic CaP precursors lead to higher percentages of cellular death, especially during the first days of culture. This fact seems to be related to the high reactivity of the biomimetically formed OCP or ACP precursors with the culture media, leading to strong microenvironmental changes in the calcium and phosphate ion concentrations. In contrast, biomimetic HA enhances the formation of extracellular matrix (ECM) and the biomineralization process by the osteoblastic cells. Moreover, when osteoblasts are seeded onto biomimetic HA there is an enhanced expression of osteocalcin and bone sialoprotein – ECM mineralization markers – compared with polystyrene substrates, especially in

Table 3.2   Different treatments for synthesizing biomimetic apatites and their effect on the osteoblastic response.[a]

Simulated fluid treatment	Biomimetic CaP precipitated	Osteoblastic cell response
SBF, 1 day + 1.5× SBF, 14 days	Conventional biomimetic CHA	Good development of anchoring elements Osteoblast elongation
5× SBF (pH 5.8) or 5× SBF (pH 6.5), 1 day	HA precursors (OCP or ACP)	High cellular death
5× SBF (pH 5.8), 1 day + SBF depleted of Mg^{2+} and HCO_3^-, 2 days	Small plate CHA	Cell viability, narrowing of anchoring elements, better spreading degree
5× SBF (pH 6.5), 1 day + SBF depleted of Mg^{2+} and HCO_3^-, 2 days	Large plate CHA	Enhanced formation of extracellular matrix and biomineralization process Enhanced cell differentiation

[a]SBF: simulated body fluid; CHA: carbonate hydroxyapatite; HA: hydroxyapatite; OCP: octacalcium phosphate; ACP: amorphous calcium phosphate.

those media depleted of inductive agents of osteoblastic gene expression, such as exogenous ascorbic acid and β-glycerol phosphate. The phosphorus presence at the microenvironment of biomimetic surfaces seems to provoke this response. Finally, biomimetic HA enhances the cell differentiation as deduced from the higher expression of osteopontin mRNA, especially large HA platelets. The mechanism is not still clear, but the better protein adsorption indicates that integrin-mediated signalling would be involved in the process.

Although SBF is a very useful fluid to mimic the "inorganic events" that occur during the bioactive process *in vivo*, the high ionic saturation makes the study of dissolution, precipitation and ionic exchange processes between the fluid and the ceramic difficult. For this reason simpler solutions such as tris(hydroxymethyl) aminomethane buffered solution at pH 7.3 are often preferred to determine the bioactive behaviour of bioceramics like bioglass,[56] especially for those studies where ion kinetic dissolution is the main focus of the research.

Table 3.1 also displays other solutions commonly used for biomimetic purposes. Hanks and Wallace's[57] balanced salt solution (HBSS) was the first successful simulated medium, containing the ions of calcium and phosphates together with other inorganic ions and glucose. HBSS is commercially available and still used in biomimetic experiments.[58,59] Homsy's pseudoextracellular fluid (PECF) is another phosphate-containing solution that also used for biomimetic apatite growth. Earl's balanced salt solution (EBSS) is a tissue culture medium that contains varying amounts of $CaCl_2$, $MgSO_4$, KCL, $NaHCO_3$, NaCl, $NaH_2PO_4 \cdot H_2O$ and glucose, according to the application and technique. It is commercially available in premixed salts or in solution. Finally, phosphate buffered saline (PBS) is a buffer solution commonly used in biochemistry. It is also a commercially available solution that only contains inorganic components and is suitable for biomimetic purposes.

3.3 Biomimetic Crystallization Methods

In principle, the biomimetic coating procedures and bioactivity tests described above involve a solution that is not renewed. Thus, the ions released from the glass remain in the container. This method is termed *static* or *integral*[60] and it is widely accepted that monitoring the formation of a CHA layer in these conditions predicts the material's bioactive behaviour. However, the use of the *static procedure* with highly reactive materials in aqueous solutions leads to remarkable variations in the ionic concentration and pH, reaching values far from physiological ones. This fact makes questionable the accuracy of these assays or the similarity of the coating with respect to biological apatites. Increases of pH of around 0.6 units from the initial 7.4 can be observed in the SBF, when bioactive sol–gel glasses are soaked for a few hours. Besides, variations in the ionic concentration of Ca(II), P(V) and Si(IV) are also detected just after a few minutes of assay.[61] Such pH increases could favour the CHA formation even in weakly bioactive materials.

Some authors have proposed the so-called *differential* method[62,63] in which the solution is renewed at predetermined intervals. However, the periodical solution exchange to eliminate such effects in bioactive glasses would require such short time intervals that the formation process of the CHA layer could be affected by the sample manipulation.

For that reason, and to simulate the continuous flux of body fluids at the implant surface, *dynamic* or *continuous in vitro* procedures have been proposed,[64] in which SBF is continuously renewed with the aid of a peristaltic pump. Figure 3.3 shows the scheme of the device used for *dynamic in vitro* assays.

Dynamic tests have been used to asses the *in vitro* bioactivity of several glasses, and compared with that without the renewal of the *in vitro* solution (*static*). A SBF flux at 1 mL min^{-1} allows the ionic concentration and pH of the solution to be maintained almost constant. As expected, the protocol modifications result in variations of the nanoapatite growth from both chemical and microstructural points of view. In *static* procedures, a faster initial formation of the amorphous phosphate coating is detected, but for higher soaking times the situation is equivalent in both cases. Under *dynamic* conditions, the apatite crystals formed are larger. Regarding the layer composition in *dynamic*, the Ca/P molar ratio is considerably lower than in *static* (1.2 *vs.* 1.6). This variation is explained by the differences in pH. The lower pH in *dynamic* (7.4) increases the HPO_4^{2-} concentration in solution compared with *static* where pH is close to 8. Thus, *dynamic* would favour the formation

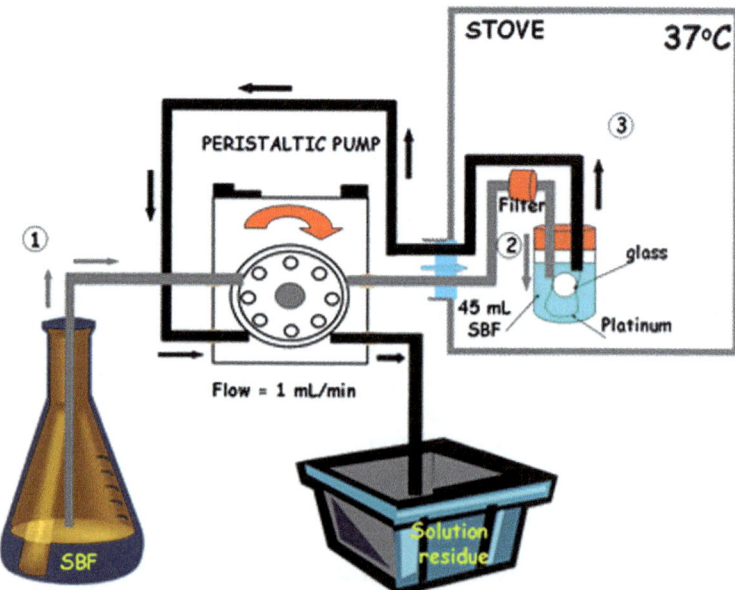

Figure 3.3   Schematic description of dynamic *in vitro* bioactivity assays. The continuous flow of the body fluids is modelled by the continuous renewal of the simulated body fluid (SBF) solution.

of calcium-deficient apatite, which might coexist with other calcium phosphates having a lower Ca/P molar ratio. In addition, the larger size of the CHA crystal aggregates formed under *dynamic* conditions is explained on the basis of the continuous supply of calcium and phosphate ions.

Other alternatives are the use of constant composition techniques such as those proposed by Nancollas and co-workers.[65–67] In these methods, multiple titrant solutions containing lattice ions are added to the reaction solution to compensate for the removal of these ions during growth. Thus, a constant thermodynamic driving force for crystal growth is maintained during the CaP growth. In order to mimic the kinetics of biological apatite crystallization, other methods such as use of a double diffusion crystallization device or crystallization within viscous gels have been proposed.[68–72] These methods are based on the restrained diffusion of calcium and phosphate ions from the opposite direction. Together with a double diffusion process, currently this is considered one of the most advanced experimental tools for mimicking biomineralization processes.

3.4 Calcium Phosphate Bioceramics

3.4.1 Bone Tissue Response to Calcium Phosphate Bioceramics

Calcium phosphates fall into the categories of biocompatible materials for bone and dental applications. Depending on their chemical composition, crystalline phase and microstructure, CaPs can slightly dissolve, promoting the formation of biological apatite before directly bonding with the tissue at the atomic level. This process results in the formation of a direct chemical bond with bone and it is named *bioactivity*.

After implantation, CaPs can act in different ways:

(1) *Osteoconduction*. Giving rise to a good stabilization through an osteoconductive mechanism, *i.e.* providing a bioactive surface on which the bone can grow without implant resorption.
(2) *Osteoinduction*. Osteoinductive materials will stimulate osteoblast proliferation and differentiation by providing biochemical signals that result in bone tissue regeneration. Osteoinduction is a property not traditionally attributed to CaP ceramics, but recent studies have demonstrated osteoblast stimulation for several CaP compositions.[73]
(3) *Bioresorption*. A bioresorbable material will dissolve and allow a newly formed tissue to grow into any surface irregularities, but may not necessarily interface directly with the material. In the field of CaP-based bioceramics, we can find examples for all the situations described above.[74,75]

Independently of the chemical composition, structure and microstructure of a bioactive ceramic, the analysis of the bone–implant interface reveals that the *presence of nanocrystalline CDHA* is one of the key features in the bonding zone.[76] In the case of CaP bioceramics, a second rule can be also established: *the implant solubility enhances the bone repair process.*[77-80] It does not mean that only highly soluble CaPs are useful for bone repair; CaPs with higher solubility are applied in those applications where the implant resorption is expected, followed by bone colonization, whereas less soluble CaPs are intended as osteoconductive materials, providing a bioactive surface that supports bone growth without dissolving, with better mechanical stability during the first stages of the repair process. Bioceramics made of dense HA would be a good example of bioactive material,[81,82] while porous scaffolds made of biphasic calcium phosphate (BCP; β-TCP/HA[83] or α-TCP/HA[84]) or bone grafts made of CDHA or ACP are examples of bioresorbable materials.[85,86]

3.4.2 Reactivity of Calcium Phosphate Bioceramics with the Biological Environment

The ability to bond to bone tissue is a unique property of bioactive materials. During this process, dissolution and precipitation reactions occur. Figure 3.4 schematically shows these phenomena, with a list of events occurring during the bioactive process. The events that constitute the bioactive process are commonly overlapped or simultaneously occurring, and the scheme displayed in Figure 3.4 should not be considered in terms of a time sequence.[87]

The scheme displayed in Figure 3.4 does not represent a mechanism in itself, but a description of observable events that occur at the interface after implantation. The mechanism must be related to physicochemical phenomena that occur in the presence or absence of cells, or are related to reactions mediated by cellular activity. An important aspect of the overall reaction sequence between these materials and tissues is that in the absence of biologically equivalent calcium-deficient carbonate apatite, dissolution,

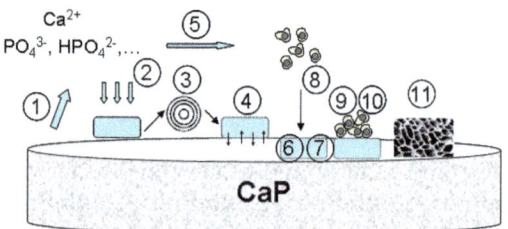

Figure 3.4   Events occurring during bone formation onto bioactive calcium phosphate (CaP) ceramics. (1) Dissolution from the ceramic. (2) Precipitation from solution onto the ceramic. (3) Ion exchange and structural rearrangement at the ceramic–tissue interface. (4) Interdiffusion from the surface boundary layer into the ceramic. (5) Solution-mediated effects on cellular activity. (6) Deposition of either the mineral phase or the organic phase, without integration into the ceramic. (7) Deposition with integration into the ceramic. (8) Chemotaxis to the ceramic surface. (9) Cell attachment and proliferation. (10) Cell differentiation. (11) Extracellular matrix formation.

precipitation and ion-exchange reactions lead to a biologically equivalent apatitic surface on the implanted material: the *in vivo* bioactivity is only strongly expressed if this new calcium-deficient carbonate apatite is formed. Under *in vitro* conditions in noncellular simulated physiological conditions, stages 1–4 are reproduced, leading to the precipitation of biomimetic CaPs.

3.4.3 Physical–Chemical Events in CaP Bioceramics During the Biomimetic Process

3.4.3.1 *CaP Dissolution During Biomimetic Processes*

The reactivity of a CaP is dependent on its composition and structure.[88] One of the mechanisms underlying the phenomenon of *in vitro* bioactivity is that dissolution from the ceramic produce solution-mediated events leading to mineral precipitation.[77,78] Under *in vivo* conditions, the process involves more complicated biological reactions affecting cellular activity and organic matrix deposition.[89,90]

Table 3.3 displays some chemical and textural characteristics of the most important CaP bioceramics in the field of dental and orthopaedic surgery. When studying the physical–chemical features of CaP compounds, we must take into account the following parameters:

(1) Type of CaP ceramic, *i.e.*, hydroxyapatite, tricalcium phosphate, tetracalcium phosphate, *etc.*
(2) Type of crystal-chemical defects, such as deviation from stoichiometry leading to calcium-deficient compounds, dehydroxylation, *etc.*
(3) Polymorph considered for a chemical compound, such as α-TCP and β-TCP.
(4) Number and type of CaP phases existing in the system, commonly monophasic or biphasic CaP systems.

Even considering all these parameters, the question about CaP solubility under the action of a physiological solution is not trivial. In order to quantify the dissolution of a CaP when soaked into a buffered fluid, two different

Table 3.3 Crystalline phases, Ca/P ratio and surface area of some calcium phosphate bioceramics.

Bioceramic	Phases	Ca/P ratio	S_{BET} ($m^2\ g^{-1}$)
HA	Stoichiometric hydroxyapatite	1.67	5.1
CDHA	Ca-deficient HA	1.61	62.9
OHA	Oxyhydroxyapatite	1.67	2.48
β-TCP	β-Tricalcium phosphate	1.5	0.64
α-TCP	α-Tricalcium phosphate	1.5	0.08
TTCP	Tetracalcium phosphate	2.0	0.24
BCP-45	45 HA/55 β-TCP	1.58	5.05
BCP-27	27 HA/73 β-TCP	1.55	4.15

approaches can be applied: the *initial dissolution rate* and *concentration of dissolved ions at the equilibrium.*[91]

3.4.3.1.1   Initial Dissolution Rate.   Determination of the initial dissolution rate must be performed with the data points experimentally obtained at the short initial immersion period, when the ionic product does not vary or does not significantly affect the undersaturation factor. Under these conditions, the initial dissolution rate is related with the following rate expression:

$$\frac{d[Ca]}{dt} = k \cdot t^m \tag{3.1}$$

with

[Ca]: Ca concentration in solution
k: constant
m: effective order of the reaction

Developing the derivative function, the initial dissolution rate can be expressed in an easy logarithmic expression as a function of soaking time, as follows:

$$\log[Ca] = A_0 + A_1 \log t \tag{3.2}$$

where

$A_0 = \log(k/m + 1)$
$A_1 = m + 1$

In this way, by measuring the Ca^{2+} concentration as a function of soaking time, the solubility of the CaP substrate can be determined by attending to some specific characteristic. For instance, the influence of crystal-chemical defects can be estimated by comparing the solubility of stoichiometric HA, partially dehydroxylated HA (OHA) and CDHA. Experimentally, it can be observed that solubility increases in the order

HA < CDHA < OHA

Following the same procedures, it was observed that factors such as high specific surface area, crystallographic defects and nonstoichiometry enhance the dissolution rate of the CDHA.[92] The general formula for CDHA is $Ca_{10-x}(HPO_4)_x(PO_4)_{6-x}(OH)_{2-x}$, where x can vary from 0 to almost 2.[93] In addition to the Ca deficiency, the low carbonate content generally contained in these compounds contributes to the structural disorder of the CDHA by replacing the tetrahedral PO_4 group by a planar CO_3. Finally, the Ca deficiency is also accompanied by hydroxyl group deficiency. This set of crystal-chemical defects lead to the higher initial dissolution rate of CDHA when compared to HA.

OHA can be presented by a formula: $Ca_{10}(PO_4)_6(OH)_{2-2x}O_x\square_x$, where $\square$ means a vacancy. In the case of OHA, one O^{2-} ion and a vacancy substitute for two monovalent OH⁻ ions. The enhanced Ca^{2+} release would be a consequence of the weak bonding interaction of Ca^{2+} ions around the vacancies, as presented in Figure 3.5.

Regarding phosphate release, CDHA also shows a higher dissolution rate compared with HA. The same factors contributing to Ca^{2+} release can explain the phosphate dissolution. Contrarily, OHA does not show a P release enhancement with respect to HA. Probably, the lower amount of OH⁻ groups decreases the hydrogen attraction on its surface. Since H⁺ governs the solid-to-solution exchange of the phosphate ions, this crystal-chemical feature could be responsible of the lower P release in the case of OHA.

Obviously, different dissolution rates are observed when comparing different CaP phases. Greater Ca^{2+} and PO_4^{3-} release rate from β-TCP than from HA is expected, since β-TCP is a known metastable member of the CaP family.[94] β-TCP cannot be precipitated from aqueous solutions, but it is a high-temperature phase of calcium orthophosphates, which only can be prepared by thermal decomposition, *e.g.* of CDHA, at temperatures >800 °C. Even exhibiting lower surface areas than HA, β-TCP shows a larger dissolution rate than HA. At temperatures >1125 °C, transformation of β-TCP to α-TCP takes place. α-TCP is more soluble in aqueous media and both the initial dissolution rate and the ionic product for α-TCP are significantly greater than those for β-TCP. Excellent and extensive information on this topic can be found in the books of Elliott[95] and LeGeros.[96]

Among the single-phase CaPs considered, tetracalcium phosphate (TTCP) displays one of the greatest initial dissolution rates. However, a rapid increase in Ca and P content is commonly followed by a decrease first in P content and then in Ca content. It indicates that the solution with immersed TTCP becomes rapidly saturated with one of the metastable CaPs phases.

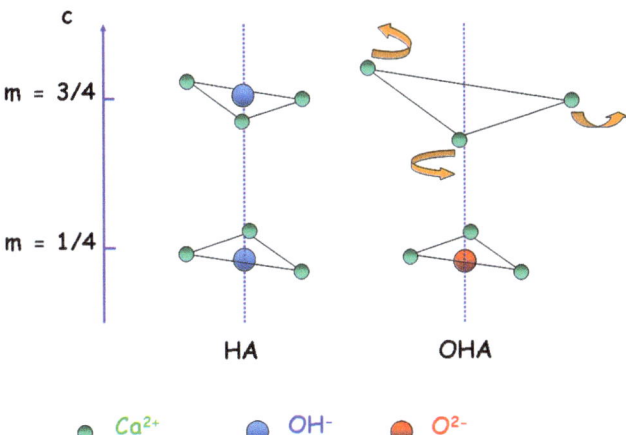

Figure 3.5 Scheme of the ions along the *c* axis in the hydroxyapatite (HA; left) and oxyhydroxyapatite (OHA; right). The vacancies at the hydroxyl sites in OHA result in weaker ionic interactions with the Ca^{2+}, facilitating the ion dissolution.

The subsequent decrease of the P and Ca content is the result of precipitation of new phase(s) on the TTCP surface.

Biphasic calcium phosphates (BCPs) consist of mixtures of HA and β-TCP. Due to the higher solubility of the β-TCP component, the reactivity increases with the β-TCP/HA ratio. Therefore, the bioreactivity of these compounds can be controlled through the phase composition.

The initial dissolution rates of the single-phase CaPs in undersaturated conditions at physiological pH increase in the order:

$$HA < CDHA < OHA < \beta\text{-}TCP < \alpha\text{-}TCP < TTCP$$

whereas BCPs solubility would fall somewhere between HA and β-TCP, depending upon the quantitative phase composition.

3.4.3.1.2 Concentration of Dissolved Ions at Equilibrium. Since dissolved ions are transported away by physiological fluids under *in vivo* conditions, the concentration of dissolved species at equilibrium is not a useful parameter to explain the bioactive behaviour of bioceramics. However, when considering the *in vitro* biomimetic synthesis of nanoapatites, it becomes an essential parameter to understand the subsequent nanoapatite precipitation, specially when *integral* methodology is applied.

Table 3.4 displays some of the CaP ceramics with applications in dental and orthopaedic surgery, together with their solubility parameters and pH stability.

The precipitation of CaPs is known to be principally determined by calcium and phosphate concentrations and condition of the nucleation site. Therefore, the amount of bioceramic dissolved at the equilibrium point strongly depends on the ionic strength of the solution.

In the case of CaP-based bioceramics, reaching the saturation points for Ca^{2+} and phosphates is very important for the biomimetic formation of CaP. Both, the crystalline phase and the amount of newly formed CaP are strongly

Table 3.4 Solubility and pH stability of some biologically relevant calcium phosphates.

Compound	Formula	Solubility $-\log(K_S)$	pH stability in aqueous solution (25 °C)
α-Tricalcium phosphate (TCP)	$\alpha\text{-}Ca_3(PO_4)_2$	25.5	NA[a]
β-Tricalcium phosphate (TCP)	$\beta\text{-}Ca_3(PO_4)_2$	28.9	NA[a]
Amorphous CaP (ACP)	$Ca_xH_y(PO_4)_z \cdot nH_2O$ $n = 3\text{–}4.5$	25.7	5–12
Ca-deficient hydroxyapatite (CDHA)	$Ca_{10-x}(HPO_4)(PO_4)_{6-x}(OH)_{2-x}$ $(0 < x < 1)$	85.1	6.5–9.5
Hydroxyapatite (HA)	$Ca_{10}(PO_4)_6(OH)_2$	116	9.5–12

[a]These compounds cannot be precipitated from aqueous solutions.

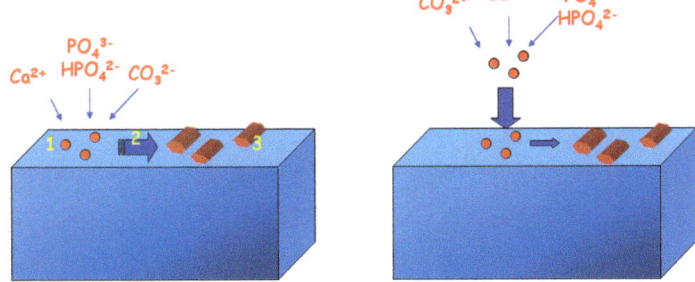

Figure 3.6 Heterogeneous (left) and homogeneous (right) precipitation of calcium phosphate nanoceramics.

dependent on the Ca^{2+} and phosphate concentration in solution. This highlights that not only heterogeneous precipitation takes place at the bioceramic surface, but also that homogeneous precipitation must occur during the biomimetic CaP formation (Figure 3.6).

3.4.3.2 Precipitation of Nanoapatites on CaP Bioceramics

Whereas it is recommended that dissolution studies are performed in simple buffered solutions such as Tris buffer, biomimetic CaP precipitation reactions are performed in SBFs with ionic compositions similar to that of physiological fluids (Table 3.1). These solutions are highly saturated in phosphates, calcium and carbonates (among other chemical species) and trend to precipitate onto the surface of bioactive ceramics as bone-like apatite phases or CaP precursors, for instance ACP and OCP.

The concept of "bone-like apatite" includes the observation that this biomimetic compound shows the apatite crystalline structure, exhibits calcium deficiency, possesses carbonate groups in the unit cell and, from the microstructural point of view, exhibits a small crystallite morphology (often needle-like). It can be said that the biomimetic apatite structure is very similar to the mineral phase of natural bone, although the kind of solution used during the biomimetic process will determine the degree of similarity. Determining the nanoapatite precipitation onto CaP bioceramics it is not a trivial issue. In fact, the apatite formation on calcium-phosphate-based ceramics has been the focus of much research for over a decade. However, convincing evidence of apatite identification does not often occur. Sometimes, researchers rely mainly on the diffraction methods to identify crystal structure. It has been a challenge to identify crystal structure of precipitates formed on surfaces of bioceramics, because the small quantity of precipitates generates very weak peak intensities in powder X-ray diffraction analysis. Identifying microcrystals formed on the surfaces of bioactive CaPs is even more difficult because the strong peaks of the substrates overlap the precipitate peaks in the powder X-ray diffraction pattern.[97] Electron

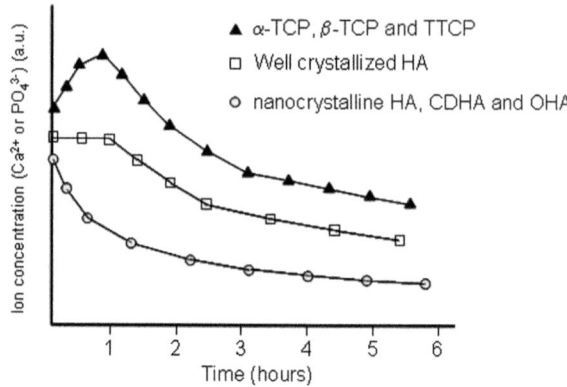

Figure 3.7 Ca^{2+} and PO_4 concentration in simulated body fluid *vs.* soaking time of different calcium phosphates.

diffraction (ED), transmission electron microscopy (TEM) and Fourier transform infrared (FTIR) spectroscopy should be considered as more effective tools than powder X-ray diffraction for identifying precipitation phases formed on bioceramics.

After soaking a CaP bioceramic in SBF, the first measurable parameter is the induction time, *i.e.* the time prior to a detectable decrease in Ca^{2+} and PO_4^{3-} concentrations of the fluid as result of precipitation[98] (see Figure 3.7). Before this point, the Ca^{2+} and PO_4^{3-} concentrations can keep constant with respect to the initial concentrations, or can increase during induction time. When a decrease is observed from the beginning, it is said that the induction time is equal to zero. As displayed in Figure 3.7, HA with a low degree of crystallinity, calcium deficient HA and oxy-hydroxyapatite shows zero induction time, whereas more soluble CaP such as α-TCP, β-TCP and TTCP lead to an increase of the Ca^{2+} and PO_4^{3-} ion concentrations during the induction times. Well-crystallized HA does not commonly show an ionic concentration increase during the induction times.

The ionic concentration of the fluid clearly determines the precipitation or not of biomimetic CaPs. For instance, when HA and β-TCP are immersed into Tris buffer or phosphate containing PECF (without Ca^{2+}), no new CaP phase is formed on the surfaces.[47] The limited solubility of HA and TCP may be the main reason of this failure of surface CaP formation in these solutions. This assumption is supported by the fact that in other biomimetic solutions, initially saturated with Ca^{2+} and phosphate, HA and TCP produced CaP layers on their surfaces. Thus, in terms of surface change, sufficient concentrations of both Ca^{2+} and phosphate are essential for low soluble HA and β-TCP. Therefore, under high Ca^{2+} and phosphate concentrations, all the CaPs bioceramics considered so far in this chapter can develop a new phase on the surface. However, kinetics, compositions and structures of the new phases are significantly different.

3.5   Biomimetic Nanoceramics on Hydroxyapatite and Advanced Apatite-Derived Bioceramics

3.5.1   Hydroxyapatite, Oxyhydroxyapatite and Ca-Deficient Hydroxyapatite

Since 1970, the beneficial effects of HA implants have been the object of study for the biomaterials scientific community. Crystalline HA is a synthetic material analogue to CaP found in bone and teeth,[99] and is a highly cytocompatible material that has been considered for coating on metallic implants,[100] porous ceramic that facilitates bone ingrowth,[101] an inorganic component in a ceramic–polymer composite,[102] a granulate to fill small bone defects[103] and for tissue engineering scaffolds.[104]

Besides its excellent biocompatibility, synthetic HA mimics many properties of natural bone.[12] HA allows a specific biological response at the tissue–implant interface, which leads to the formation of bonds between the bone and the material.[105] As described above, this response is mediated by solution, precipitation and ionic exchange reactions that result in the surface transformation into a biomimetic nanoceramic formed surface.

The data indicate that the behaviour of the HA family upon immersion in most simulated physiological solutions was structure- and composition-dependent. The structural effect is a combination of crystallinity and specific surface area, since these structural properties varied in parallel.

When apatite is soaked in any Ca^{2+} and PO_4-containing SBF, the variations observed within the fluid are essential to understand the biomimetism of these compounds. The degree of crystallinity and dehydroxylation, stoichiometry, *etc.* affect the reactivity of the apatite. In this sense, low crystalline HA (for instance synthesized by wet methods and treated <700 °C) incorporates Ca^{2+} and phosphate ions from the solution immediately after contact with the fluid. Therefore, low crystalline HA exhibits induction time equal to zero. Similar behaviour is shown by CDHA and those highly dehydroxylated HA. In fact, the Ca^{2+} and PO_4 incorporation is initially as intense as the precipitation occurring in supersaturated solutions when CaP seeds are soaked within. The Ca^{2+} and PO_4 incorporation gradually decreases insofar as the solid/solution equilibrium is reached.

In contrast, the reaction that takes place on crystalline HA is significantly different. In the absence of measurable dissolution processes, crystalline HA shows induction times of ~1 hour (at pH 7.4 and 37 °C). From then on, a decrease in Ca^{2+} and PO_4 can be measured in the SBF.

In addition to the induction time, the Ca/P molar ratio of the newly formed CaP provides essential information for elucidating its crystal-chemical characteristics. For instance, Ca/P molar ratios of 1.75–1.79 are commonly calculated for the biomimetic CaP precipitated onto CDHA. In this case, Ca/P ratio is higher than 1.67, which indicates that the newly formed CaP is a type B carbonate apatite, in which PO_4^{3-} substitutes for CO_3^{2-}. This type of apatite commonly occurs during biological mineralization processes and consists of solid solutions, whose compositions can vary between $Ca_{10}(PO_4)_6(OH)_2$ and $Ca_8(PO_4)_4(CO_3)_2$. In the case of biomimetic CaP precipitated onto low crystalline HA, the Ca/P molar ratio is around 1.66, that is, very close to stoichiometric HA, whereas those precipitated onto crystalline HA and OHA are 1.34–1.40 and 1.45, respectively, far from Ca/P ratio of 1.67 or the stoichiometric HA.

As mentioned before, the characterization of biomimetic CaPs by X-ray diffraction is very difficult when they are precipitated onto synthetic apatites. Therefore, techniques such as FTIR spectroscopy and TEM play a fundamental role in this kind of study to follow the biomimetic process on the surface of CaP-based bioceramics. Table 3.5 shows the FTIR characteristic data of the biomimetic evolution for several CaP bioceramics.

Crystalline HA does not exhibit significant changes in the FTIR spectra after being soaked in SBF. The slight formation of an amorphous phase that incorporates a small amount of carbonates is observed. Crystalline HA shows a very slow kinetic for the reactions that constitute the bioactive process (dissolution, precipitation and ionic exchange) and, consequently, several strategies have been proposed to upgrade their biomimetic capacity.

3.5.2 Silicon-Substituted Apatites

The biomimetic behaviour of HA can be improved by introducing some substitutions into the structure.[106] The apatite structure can incorporate a wide variety of ions, which affect both its cationic and anionic sublattices.

Table 3.5 Fourier transform infrared (FTIR) spectroscopy absorption band modifications for several apatites during the first week soaked in simulated body fluid.[a]

Bioceramic	FTIR spectra evolution
CDHA	Appearance/increase at 875 cm^{-1} and 1418–1460 cm^{-1} regions of C–O characteristic bands
	Gradual reduction of the splitting of the PO_4^{3-} absorption bands at 600 cm^{-1}, 550 cm^{-1} and 1100–1000 cm^{-1} corresponding to the formation of amorphous or low crystalline CaP phases
Nano-HA	Appearance/increase at 875 cm^{-1} and 1418–1460 cm^{-1} regions of C–O characteristic bands
	Gradual reduction of the splitting of the PO_4^{3-} absorption bands at 600 cm^{-1}, 550 cm^{-1} and 1100–1000 cm^{-1}, corresponding to the formation of amorphous or low crystalline CaP phases
Crystalline HA	Gradual reduction of the splitting of the PO_4^{3-} absorption bands at 600 cm^{-1}, 550 cm^{-1} and 1100–1000 cm^{-1}, corresponding to the formation of amorphous or low crystalline CaP phases
OHA	Appearance/increase at 875 cm^{-1} and 1400–1500 cm^{-1} regions of C–O characteristic bands
	Appearance at 632 cm^{-1}, corresponding to the librational mode of OH. Occasionally, appearance at 559 cm^{-1} and 525 cm^{-1}, corresponding to octacalcium phosphate formation

[a]CDHA: calcium-deficient hydroxyapatite; CaP: calcium phosphate; HA: hydroxyapatite; OHA: partially dehydroxylated HA.

For example, in biological apatites, substitutions of CO_3^{2-} by PO_4^{3-} (type B) or OH^- (type A) are likely.[107,108] In the case of B-type carbonated apatites, the neutrality is usually reached by the incorporation of single-valence cations (Na^+ or K^+) in the Ca^{2+} positions.[109,110]

Studies performed by Carlisle[111,112] have shown the importance of silicon in bone formation and mineralization. This author reported detection of silicon *in vivo* within the unmineralized osteoid region (active calcification regions) of the young bone of mice and rats. Silicon levels up to 0.5 wt% were observed in these areas, suggesting that Si has an important role in the bone calcification process. Moreover, the highest bioactivity of silica-based glasses and glass-ceramics (and the mechanism proposed for the bioactive behaviour),[113,114] suggested that the incorporation of silicon into apatites would improve the *in vivo* bioactive performance. New apatite layers are formed on the surface of bioactive silica-based glasses and glass-ceramics after a few hours in SBFs. The formation of silanol groups (Si–OH) has been proposed as a catalyst of the apatite phase nucleation, and the silicon dissolution rate is considered to have a major role on the kinetics of this process.[115,116] These events suggested the idea of incorporating Si or silicates into the HA structure.

Si-substituted hydroxyapatites (SiHA) are among the most interesting bioceramics from the biomimetic point of view. *In vitro* and *in vivo* experiments have evidenced an important improvement of the bioactive behaviour respect to nonsubstituted apatites.[117,118] Figure 3.8 shows the scanning electron micrographs (SEM) of pure HA and SiHA after 5 weeks soaked in SBF. The surface of pure HA remains almost unaltered at the SEM observation, since the slow surface reactivity does not allow the observation of significant changes under these conditions. In contrast, SiHA develops a new apatite phase with a different morphology with respect to the substrate. The surface of SiHA appears covered by a new material with acicular and plate-like morphology, characteristic of new apatite phases grown on bioactive ceramics.

The term silicon-substituted means that silicon is substituted into the apatite crystal lattice and is not simply added. Silicon or silicates are supposed

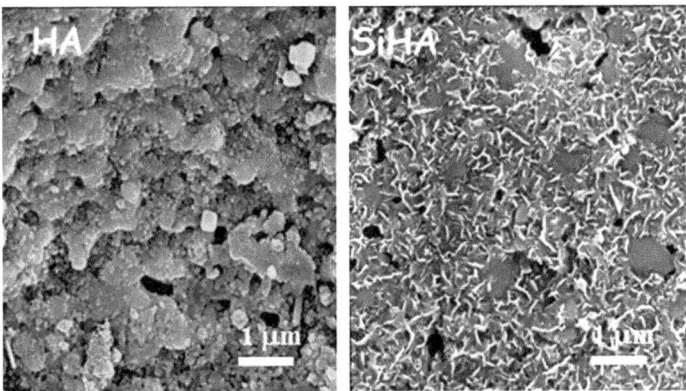

Figure 3.8 Scanning electron micrographs of hydroxyapatite (HA) and silicon-substituted HA after 5 weeks in simulated body fluid.

to substitute for phosphorus, or phosphates, with the subsequent charge imbalance.[119] The amount of silicon which can be incorporated seems to be limited. The literature includes values ranging from 0.1 to 5% by weight in silicon.[120–122] Small amounts of 0.5% and 1% are enough to yield important biomimetic improvements.

The controlled crystallization method is by far the most common synthesis route to obtain SiHA found in the scientific literature.[117–120,123] This process comprises the reaction of a calcium salt or calcium hydroxide with orthophosphoric acid or a salt of orthophosphoric acid in the presence of a silicon-containing compound. Under these conditions it is believed that the silicon-containing compound yields silicon-containing ions, such as silicon ions and/or silicate ions, which substitute into the apatite lattice. There are several synthetic routes to incorporate Si into the hydroxyapatite structure[119,121,124–126] and the kind of silicon precursor, as well as the synthesis method, can lead to different SiHA with different chemical and physical properties. This is clear in the case of the thermal stability of these compounds.

The amount of silicon incorporated also has an important influence on the thermal stability. For instance, when series of SiHA with the nominal formula $Ca_{10}(PO_4)_{6-x}(SiO_4)_x(OH)_{2-x}$ (for $x = 0$, 0.25, 0.33, 0.5 and 1) are prepared, using tetraethyl orthosilicate (TEOS) as silicon source, the as-precipitated samples are always a single nanocrystalline apatite phase (Figure 3.9). After heating at 900 °C, samples with Si content up to 0.33 remained a single apatite phase, whereas higher Si content led to decomposition into HA and α-TCP.[127,128] In fact, this is an appropriate method to obtain biphasic material α-TCP–HA at relatively low temperature. α-TCP is a high-temperature phase that appears when HA or β-TCP is treated over 1200 °C. The presence of silicon seems to stabilize the α-TCP at lower temperatures.

3.5.2.1   Crystal-Chemical Considerations of SiHA

Silicon (or SiO_4^{4-}) for P (or PO_4^{3-}) is a non-isoelectronic substitution. It means that the extra negative charge introduced by SiO_4^{4-} must be compensated by means of some mechanism, for example creating new anionic vacancies. The Si, or SiO_4^{4-} incorporation into the apatite structure at the P, or PO_4^{3-} position has been studied by several authors. Gibson *et al.*[119] have reported on the structure of aqueous precipitated SiHA. The main structural evidences reported were the decrease and increase of *a* and *c* parameters, respectively, absence of secondary phases and the increase of tetrahedral distortion. These authors have proposed a mechanism to compensate the negative charge introduced by the SiO_4^{4-} incorporation in apatites obtained by the aqueous precipitation method. They state the formation of vacancies at the OH⁻ site, in a mechanism that can be summarized as follows:

$$PO_4^{3-} + OH^- \leftrightarrow SiO_4^{4-} + \square$$

obtaining   Si   substituted   apatites   with   the   general   formula $Ca_{10}(PO_4)_{6-x}(SiO_4)_x(OH)_{2-x}\square_x$.

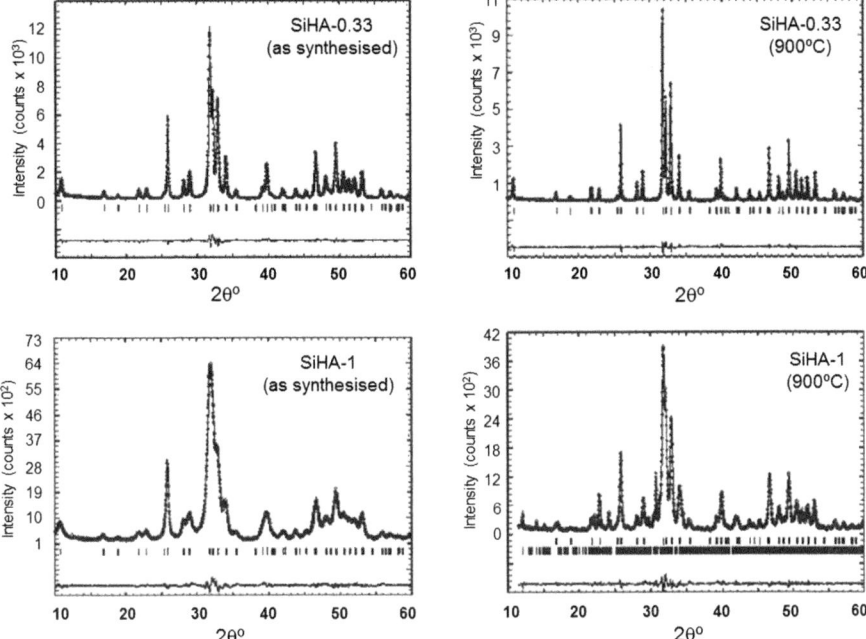

Figure 3.9 Powder X-ray diffraction pattern of $Ca_{10}(PO_4)_{6-x}(SiO_4)_x(OH)_{2-x}$, for x = 0.33 and 1. As-synthesised samples (left) and treated at 900 °C (right). The vertical lines mark the positions of Bragg peaks for an apatite-like phase and α-tricalcium phosphate (only for SiHA-1 treated at 900 °C).

The structural analysis of SiHA has been performed by X-ray diffraction studies. However this technique does not distinguish between P and Si, since they are almost isoelectronic, and the presence of H atoms cannot be determined by this technique. Neutron diffraction data seem to be an appropriate method for the structural study of Si substituted HA.[129-131] In order to explain the higher reactivity of SiHA, the neutron diffraction studies haven been focused on the presence of H^+ into the apatite structure, specially taking part of OH^-. This group has great importance in the reactivity of these compounds. As can be seen in Figure 2.14 the thermal displacement of the H atom along the *c*-axis is more than twice that for SiHA. This disorder, together with the tetrahedral distortion resulting from the substitution of PO_4^{3-} by SiO_4^{4-}, could contribute to the higher reactivity of SiHA. However, a crystal-chemical explanation of the SiHA-improved biomimetism would be clearly insufficient. The biomimetic process is a surface process, which is enhanced by the material reactivity. The sum of the different factors may justify the enhanced reactivity. From the point of view of *crystalline structure*, silicon yields tetrahedral distortion and disorder at the hydroxyl site, which could decrease the stability of the apatite structure and therefore increase the reactivity. At the *microstructural level*, the changes are even more evident. Grain boundaries defects are the starting points of dissolution under *in vivo* conditions. There is a close relationship between the amount of silicon, the number of sintering defects at the grain boundaries and the dissolution rate. In particular, the number of triple junctions in SiHA may have an important role in the material reactivity and consequently, in the rate at which the ceramic reacts with the bone. Finally, the *surface charge* undergone by the ceramic due to the presence of SiO_4^{4-} would also play an important role for the Ca^{2+} incorporation at the new biomimetic layer. This effect could be also responsible in part for the alteration in its biological response. In summary, to understand the improved biomimetic behaviour in SiHA requires its consideration as a sum of different factors at different levels.

3.6   Biphasic Calcium Phosphates (BCPs)

3.6.1   An Introduction to BCPs

Nowadays, the general requirements established for ideal implants aimed at bone regeneration should exhibit pores of several hundred micrometres, a biodegradation rate comparable to the formation of bone tissue (*i.e.* between a few months and about 2 years) and sufficient mechanical stability.[3] HA and TCP (both, α and β polymorphs) do not fulfil these requirements and some clinical failures have occurred as a consequence of a non-appropriate biodegradability kinetic, which eventually will involve a disadvantage to the host tissue surrounding the implant. For instance, some implants made of calcined HA to reconstruct mandibular ridge defects have resulted in high failure rate in human clinical applications.[132] In order to avoid this problem, the use of granular forms of HA instead of block forms was suggested,[133] although HA exhibited some drawbacks due to lack of biodegradability, independently of the implant form. On the other hand, β-TCP ceramics have been developed as a biodegradable bone replacement and are commercially available as, for instance, ChronOS™, Vitoss™, *etc.*[134] However, when used as a biomaterial for bone replacement, the rate of biodegradation of TCP has been shown to be too fast. In 1988, Daculsi *et al.*[83] thought that the presence of an optimum balance of stable HA and more soluble β-TCP should be more favourable than pure HA and β-TCP. Due to the biodegradability of the β-TCP component, reactivity increases with the β-TCP/HA ratio. Therefore, the bioreactivity of these compounds could be controlled through the phase composition. The main advantage with respect to other nonsoluble CaPs is that the mixture is gradually dissolved in the human body, acting as a stem for newly formed bone and releasing Ca^{2+} and PO_4^{3-} to the local environment.[135] *In vivo* tests have confirmed the excellent behaviour of BCP concerning the biodegradability rate.[136-139]

Since Ellinger *et al.*[136] used the term BCP for the first time to describe a mixture of β-TCP and HA, many advances have occurred in the BCP field. The work undertaken by Daculsi and co-workers[138,140] impelled the commercialization of BCP and currently can be found as trademarked products such as Triosite™, HATRIC™, Tribone™, *etc.* Nowadays, BCPs are clinically used as an alternative or as an additive to autogenous bone for dental and orthopaedic applications. Implants shaped as particles, dense or porous blocks, customized pieces and injectable polymer–BCP mixtures are common BCP-based medical devices. Moreover, research is in progress to enlarge the clinical applications to the field of scaffolding for tissue engineering[141–143] and carriers loading biotech products.[144,145]

HA chemistry and structure have been widely explained in Chapters 1 and 2. β-TCP is a phase that crystallizes in the rhombohedral system, with a unit cell described by the space group *R3Ch* and unit cell parameters a = 10.41 Å, c = 37.35 Å, γ = 120°. At temperatures above 1125 °C it transforms into the high-temperature phase α-TCP. Being the stable phase at room temperature, β-TCP is less soluble in water than α-TCP. Pure β-TCP never occurs in biological calcifications, *i.e.* there is no biomimetic process that results in β-TCP. Only the Mg-substituted form (whitlockite) is found in some pathological calcifications (dental calculi, urinary stones, dentinal caries, *etc.*).

α-TCP is usually prepared from β-TCP by heating >1125 °C, and it might be considered as a high-temperature phase of β-TCP. α-TCP crystallizes in the monoclinic system, with a unit cell described by the space group $P2_1/a$ and unit cell parameters of a = 12.89 Å, b = 27.28 Å, c = 15.21 Å and β = 126.2°. Therefore, α-TCP and β-TCP have exactly the same chemical composition, but they differ in the crystal structure. This structural difference determines that β-TCP is more stable than the α-phase. Actually, α-TCP is more reactive in aqueous systems, has a higher specific energy and it can be hydrolyzed to a mixture of other CaPs. Similarly to the β-phase, α-TCP never occurs in biological calcifications, and it is occasionally used in CaP cements.[86,146,147] In recent years, α-TCP is being used as a component of biphasic HA–α-TCP bioresorbable scaffolds. This material is obtained by heating SiHA at temperatures ~1000 °C, obtaining the so-called silicon-stabilized α-TCP.[84,148–150]

3.6.2 Biomimetic Nanoceramics on BCP Biomaterials

As described above, the biomimetic process in CaPs is based on dissolution, precipitation and ion exchange processes. The dissolution rate of BCPs depends on the ratio of TCP to HA in the compound.[151] Under *in vivo* conditions, the CaP ceramics containing a greater amount of TCP phase also show greater biodegradation. Many factors influence both dissolution and biodegradation, including the size and the conditions under which HA and TCP are synthesized. Interfacial aspects include stability when subjected to body fluid, porosity of surface and grain boundaries condition. However, the most important factor determining the dissolution and biodegradability is the TCP to HA ratio.[152] At a pH range of 4.2–8.0, and therefore at the physiological pH 7.4, HA is less soluble than other tricalcium phosphates. In fact, the tricalcium phosphate dissolves 12.3 times faster than HA in acidic medium and 22.3 times faster than HA in basic medium.

When β-TCP/HA biphasic materials are soaked in a simulated physiological solution, SBF for instance, the pH values of any experimental solution decreases to values between 4.6 and 6.0 after several weeks of immersion. This fact is consistent with most biomimetic processes, which evidence a pH decrease of the solution during the CaP precipitation. HA does not dissolve, but β-TCP is subject to dissolution. The interaction of the TCP phase with the solution takes place in a very short time after soaking.

Following the phase content patterns as a function of soaking time collected using X-ray diffraction, it can be seen that depending on the TCP amount contained in the BCP, 25–100% of the β-TCP or α-TCP contained can be dissolved after 4 weeks of soaking. However, BCPs do not only degrade under the action of physiological solutions. In fact, the changes in weight of most BCP materials tested after immersion in SBF are negligible, which means that the precipitation of new CaP phases (biomimetic ones) and/or hydrolysis of the TCP phase also occur on the surface of the materials.

Whereas the dissolution process seems to be an easy question to resolve in BCPs, the precipitation process becomes more difficult to understand. In fact, the newly formed phases are different when the biomimetic process is performed under static or dynamic conditions and, of course, completely different for *in vivo* experiments.[97] In static biomimetic conditions, calcium and phosphate ions from the TCP phase in the BCP dissolves into solution and reprecipitates onto the BCP as CDHA. Due to its greater stability, the HA of the BCP acts as a seed material in physiological solutions. However, a deep study of the biomimetic CaP formed under dynamic conditions can show a different scenario for the BCPs. Single crystalline precipitates of CaPs on porous BCP bioceramics obtained after immersion in dynamic SBF and after implantation in pig muscle were examined using TEM electron diffraction. The crystals formed *in vitro* in dynamic SBF were identified as OCP, instead of apatite. The hard evidence provided by single-crystal diffraction indicates that the precipitation of BCP in SBF may be neither "bone-like" nor "apatite".

3.7 Biomimetic Nanoceramics on Bioactive Glasses

3.7.1 An Introduction to Bioactive Glasses

Bioactive glasses were discovered by Professor Hench in 1971 and nowadays are considered as the first expression of bioactive ceramics. Due to the high bioactivity level and their brittleness, these materials find clinical application in those cases where high tissue regeneration is required without supporting high loads or stresses. Currently they are used for the replacement of ear bones and as powders for periodontal surgery and bone repair.[56]

The starting point for the first bioglass synthesis was based upon the following simple hypothesis:[153]

> *The human body rejects metallic and synthetic polymeric materials by forming scar tissue because living tissues are not composed of such materials. Bone contains a hydrated calcium phosphate component, hydroxyapatite and therefore if a material is able to form a hydroxyapatite layer in vivo it may not be rejected by the body*

Actually, the apatite phase formed on the surface of bioactive glasses is calcium deficient, carbonate containing, nanocrystalline and therefore very

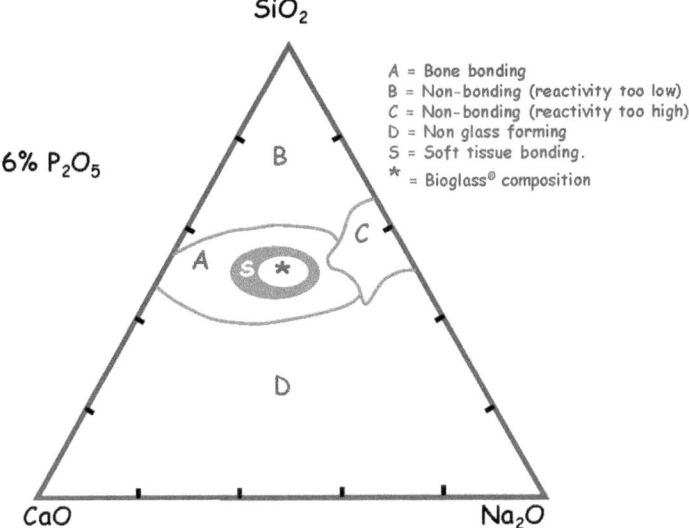

Figure 3.10 Compositional diagram for bone-bonding ability of melt-derived glasses.

similar to the biological ones. The first *in vivo* experiments carried out with the so-called 45S5 Bioglass® (see Figure 3.10) demonstrated that these apatite crystals were bonded to layers of collagen fibrils produced at the interface by osteoblasts. This chemical interaction between the newly formed apatite layer and the collagen fibrils constitute a strong chemical bond denoted a "bioactive bond".[154,155]

Bioactive glasses exhibit class A bioactivity, *i.e.* they are *osteoproductive*[†] materials,[156] instead of those ceramics such as HA that behave as *osteoconductive*[‡] materials and are classified as class B bioactive materials. Since both kinds of materials are bioactive, they form a mechanically strong bond to bone. However, as a class A bioactive material, bioactive glasses exhibit a higher rate of bonding to hard tissues (although they also bond to soft tissues). Together with a rapid bonding to bone, bioactive glasses also show enhanced proliferation compared to calcium phosphate bioceramics or any other class B bioactive ceramic.

3.7.2 Composition and Structure of Melt-Derived Bioactive Glasses

The first bioactive glass reported in 1971 was synthesized in the system SiO_2–P_2O_5–CaO–Na_2O.[7] The glass composition of 45% SiO_2 – 24.5% Na_2O – 24.5% CaO – 6% P_2O_5 was selected. This composition provides a large amount of

---

[†]Osteoproduction is the process whereby a bioactive surface is colonized by osteogenic stem cells free in the bone defect environment as a result of surgical intervention.
[‡]Osteoconduction is the process of bone migration along a biocompatible surface.

CaO with some P_2O_5 in a Na_2O–SiO_2 matrix, and it was very close to a ternary eutectic and therefore easy to melt. Actually, the synthesis process consisted of melting the precursor mixture and quenching. In the following years, several compositions contained in the phase equilibrium diagram of such system were studied.[157-159]

Silicate glasses can be considered as inorganic polymers, whose monomer units are SiO_4 tetrahedra. These units are linked through the O placed at the tetrahedral apexes (bonding oxygen atoms) and the polymeric network is disrupted when the oxygen atoms are not shared with another SiO_4 tetrahedron (nonbonding oxygen). The presence of cations such as Na^+ and Ca^{2+} in the bioglass composition causes a discontinuity of the glassy network through the disruption of some Si–O–Si bonds. As a consequence, nonbridging oxygens are created. The network modifiers are in this case MO and M_2O-type oxides such as CaO and Na_2O, respectively. The properties of such glasses may be explained on the basis of the crosslink density of the glass network using concepts taken from polymer science that are normally used to predict the behaviour of organic polymers. The network connectivity (NC) or the crosslink density of a glass can be used to predict its surface reactivity and solubility among other physical–chemical properties.[160,161] In general terms, the lower the crosslink density of the glass, the greater the reactivity and solubility.

The crosslink density is defined as the average number of additional crosslinking bonds above 2 for the elements other than oxygen forming the glass network backbone. In bioglasses these elements are silicon, phosphorus, boron and occasionally aluminium. Thus, a glass with a network connectivity of 2, equivalent to a crosslink density of 0, corresponds to a linear polymer chain, while a pure silica glass has a network connectivity of 4. The calculation of the network connectivity is a very easy operation defined by eqn (3.3).

$$NC = 8 - 2R \tag{3.3}$$

R = number of O atoms/number of network former atoms (Si, P, B or Al).

New components were added to the system almost simultaneously in order to act as network formers and/or modifiers and to decrease the synthesis temperature of bioglasses. But the main purpose of their inclusion was to improve their properties focused on clinical applications, *i.e.* to increase their bioreactivity or at least to preserve or increase their bioactivity, while adding new properties to the materials. In this sense, the addition of K_2O, MgO, CaF_2, Al_2O_3, B_2O_3 or Fe_2O_3 was tested.[162] However, all these efforts did not always lead to positive results, since the addition of some of these oxides degraded or totally avoided the bioactive behaviour of bioglass. For instance, 3% Al_2O_3 added to the initial composition of Greenspan and Hench,[163] in order to improve its mechanical properties eliminated its bioactivity, and the addition of Fe_2O_3 to obtain glass-ceramics for hyperthermia treatment of cancer[164] decreased the bioactivity. In 1997, Brink and co-workers[165,166] studied the *in vivo* bone-bonding ability of 26 melt glasses in the system Na_2O–K_2O–CaO–MgO–B_2O_3–P_2O_5–SiO_2, concluding that the compositional limits for bioactivity were 14–30 mol% of alkali oxides (Na_2O + K_2O), 14–30 mol% of alkaline earth oxides (CaO + MgO), and <59 mol% of SiO_2.

3.7.3 Sol–Gel Bioactive Glasses

The sol–gel method is a synthesis strategy that consists of obtaining a sol by hydrolysis and condensation of the precursors, commonly metal alkoxides and inorganic salts, and the subsequent gelation of the sol. The sol–gel method presents some advantages with respect to melting for glass processing. Glasses are obtained with a higher degree of purity and homogeneity. However, the real potential of sol–gel bioactive glasses is based on two aspects:

(1) Sol–gel synthesis offers a potential processing method for molecular and textural tailoring of the biological behaviour of bioactive materials. The inherent features of this method allow bioactive compositions to be obtained in the form of particles, fibres, foams, porous scaffolds, coatings and, of course, monoliths.

(2) The sol–gel method of glass processing provides materials with high mesoporosity and high surface area, enhancing the kinetics of the apatite formation and expanding the composition range for which these materials show bioactive behaviour.[167–170] It must be taken into account that the reactions that trigger the bioactive behaviour take place on the surface. Therefore not only chemical composition but also the textural properties (pore size and shape, pore volume, *etc.*) play a fundamental role in the development of the biomimetic CHA layer.[171–176]

Among the different bioactive sol–gel glasses, $SiO_2 \cdot CaO \cdot P_2O_5$ is the most widely studied system.[177–180] Each component contributes to the structure-reactivity relationship, so providing the different *in vivo* response for each composition. Silica is a network former and constitutes the basic component of the glasses. Higher amounts of SiO_2 result in more stable glasses. CaO is a network modifier, *i.e.* its presence partially avoids the Si–O–Si link formation, resulting in more reactive glasses. As a general trend, the higher the CaO content, the higher the bioactive behaviour of the glass. Finally, the role of P_2O_5 is not clear from the structural point of view. It can be found as tetrahedral units that contribute to the network formation, or as orthophosphates grouped into clusters. The presence of P_2O_5 contributes to the CHA crystallization on the glass surface during the bioactive process, although amounts greater than 12% in weight inhibit the bioactivity.

The sol–gel method enables the expansion of the bioactive compositional range studied in the phase equilibrium diagram of melted glasses, and the glasses so obtained exhibit higher surface area and porosity values, critical factors in their bioactivity.[181,182] This feature allows the simplification of the chemical systems, thus obtaining bioactive compositions in the diagram SiO_2–CaO. This binary system was tested by Kokubo and co-workers,[183,184] producing glasses of the binary system SiO_2–CaO with a SiO_2 content ≤65%, prepared by melting; Vallet-Regí and co-workers also prepared glasses in this system, with SiO_2 contents of up to 90% (50–90% SiO_2), prepared by the sol–gel technique.[61,169,170]

The application of sol–gel chemistry to the synthesis of bioactive glasses opened new perspectives in the chemistry of these compounds. For the same silica content, the rate of CHA formation is higher in sol–gel-derived glasses than in melt-derived ones. The higher bioactivity of the sol–gel glasses is attributed to the high surface area and concentration of silanol groups on the surface of these materials. These features come from the sol–gel processing that allows the production of glasses and ceramics at much lower temperatures compared with conventional methods.

3.7.4   The Bioactive Process in SiO_2-Based Glasses

The cascade of event that leads to the growth of a nanoapatite phase and the subsequent bonding between glass and bone has been described by Hench *et al.*[185] The basis for bone bonding is the reaction of the glass with the surrounding solution. A sequence of interfacial reactions, which begin immediately after the bioactive material is implanted, leads to the formation of a CHA layer and the establishment of an interfacial bonding. Hench summarizes the sequence of interfacial reactions as follows:

(1) Rapid exchange of Na^+ or Ca^{2+} with H^+ or H_3O^+ from solution and formation of silanols (Si–OH) at the glass surface:

$$Si-O-Na^+ + H^+ + OH^- \rightarrow Si-OH + Na^+ + OH^-$$

(2) Loss of soluble silica, in the form of $Si(OH)_4$ resulting from breaking of Si–O–Si bonds and formation of silanols:

$$2(Si-O-Si) + 2(OH^-) \rightarrow Si-OH + OH-Si$$

(3) Condensation of silanols to form a hydrated silica gel layer:

$$2(Si-OH) + 2(OH-Si) \rightarrow -Si-O-Si-O-Si-O-Si-O-$$

(4) Migration of Ca^{2+} and PO_4^{3-} groups to the surface through the silica layer, forming a $CaO-P_2O_5$-rich film on the top of the silica-richer layer.

(5) Crystallization of the ACP layer by incorporation of OH^-, CO_3^{2-} or F^- from solution to form a mixed hydroxyl-carbonate apatite layer (CHA) or hydroxyl-carbonate fluorapatite (HCFA) layer from the solution.

(6) Adsorption of biological moieties in the CHA layer.

(7) Action of macrophages.

(8) Attachment of stem cells.

(9) Differentiation of stem cells.
(10) Generation of collagen matrix.
(11) Crystallization of mineral matrix.

Stages 1–5 occur under *in vitro* conditions and do not require any biological or organic entity. Therefore, these stages constitute the mechanism that rules the synthesis of biomimetic apatites on bioactive glasses.

3.7.5 Biomimetic Nanoapatite Formation on SiO₂-Based Bioactive Glasses

The problem of the mechanism of apatite formation on the surfaces of glasses and glass-ceramics was a controversial topic during the 1990s. The body fluid and artificial SBFs are supersaturated with respect to the apatite under normal conditions. In such an environment, once the apatite nuclei are formed on the surfaces of glasses and glass-ceramics, they can grow spontaneously by consuming the calcium and phosphate ions from the surrounding solution. The problem is therefore reduced to the mechanism of the apatite nucleation on the surfaces of glasses.

SiO_2–CaO and SiO_2–CaO–Na_2O-based glasses, including those with and without P_2O_5, form the apatite layer on their surface *in vivo* as well as *in vitro* by biomimetic processes. In contrast, CaO–P_2O_5-based glasses do not develop such a phase,[186] indicating that the SiO_2 presence is mandatory to set off the bioactive process. Calcium ions dissolve from the glass and increase the degree of the supersaturation of the surrounding body fluid with respect to the apatite, and the hydrated silicate ion formed on their surfaces might provide favourable sites for the apatite nucleation. The importance of the hydrated silicate ion in forming the apatite layer had been also proposed by Hench, as mentioned earlier.[187,188]

SiO_2–CaO glasses containing a small amount of P_2O_5, for example SiO_2 50-CaO 45-P_2O_5 5 (mol%), develop an apatite layer on the surface in SBF faster (~6 hours) than those compositions without P_2O_5 (~3 days). They succeed in developing biomimetic nanoapatites, in contrast to CaO–P_2O_5-based glasses, which do not form them.

Since body fluid is already supersaturated with respect to the apatite under normal conditions, once the apatite nuclei are formed they can grow spontaneously by consuming the calcium and phosphate ions from the surrounding body fluid. In view of these factors, Ohtsuki *et al.*[114] established that the rate of apatite nucleation on glasses in SBF increases in the order:

$$CaO–P_2O_5 << SiO_2–CaO < SiO_2–CaO–P_2O_5$$

The rate, *I*, of nucleation of a crystal on a substrate in a solution at the temperature, *T*, is generally given by eqn (3.4).[189]

$$I = I_0 \exp\left(\frac{-\Delta G^*}{kT}\right) \exp\left(\frac{-\Delta G_m}{kT}\right) \qquad (3.4)$$

where $\Delta G*$ is the free energy for formation of an embryo of critical size, ΔG_m is the activation energy for transport across the nucleus–solution interface. Among them, ΔG_m is independent of the substrate. $\Delta G*$ is given by the eqn (3.5).

$$\Delta G* = \frac{16\sigma^3 f(\theta)}{3\left(\frac{kT}{V_\beta}\ln\left(\frac{IP}{K_0}\right)\right)^2} \tag{3.5}$$

where σ is interface energy between the nucleus and the solution, IP is ionic activity product of the crystal in the solution, K_0 is the value of IP at equilibrium, *i.e.* solubility product of the crystal; $f(\theta)$ is a function of contact angle between the nucleus and the substrate, and V_β is the molecular volume of the crystal phase. Among them, $f(\theta)$ depends upon the substrate, and IP/K_0 a measure of the degree of supersaturation, also depends upon the substrate when the substrate releases some constituent ions of the crystal, while others are independent of the substrate.

Experimental results have demonstrated that SiO_2–CaO-based glasses dissolve significant amounts of calcium ions, whereas CaO–P_2O_5-based glasses dissolve important amounts of phosphate ions. Consequently, the changes of IP in the SBF for both cases are very similar and, therefore, the different biomimetic behaviour cannot be attributed to larger increase in the degree of the supersaturation due to the dissolution of the calcium ion.

The term $f(\theta)$, generally given by eqn (3.6) decreases with decreasing interface energy between the crystal and the substrate:

$$f(\theta) = \frac{(2+\cos\theta)(1-\cos\theta)^2}{4} \tag{3.6}$$

This indicates that the SiO_2–CaO-based glasses provide a specific surface with lower interface energy against the apatite. Bioactive glasses form a silica hydrogel layer prior to the formation of the apatite layer. This layer is responsible for the decrease of $f(\theta)$, decreasing the contact angle and providing specific favourable sites for apatite nucleation.

The works carried out by Li *et al.*[167] on bioactive sol–gel glasses revealed the importance of surface area and porosity in the formation of biomimetic nanoapatites. The apatite growth in SBF was demonstrated for sol–gel glasses composition with nearly 90% of SiO_2. The rate of surface CHA formation for 58S composition (see Table 3.6) was even more rapid than for melt derived 45S5 Bioglass. Table 3.6 shows some of the more-often tested compositions with their corresponding nomenclature. More information about the numerous sol–gel glasses compositions can be found in Vallet-Regí *et al.*[190]

High surface area seems to be very important for SiO_2-based bioactive glasses, both melt-derived and sol–gel glasses. Melt-derived glasses initially exhibit surface area values <1 m^2 g^{-1}. However, they develop to >100 m^2 g^{-1}

Table 3.6 Chemical composition (wt%) for some melt-derived and sol–gel glasses.[a]

	SiO_2	P_2O_5	CaO	Na_2O
45S5 melt(+)	45	6	24.5	24.5
60S melt(−)	60	6	17	17
58S sol–gel(+)	48	9	33	
68S sol–gel(+)	68	9	23	
77S sol–gel(+)	77	9	14	
91S sol–gel(−)	91	9		

[a](+): bioactive glasses; (−): nonbioactive glasses.

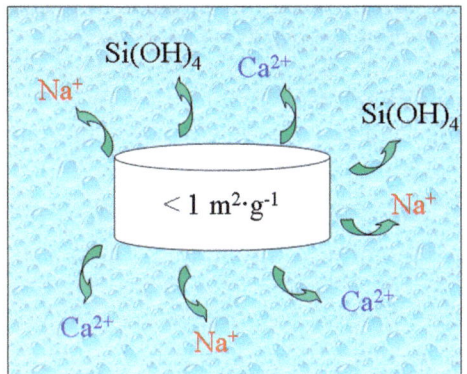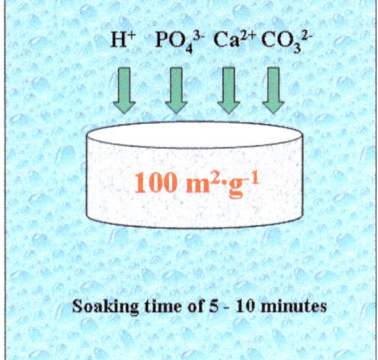

Figure 3.11 Ionic exchange and surface area evolution in bioactive melt-derived glasses after being soaked in simulated body fluid.

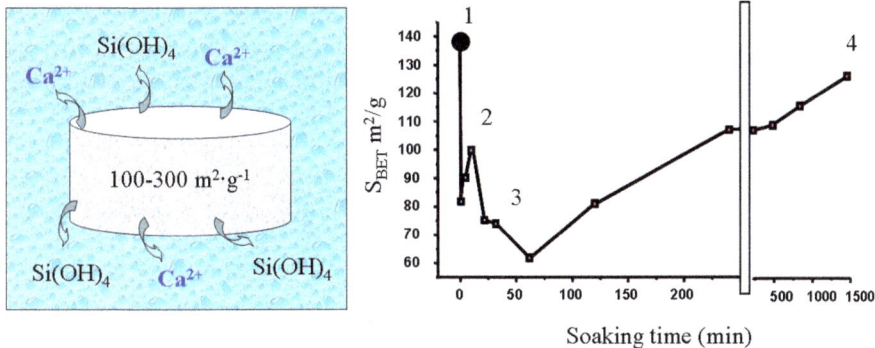

Figure 3.12 Evolution of S_{BET} as a function of soaking time for 58S sol–gel glass.

when they get in contact with fluids at physiological pH, as was demonstrated by Greenspan *et al.*[191] Once this surface area is developed, the melt-derived bioglass is suitable to be coated by biomimetic nanoapatites (Figure 3.11).

In the case of sol–gel glasses, the surface evolution is very different compared with 45S5 melt-derived bioglass, for instance.[192] Figure 3.12 shows the S_{BET} evolution of the glass as a function of the soaking time in SBF. Four stages can be clearly differentiated. During the first minute (stage 1), the glass undergoes a drastic surface decrease from 138 $m^2 g^{-1}$ (original value) to 82 $m^2 g^{-1}$, which means a 40% surface reduction in a very short time. This is a very different behaviour in comparison to melt-derived glasses, which have a very low surface area but develop surfaces of ~100 $m^2 g^{-1}$ after being soaked in physiological simulated solutions. Afterwards, a partial surface recovering occurs between 1 minute and 10 minutes (stage 2), reaching a surface value of 100 $m^2 g^{-1}$. From this point the glass begins to lose surface gradually (stage 3), and after 1 hour it has lost ~55% of the initial surface, showing values of 62 $m^2 g^{-1}$. Finally, stage 4 involves the progressive recovery of the surface area from 1 hour until the end of the experiment, reaching values of 127 $m^2 g^{-1}$ after 24 hours in SBF. These four stages can be explained in terms of the bioactivity theory of glasses:

(1) Loss of surface area due to fast Ca^{2+} release.
(2) Partial surface area restoring due to the Si–OH formation and CO_3^{2-} incorporation.
(3) Second surface area loss as a consequence of the amorphous CaP formation.
(4) Surface area restoring during the CaP crystallization into hydroxycarbonate apatite.

In the case of sol–gel glasses, the values of textural parameters depend on the chemical composition of the glass and the stabilization temperature used.[193,194] Moreover, the changes of surface area and porosity depend on the kinetics of the bioactive process for each glass composition. The textural properties of SiO_2–CaO–P_2O_5-glasses have been studied by varying the SiO_2/CaO ratio.[195] This systematic study confirmed that higher presence of SiO_2 results in higher surface area, whereas higher CaO content provides more mesopore volume and larger pore diameter. The morphology of mesopores is modified as function of SiO_2 (or CaO) content. While the glasses with larger SiO_2 content (80 mol% and 75 mol%) have inkbottle-type pores with narrow necks, glasses with lower SiO_2 content (58 mol%, 60 mol% and 65 mol%) have cylindrical pores open at both ends with occasional necks along the pores. The pore morphology parallels the variations of pore diameter and volume. The transition from narrow-neck inkbottle-type pores to open-ended cylindrical pores apparently takes place when the pore diameter increases. The higher Ca content leads to the increase of the pore size and volume and causes a change of morphology from inkbottle pores to cylindrical ones. Since the higher ionic concentration occurs in the mesopores, the apatite growth (nucleation and crystallization) will depend on this porosity. This model is schematically plotted in Figure 3.13.

Although the influence of the texture of the substrate on the formation of apatite is generally admitted, the detailed nature of the nucleation process of the apatite is still a matter of debate. Almost all authors focus the discussion on apatite nucleation upon the role of the silanol groups existing on the glass surface under the environmental conditions under which the assays are conducted.[196–198] Wang and Chaki[199] show an epitaxial relationship

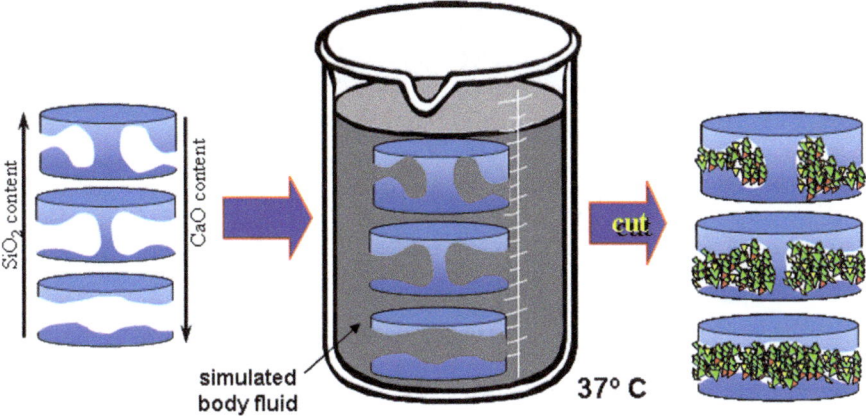

Figure 3.13 Schematic model of the mesopore morphology as a function of SiO_2:CaO ratio. The figure also shows a scheme of apatite formation within the mesopores after soaking in simulated body fluid. For the sake of clarity, the apatite layer grown all over the particle free surface is not shown.

between Si(111) and apatite in [102] orientation. Interestingly, the phosphorus and calcium of the substrate are generally considered to be a mere reservoir that influences the supersaturation of the solution as they are leached from the glass. Nevertheless, phosphorus and calcium as components of bioactive glasses could in fact be potential nucleation centres for apatite crystallization, although the role of P_2O_5 is controversial, as explained in the next section.

3.7.6 Role of P_2O_5 *In Vitro* Bioactivity of Sol–Gel Glasses

From the early 1990s, when the first bioactive sol–gel glasses were prepared in the $CaO–P_2O_5–SiO_2$ system, diverse studies were performed to understand the role of the gel glass constituents in the surface properties and the *in vitro* formation of a CHA phase. That way, the role of SiO_2 and CaO was reported, but the effect of P_2O_5 was not fully understood.

The bioactive behaviour of $CaO–SiO_2$ glasses demonstrates that P_2O_5 is not an essential requirement for bioactivity, even for high SiO_2 content.[170] However, if not essential, P_2O_5 plays an important role on the kinetic formation and final features of the biomimetic apatite growth on glass surfaces. Two series of $CaO–P_2O_5–SiO_2$ glasses were prepared, first with SiO_2 constant (80%),[200] the second with CaO constant (25%) (in mol%).[201] Finally, the nano-structural characterization of glasses by high-resolution TEM[202] allowed the determination of calcium and phosphorus location in the silica network.

Regarding the *in vitro* bioactivity, it was concluded that P_2O_5 retards the initial *in vitro* reactivity of glasses, defined as the time required for the formation of a layer of amorphous calcium phosphate. However, once some nuclei are formed, for contents of P_2O_5 up to 5%, the growth of CHA crystals in the layer is quicker and yields larger crystals. With respect to the textural

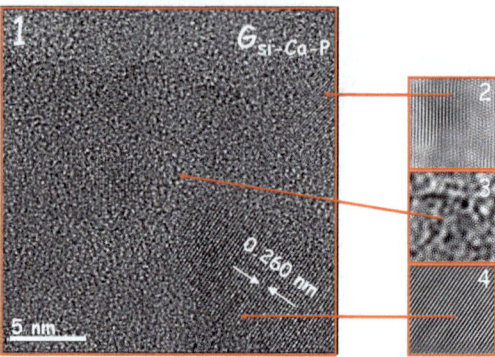

Figure 3.14   High-resolution transmission electron microscopy (TEM) study of a gel glass of composition 17% CaO–3% P_2O_5–80% SiO_2. (1) High-resolution TEM image and (3) filtered high-resolution TEM image of the amorphous matrix. (2) and (4) P-rich crystalline areas oriented along different directions with interplanar spacings close to 0.26 nm.

characterization, it was shown that the surface area increases and the diameter and volume of pores decrease when increasing the P_2O_5 content in glasses with 25% CaO, pointing out that P_2O_5 bonds to CaO, given that increasing the P_2O_5 content produces similar textural effects as decreasing the CaO content.

This assumption has been confirmed by high-resolution TEM, since distances of 0.53 nm were found between the $[SiO_4{}^{4-}]$ tetrahedra in a P-free glass of composition SiO_2 80–CaO 20 mol%, but only 0.36 nm were measured in a P-containing glass (SiO_2 80–CaO 17–P_2O_5), indicating that in the latter the calcium was out of the glass network. In addition, in P-containing glasses small crystalline clusters (<10 nm), identified as silicon-doped calcium phosphate nuclei were detected (Figure 3.14).

In P-free glasses bioactivity is controlled by the rapid exchange of calcium in the glass network by protons in solution forming silanol (Si–OH) groups, which attract calcium and phosphorous in SBF to form an amorphous calcium phosphate.

Afterwards, a relatively long period is required for the *in vitro* crystallization of CHA. However, for P-containing glasses the silanol concentration is lower, retarding the ACP formation, but the presence of the mentioned nanocrystals that could act as nucleation centres increasing the CHA crystallization rate.

3.7.7   Biomimetism Evaluation on Silica-Based Bioactive Glasses

Once a protocol (biomimetic solution, dynamic or static test, *etc.*) has been established, the evolution of the bioglass surfaces can be verified by several techniques. In the same way that CaP-derived bioceramics are studied, FTIR spectroscopy is one of the most widely used methods to evaluate the biomimetic growth on silica-based glasses. Figure 3.15 indicates the formation of

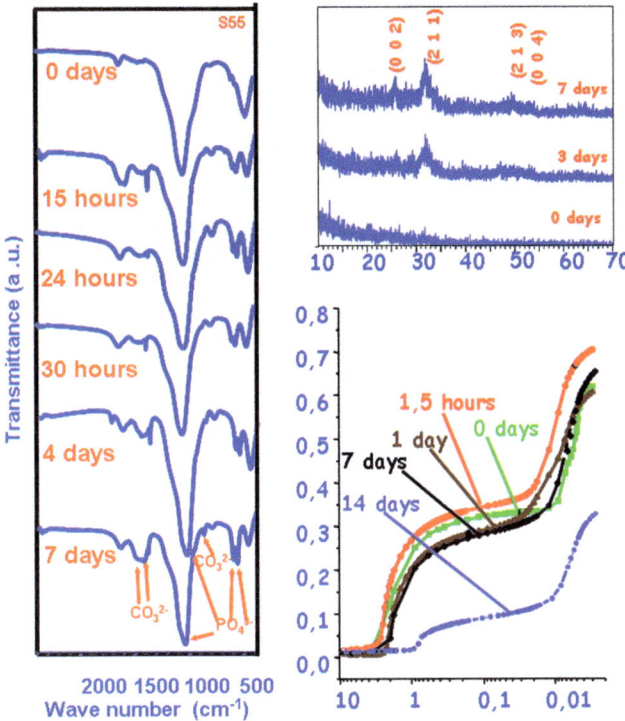

Figure 3.15 Study of the nucleation and growth of an apatite-like layer on the surface of a bioactive sol–gel glass as a function of soaking time in simulated body fluid. Left: Fourier transform infrared spectroscopy; right: X-ray diffraction patterns (top), Hg intrusion porosimetry (bottom).

an apatite-like layer on the glass surface after soaking in SBF. Silicate absorption bands at approximately 1085, 606 and 462 cm^{-1} are observed on the glass spectra before soaking. Phosphate absorption bands at approximately 1043, 963, 603, 566 and 469 cm^{-1} and carbonate absorption bands at approximately 1490, 1423 and 874 cm^{-1} can be observed on the spectra of materials scrapped from the surfaces of soaked glass disks. The increase in the intensity of the carbonate bands is associated with the soaking period in SBF solution. The phosphate and carbonate absorption bands observed on the glass surfaces after soaking are similar to those observed in synthetic carbonate hydroxyapatite.[180] These bands confirm the formation of an apatite-like layer, but also indicate that the apatite-like layer material is a carbonate hydroxyapatite similar to biological apatites, in which a coupled substitution of Na^{+} by Ca^{2+} and CO$_3^{2-}$ by PO$_4^{3-}$ is observed.[203,204]

The changes in the bioglass surface can also be monitored by X-ray diffraction. Given the amorphous nature of the glass, and its evolution towards an apatite of very low crystallinity, at first it does not seem a very adequate technique for such study. However, it is a very useful tool to visualize the transformations on the glass when in contact with SBF, following the evolution

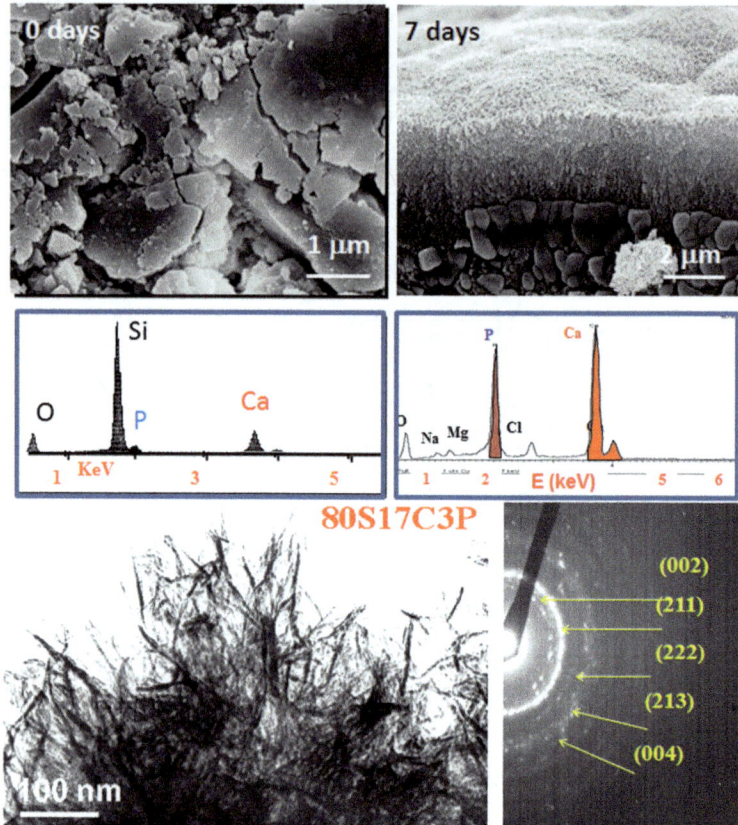

Figure 3.16   Top: scanning electron micrographs of a bioactive glass before and after soaking in simulated body fluid (SBF) for 7 days; middle: energy dispersive X-ray spectrum of the glass surface before and after 7 days soaked in SBF and bottom: high-resolution transmission electron microscope image of the newly formed biomimetic apatite phase and electron diffraction pattern.

with soaking time. It also allows the comparison of the diffraction patterns obtained with those of natural bone; for soaking times ≥7 days, clear similarities can be observed (Figure 3.16). As can be observed, the diffraction patterns of bioactive glasses show two diffuse reflections centred at 2θ values of 26° and 32°, which correspond to the hydroxyapatite (002) and (211) reflections, respectively. Even after 7 days of soaking in SBF, the X-ray diffraction patterns correspond to a material with a very low degree of crystallinity.

The biomimetic growth also induces changes in the textural properties of the substrate. In the case of bioactive pieces these changes can be observed at the macroporous level, since the intergranular spaces are filled insofar as the new apatite phase grows. This evolution can be followed by Hg intrusion porosimetry (Figure 3.15) since the volume of Hg intruded is drastically reduced after 2 weeks in SBF.

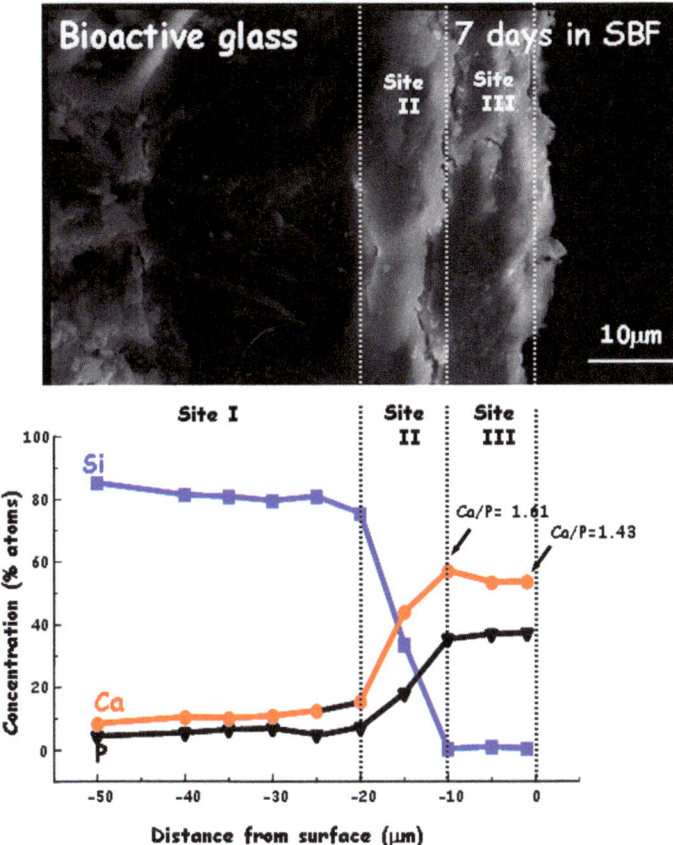

Figure 3.17 Scanning electron micrograph of a cross-section of a bioactive glass after being soaked in simulated body fluid (SBF) for 7 days (top). Element distribution obtained by energy dispersive X-ray spectroscopy (bottom).

The changes on the bioglass surface can be clearly observed using SEM techniques. Thus, Figure 3.16 shows images of the glass surface after soaking in SBF for 1 week. These images confirm the formation of a layer constituted of spherical particles, which coats the whole surface of the initial glass. It can be observed that the particles are formed by small crystalline aggregates. The combination of SEM and energy dispersive X-ray spectroscopy (EDX) techniques yields additional information about the nature of this newly formed layer. In fact, the EDX profiles of the glass surface after 1 week of soaking in SBF reveals the presence of P and Ca only, with a Ca/P ratio of ~1.25. These results support the growth of a layer with similar composition to that of biological apatites.

In turn, the particles observed by SEM can be further studied by TEM and ED, analysing their composition with EDX equipment connected to the TEM microscope. Figure 3.17 shows the high-magnification image, ED pattern and

EDX spectrum of particles at the apatite-like layer grown onto the bioglass surface upon 1 week of soaking in SBF. A small area was selected using the microdiffraction technique. The ED pattern obtained showed the presence of diffuse diffraction rings in which the interplanar spacings agreed with those of an apatite-like structure, indicating that crystalline nuclei were embedded in a glassy matrix. In the correspondent micrograph, the needle-like shape of the aggregated crystals forming the spherical particles may be observed. Taking into account the hydroxyapatite lattice parameters ($a = 9.5$ Å and $c = 6.8$ Å), and its symmetry (hexagonal, S.G. $P6_3/m$), most likely its unit cells will be arranged along the c axis. This would justify a preferred orientation that gives rise to an oriented growth along the c axis and a needle-like morphology, which agrees with the morphology observed by TEM. Conversely, the EDX spectrum obtained using TEM showed that the crystals were composed of Ca, P and O, corresponding to biological apatites.

Another interesting aspect of the study of the apatite-like layer is to ascertain its thickness. The combination of SEM and EDX techniques can be very useful in this question. In Figure 3.17, the cross-section of 55S glass (55: SiO_2 percentage; S: sol–gel) after 15 hours of soaking is shown. The EDX spectra inside the glass and on the layer are also included. As observed, the obtained analysis of the inner region agrees with the nominal glass composition, that is 55% SiO_2–41% CaO–4% P_2O_5 (mol%). However, in the EDX spectrum of the layer, a remarkable increase of Ca and P concentrations, together with a significant decrease of Si, was observed. The decrease of Si with increasing Ca and P concentrations indicates the formation of an apatite-like material. The SEM study of the cross-section of samples after different soaking times allows monitoring the evolution of the layer thickness with the soaking time in SBF. Layer thickness grew from 2 µm after 15 hours of immersion up to 10 µm after 5 days of assay. It is also observed that there is no difference in layer thickness between 5 and 7 days, which suggests that, at least under *in vitro* conditions, the apatite-like layer does not keep growing indefinitely.

3.8   Mesoporous Bioactive Glasses (MBGs)

In 2004, the research group of Professor Vallet-Regí proposed for the first time the possibility of using silica-based mesoporous materials for bone regenerative purposes.[205] Silica-based mesoporous materials are ordered porous structures of SiO_2, characterized by high pore volume, narrow pore size distribution and high surface area. They demonstrated that under specific conditions some SiO_2 structures, such as those depicted in Figure 3.18, could develop biomimetic apatites onto the surface. However, large surface areas and porosities are not enough to achieve satisfactory biomimetic behaviour. For instance, MCM-41 is not bioactive and requires doping to show bioactivity.[206] Other phases such as MCM-48 or SBA-15 must be soaked in SBF for 60 and 30 days, respectively, before developing an apatite-like phase.[207,208] Only SBA-15 obtained as coating shows bioactivity after 1 week, which can be considered reasonable for clinical applications.[209] Although the highly ordered porosity means an added value over conventional bioactive sol–gel glasses,

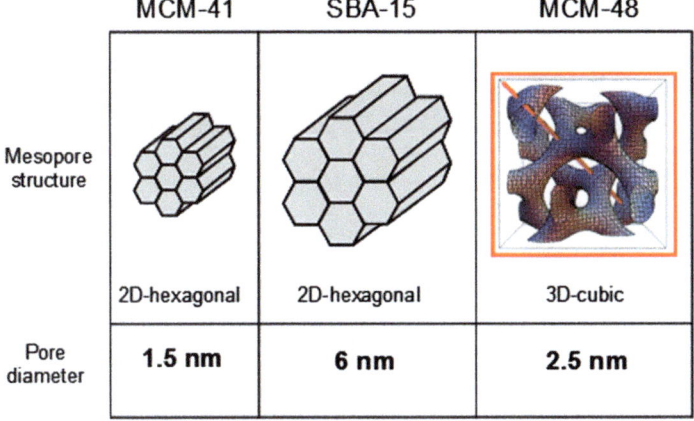

Figure 3.18 Different mesoporous structures.

none of the mesoporous materials described until now improve the bioactive behaviour of the conventional sol–gel glasses.

The real challenge was to obtain bioactive multicomponent sol–gel glasses, with the textural properties of ordered mesoporous silica. In 2004 Yan *et al.* synthesized the first MBGs.[210] This group combined the sol–gel chemistry of the multicomponent system SiO_2–P_2O_5–CaO, with the benefits of adding a structure-directing agent (SDA). They obtained mesoporous materials with chemical compositions analogous to the conventional SiO_2–P_2O_5–CaO sol–gel glasses, but exhibiting porosities and surface areas very similar to the mesoporous materials MCM-41, SBA-15, *etc.*

MBGs cannot be prepared following the conventional hydrothermal synthesis, used for most silica mesoporous materials. MBGs commonly contain CaO and P_2O_5, in addition to SiO_2. The multicomponent nature of MBGs requires a synthesis method that has led to more robust mesoporous structures. Otherwise, the presence of network modifiers such as Ca^{2+} cations interferes in the SDA–silica interaction, leading to very defective structures. For this reason, Yan *et al.* turned to the evaporation-induced self-assembly (EISA) route, developed by Brinker *et al.*[211] This method is illustrated in Figure 3.19 and is based on the preparation of diluted solutions containing the precursors and the SDA in a volatile solvent, for instance ethanol. As the solvent slowly evaporates, the critical micelle concentration is reached, resulting in the self-assembly of the micelles into ordered phases.

The mesoporous structures for the firsts MBGs were limited to 2D-hexagonal *p6m* ordering. These MBGs were prepared by using Pluronic 123 (P123) as structure directing agent. P123 is an amphiphilic triblock polymer with composition $EO_{20}PO_{70}EO_{20}$, where EO is poly(ethylene oxide) and PO is poly(propylene oxide). Tetraethyl orthosilicate (TEOS), triethyl phosphate (TEP) and calcium nitrate tetrahydrate $(Ca(NO_3)_2 \cdot 4H_2O)$ are commonly used as SiO_2, P_2O_5 and CaO precursors, respectively. In a typical synthesis P123 is dissolved in ethanol/acid solution at room temperature. Afterward the

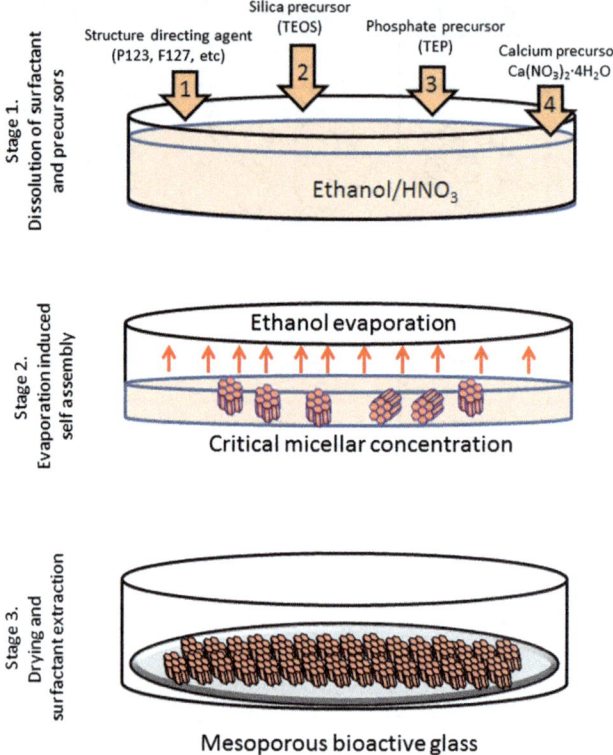

Figure 3.19   Scheme of the synthesis of a mesoporous bioactive glass by the evaporation-induced self-assembly (EISA) process.

appropriate amounts of TEOS, TEP and/or $Ca(NO_3)_2 \cdot 4H_2O$ are added under continuous stirring. The resulting sols are stirred at room temperature for 24 hours and then they must be transferred into dishes with high surface contact with atmosphere. The EISA process takes place at room temperature for about 7 days, until homogeneous and flexible membranes of surfactant–inorganic components composite are obtained. Finally, the membranes are heated in air at 700 °C for 6 hours, thus obtaining ordered mesoporous powders of SiO_2–P_2O_5–CaO composition.

3.8.1   Structures and Compositions

The mesoporous structure and textural properties are strongly dependent upon several factors, such as:

- type of SDA and concentration;
- chemical composition of the MBG; and
- temperature of evaporation during the EISA process, *etc.* (See Table 3.7).

Table 3.7 Structural and textural properties of mesoporous bioactive glasses, as a function of the type of surfactant, chemical composition and evaporation temperature.

Composition (%mol)	Surfactant	Temp. (°C)	Symmetry	S_{BET} (m² g⁻¹)	V_p (cm³ g⁻¹)	D_P (nm)	Reference
$58SiO_2$–$37CaO$–$5P_2O_5$	P123	40	2D-hexagonal (*p6mm*)/wormlike	195	0.46	9.45	212
$75SiO_2$–$20CaO$–$5P_2O_5$	P123	40	2D-hexagonal (*p6mm*)/orthorombic (*p2mm*)	393	0.59	6.0	217
$85SiO_2$–$10CaO$–$5P_2O_5$	P123	40	3D-cubic (*Ia3d*)	427	0.61	5.73	217
$85SiO_2$–$10CaO$–$5P_2O_5$	P123	20	2D-hexagonal (*p6mm*)	473	0.63	5.37	213
$90SiO_2$–$10CaO$	P123	40	2D-hexagonal (*p6mm*)	468	0.63	5.37	218
$75SiO_2$–$21CaO$–$4P_2O_5$	F127	40	Cubic (*Im3̄m*)	400–506	0.25–0.53	3.2–5.4	214
$90SiO_2$–$5CaO$–$5P_2O_5$	P123	r.t	2D-hexagonal (*p6mm*)	338	0.46	5.5	215
$80SiO_2$–$15CaO$–$5P_2O_5$	P123	r.t	2D-hexagonal (*p6mm*)	229	0.31	5.2	220
$80SiO_2$–$15CaO$–$5P_2O_5$	P123	r.t	2D-hexagonal (*p6mm*)	351	0.49	4.6	215
$70SiO_2$–$25CaO$–$5P_2O_5$	P123	r.t	2D-hexagonal (*p6mm*)	319	0.49	4.6	215
$60SiO_2$–$35CaO$–$5P_2O_5$	P123	r.t	2D-hexagonal (*p6mm*)	310	0.43	4.3	215
$70SiO_2$–$25CaO$–$5P_2O_5$	F127	r.t	Wormlike	300	0.36	5.0	215
$60SiO_2$–$35CaO$–$5P_2O_5$	B50-6600	r.t	Wormlike	228	0.42	7.1	215
$80SiO_2$–$15CaO$–$5P_2O_5$	PN-430 + Brij 70	r.t	2D-hexagonal (*p6mm*)	485	0.27	<2.0	216
$80SiO_2$–$15CaO$–$5P_2O_5$	P85	r.t	2D-hexagonal (*p6mm*)	328	0.36	3.4	221
$80SiO_2$–$15CaO$–$5P_2O_5$	P123	r.t	2D-hexagonal (*p6mm*)	325	0.40	5.0	221
$80SiO_2$–$15CaO$–$5P_2O_5$	B50-6600	r.t	2D-hexagonal (*p6mm*)	301	0.41	6.4	221

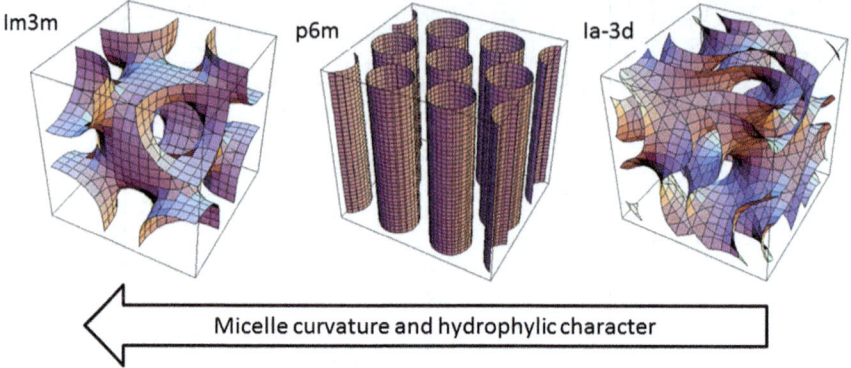

Figure 3.20 Reconstruction of *Im3m*, *p6m* and *Ia3d* structures and their relative hydrophilic characters.

In 2006, the research team of Professor Vallet-Regí prepared SiO_2–P_2O_5–CaO-based MBGs with different mesoporous structures, using the same SDA (P123) in all cases.[212] Keeping constant the molar ratio between the network formers (SiO_2 and P_2O_5) and the SDA (P123), the dependence of the mesoporous structure on the CaO content was demonstrated. For high CaO contents (~36 mol%), the already-known 2D hexagonal *p6m* phase was obtained. However, for lower CaO contents, the formation of a cubic 3D bicontinuous *Ia3d* phase takes place. This cubic phase is related to more hydrophobic systems, as it is formed from micelles with less curvature. In these cases, the evaporation temperature during the EISA process is extremely important. For instance, the material MBG-85 with composition 85 SiO_2–5 P_2O_5–10 CaO (mol%) exhibits *p6m* or *Ia3d* structures when the EISA process is carried out at 20 °C or 40 °C, respectively.[213] These observations indicate that high CaO contents and low temperatures favour the formation of hydrophilic mesoporous systems, derived from micelles with high curvature. The micelles exhibit cylindrical rod shapes resulting in 2D hexagonal structures (*p6m*) after the self-assembly stage. In contrast, low CaO contents and higher evaporation temperatures favour the formation of less hydrophilic systems, derived from micelles with lower curvature. In this case, the micelles exhibit 3D bicontinuous structures resulting in 3D cubic *Ia3d* after the self-assembly process.

New bioactive mesoporous structures were prepared by adding a surfactant with higher hydrophilic character.[214-216] By using F127, which contains longer chains of ethylene oxide (hydrophilic head) compared with P123, MBGs with cubic *Im3m* cage-type structures were prepared. This structure derives from spherical micelles characteristic of highly hydrophilic systems; that is, when the volume of the hydrophilic part (inorganic components and polar chains of surfactant) clearly predominates over the hydrophobic part (in this case polypropylene oxide chain of F127). In this way, this group prepared for the first-time MBGs with spherical pores instead of channel-like structures prepared so far. Figure 3.20 shows the reconstruction of *Im3m*, *p6m* and *Ia3¯d* structures exhibited for the different MBGs prepared so far.

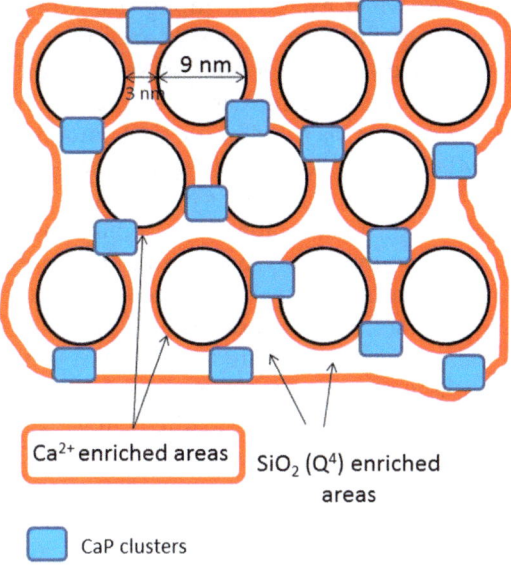

Figure 3.21 Scheme of the local environment in mesoporous bioactive glass (MBG)-58 with calcium phosphate (CaP) clusters and Ca^{2+} cations close to the MBG surface.

The structure of MBGs composed of glassy SiO_2–P_2O_5–CaO can be described as an ordered arrangement of mesopores with sizes between 5 and 9 nm. The wall thickness has been calculated to be around 3 nm, which means very thin walls material. Such a small volume of the walls initially led to think that the three components would be homogeneously distributed that is, without CaP clustering as was observed in conventional sol–gel glasses of identical compositions. However, the local environment of MBGs is also very heterogeneous even in such narrow walls.

This is clearly indicative of the presence of CaP clusters located at the wall surface, which are consequently very accessible to the surrounding fluids when implanted. A similar scenario has been observed for MBGs with high CaO contents, for instance MBG-58S with 37 mol% of CaO. CaP clusters are also formed in these compositions, but the remaining of Ca^{2+} is not entrapped by phosphates and remains within the silica network acting as a network modifier. Figure 3.21 represents a model of the local environment in this MBG.

3.8.2 The Bioactive Behaviour of MBGs

We have seen in previous sections that the bioactive process comprises a set of reactions with the living tissues, which start at the interface between the implant surface and the bone. The amount of material exchanged between the MBGs and the bone determines the kinetic and the type of reaction that will take place between both surfaces. The ionic exchange is higher in those structures with

3D interconnected pores, which extend all through the MBG volume, compared with bidimensional hexagonal structures with nonconnected channels.[217]

Conventional nonmesoporous sol–gel glasses exhibit better bioactive behaviors insofar as the CaO content increases. In this sense, the bioactivity of conventional sol–gel glasses is ruled by the CaO amount. However, the bioactive behavior of MBGs seems to be ruled by the mesoporous ordering and the textural properties. Low CaO contents (~10 mol%) lead to bicontinuous 3D cubic structures with the fastest bioactive response observed so far. This is explained by the higher connectivity, surface area and pore volume exhibited by the structure in comparison to that presented by MBGs with greater CaO content (~36 mol%). These compositions result in *p6m* bidimensional hexagonal structures. In any case, a minimum CaO amount is required to observe acceptable bioactive behaviours independently of the mesoporous structure.

In vitro apatite-forming capability tests in SBF demonstrate that MBG-85 (85 SiO_2–10 CaO–5 P_2O_5, mol%) material develops an amorphous calcium phosphate after a few minutes of soaking, which evolves to HA after 30 minutes. Although this composition comprises a low CaO amount (10 mol%), the outstanding textural properties of the mesoporous glass greatly accelerates the surface reactions, such as ionic exchange and CaP nucleation and growth. Besides, MBGs with high calcium content, such as MBG-58 (58 SiO_2–37 CaO–5 P_2O_5, mol%) together with the good textural properties lead to a new mechanism of bioactivity that mimics the natural mineral maturation of bone in mammals. Figure 3.22 shows the TEM images obtained from the MBG-58 surface after soaking in SBF at different times. After 1 hour, the surface is coated by a thick layer of ACP. This ACP does not maturate directly to HA, but evolves, forming OCP that transforms into HA a few hours later. The ACP–OCP–HA mineral maturation pathway is equivalent to that proposed *in vivo* in mammals. No other bioactive bioceramic is able to reproduce this mineral maturation.

The unique characteristics of MBGs with high CaO content explain this fact. MBG-58 has a high amount of Ca^{2+} cations very close to the material surface and accessible to the surrounding fluids. When MBG-58 makes contact with SBF, an ionic Ca^{2+}–H^+ exchange occurs, increasing the Ca^{2+} concentration in SBF while incorporating H^+ into the MBG surface. This exchange is so intense that the MBG surface is transiently acidified, thus falling into the pH range of stability for OCP nucleation. Thereafter, the presence of preformed CaP clusters facilitates the fast nucleation of nanocrystalline HA.

3.9 Biomimetism in Organic–Inorganic Hybrid Materials

3.9.1 An Introduction to Organic–Inorganic Hybrid Materials

Organic–inorganic hybrid materials have the unique feature of combining the properties of traditional materials, such as ceramics and organic polymers, on the nanoscopic scale.[218–226] Nowadays, these materials represent the

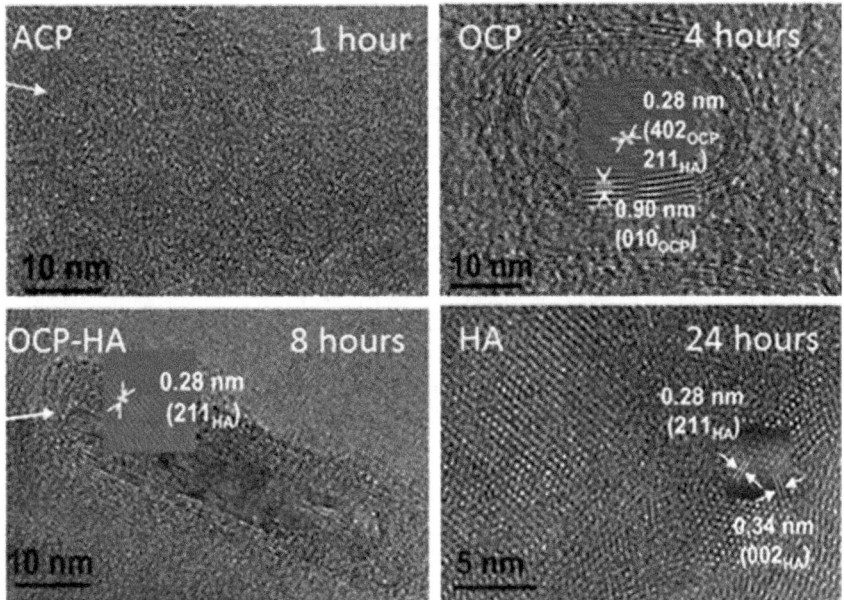

Figure 3.22 Transmission electron microscope images of mesoporous bioactive glass (MBG)-58 after being soaked in simulated body fluid at different times. 1 hour: a large amount of amorphous calcium phosphate (ACP) is observed next to MBG grain; 4 hours: nanometrical octacalcium phosphate (OCP) oval nuclei are observed within the ACP matrix; 8 hours: needle-shaped apatite crystallizes within the ACP matrix; 24 hours: evidence of the microstructural evolution of OCP nuclei to apatite crystallites.

most direct approach towards the development of grafts that mimic the composition and structure of natural bone.

The synthesis methodology is closely related to the development of sol–gel science.[227,228] The general behaviour of these organic–inorganic nanocomposites is dependent on the nature and relative content of the constitutive inorganic and organic components, although other parameters, such as synthesis conditions also determine the properties of the final materials. The final product must be an intimate "mixture" where at least one of the domains (inorganic or organic) has a dimension ranging from a few angstroms to a few tens of nanometers. In this section, we review the behaviour of these implants able to mimic some of the functional properties of bone, especially that concerning the production of nanoapatites in contact with physiological fluids.

The main goal when synthesizing a silicate-containing hybrid material for any application, including biomedical ones, is to take advantage of both domains to improve the final properties of the material. In previous sections, we could see how the silica-based bioactive glasses are able to promote the

Table 3.8 Properties provided by the organic and inorganic domains, expected to be combined in hybrid materials.

Inorganic	Organic
Hardness, brittleness	Elasticity, plasticity
Strength	Low density
Thermal stability	Gas permeability
High density	Hydrophobicity
High refractive index	Selective complexation
Mixed valence state (red-ox)	Chemical reactivity
Bioactivity	...

formation of nanoapatites in contact with physiological fluids. The high bioactivity of silicate-based glasses suggests that the incorporation of silicate as inorganic component would supply bioactivity to the organic component through the synthesis of the hybrid material.

The final properties are not only the sum of the properties of the individual components, but synergetic effects can be expected according to the high interfacial area. Table 3.8 summarizes some of the features that each domain can supply to the hybrid.

Based on the nature of the interactions exchanged by both components, organic–inorganic hybrid materials can be classified as class I and II.[229] Class I hybrid materials show weak interactions between both domains, such as van der Waals, hydrogen bonds and electrostatic interactions. No chemical links (covalent or iono-covalent) are present between the components. In these cases, silica is considered as inorganic nanofiller incorporated to the organic component. In contrast, class II organic–inorganic hybrid materials show chemical links between the components and consequently strong interactions are produced. In this last case, the silicates are considered to be organically modified and are usually referred as *ormosils*.

3.9.2 Synthesis of Biomimetic Nanoapatites on Class I Hybrid Materials

The possibility to design class I hybrid materials associating biopolymers with mineral phases relies on the understanding and control of their mutual interaction. An interesting approach is the synthesis of organic–inorganic hybrids based on bioactive gel glasses (BG) and a biocompatible hydrophilic organic polymer such as poly(vinyl alcohol) (PVAL). The synthesis of BG–PVAL-based hybrid materials is aimed to obtain a new family of compounds, which exhibits the bioactive behaviour of sol–gel glasses together with the mechanical properties and biodegradability of PVAL. The bioactive glass component can belong to the SiO_2–CaO–P_2O_5 or SiO_2–CaO systems. The presence of this kind of component not only ensures the integration of the implant, but also stimulates the new bone formation due to the action of their degradation products (soluble silica, Ca^{2+} cations, *etc.*) on the gene expression of bone growth factors.

This system can be synthesised as monoliths, being potentially applicable for the treatment of medium and large bone defects. When the biodegradability and bioactivity of these hybrids were studied after being soaked in SBF, it could be observed that the addition of PVAL helped the synthesis of crack-free monoliths able to develop an apatite-like phase.[230,231] In contrast, higher amounts of P_2O_5 made the hybrids synthesis difficult and decreased their *in vitro* bioactivity, although it also contributes to the material degradability. Thus, hybrids with very high amounts of both PVAL and P_2O_5 showed such a fast degradation that apatite formation was impeded.

3.9.3 Synthesis of Biomimetic Nanoapatites on Class II Hybrid Materials

The strategy to synthesize class II hybrid materials consists of making intentionally strong bonds (covalent or iono-covalent) between the organic and inorganic components. Organically modified metal alkoxides are hybrid molecular precursors that can be used for this purpose,[232] but the chemistry of hybrid organic–inorganic networks is mainly developed around silicon-containing materials. Currently, the most common way to introduce an organic group into an inorganic silica network is to use organo-alkoxysilane molecular precursors or oligomers of general formula $R'_nSi(OR)_{4-n}$ or $(OR)_{4-n}Si-R''-Si(OR)_{4-n}$ with $n = 1, 2, 3$. The sol–gel synthesis of siloxane-based hybrid organic–inorganic implants usually involves di- or trifunctional organosilanes co-condensed with metal alkoxides, mainly $Si(OR)_4$ and $Ti(OR)_4$. Finally, the incorporation of Ca salts is a common strategy to provide bioactivity at the systems.

3.9.3.1 PMMA–Silica Ormosils

Polymethyl methacrylate (PMMA)–silica hybrid composites have been prepared for dental restorative and bone replacement applications.[233,234] This hybrid material exhibits growth of a low-crystalline CHA layer on the surface when soaked in SBF, demonstrating the bioactive behaviour of this hybrid. Biocompatibility tests have been performed with this kind of material.[235] Mouse calvarial osteoblast cell cultures showed better biological response when seeded on PMMA–SiO_2 hybrid materials than on PMMA in terms of cell attachment, proliferation and differentiation. The enhanced biocompatibility of the PMMA–SiO_2 hybrid was explained by two possible interrelated mechanisms: (1) the capability of inducing a CaP layer formation on the surface of the PMMA–SiO_2 in cell culture media and (2) the capability to release silica (as silicic acid), which induces osteoblast early mineralization.

3.9.3.2 PEG–SiO₂ Ormosils

Poly(oxyethylene)–SiO_2 ormosils have been prepared as an approach to the preparation of biologically active polymer–apatite composites. For this purpose, Yamamoto *et al.*[236] obtained class II hybrids from triethoxysilyl-terminated poly(oxyethylene) (PEG) and TEOS by using the *in situ* sol–gel process. After being subjected to the biomimetic process for forming the bone-like apatite layer, it was found that a dense apatite layer could be prepared on the hybrid materials, indicating that the formed silanol groups provide effective sites for CHA nucleation and growth.

3.9.3.3   PDMS–CaO–SiO₂–TiO₂ Ormosils

One of the more thoroughly studied organic–inorganic hybrid systems for bone and dental repair is that including poly(dimethylsiloxane) (PDMS) as a precursor, together with titanium or silicon alkoxides such as tetraethylorthotitanate (TEOT) or TEOS, respectively. These hybrid materials show properties comparable to those of organic rubbers.[237,238]

Chen and co-workers[239,240] have worked extensively on the PDMS-modified CaO–SiO₂–TiO₂ system, obtaining dense and homogeneous monoliths composed of a silica and titania network incorporated with PDMS and the calcium ion ionically bonded to the network. The hybrids show relatively large amounts of calcium in their surfaces and an apatite like phase is developed within 12–24 hours in SBF. Together with this fairly high apatite-forming ability, some compositions of PDMS–CaO–SiO₂–TiO₂ ormosils exhibit high extensibilities and Young's modulus almost equal to that of the human cancellous bone, although all these features also depend on synthesis parameters, such as the thermal treatment.

3.9.3.4   PDMS–CaO–SiO₂ Ormosils

PDMS–CaO–SiO₂ ormosils combine in a single material the excellent bioactivity of the inorganic component, CaO–SiO₂, and the rubber-like mechanical properties induced by the organic constituent, PDMS. As expected, the bioactive behaviour is strongly dependent on the CaO content.[241] The apatite-forming ability of the hybrids appears when the calcium content in the CaO/SiO_2 molar ratio falls into the range of 0–0.1. The hybrids with a CaO/SiO_2 molar ratio between 0.1 and 0.2 formed apatite on their surfaces in SBF within 12 hours. These ormosils also showed mechanical properties analogous to those of human cancellous bone.

High-resolution TEM of this hybrid material (Figure 3.23) shows the characteristic contrast distribution observed for amorphous materials, suggesting similar structural features to those of conventional glasses. EDX microanalysis results show the incorporation of Ca atoms randomly distributed into the SiO_2 cluster network. The nanostructural analysis revealed distances of 0.53 nm between the $[SiO_4^{4-}]$ units. Besides, nonbioactive $CaO–SiO_2$–PDMS materials were also synthesized. For this synthesis, the same amounts of reactants and catalyst as for the bioactive one were used, but in this case twice the amount of H_2O was used. The corresponding Fourier-filtered high-resolution TEM image showed an average distance of 0.39 nm between $[SiO_4^{4-}]$ units. This distance is clearly lower than 0.53 nm

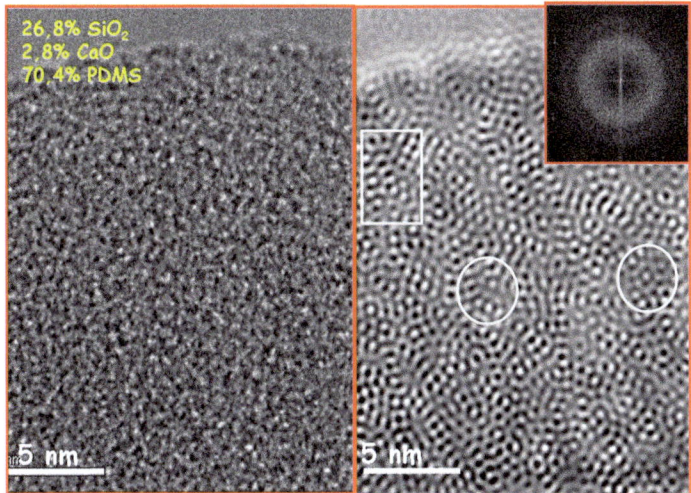

Figure 3.23   Electron microscopy study of a poly(dimethylsiloxane) (PDMS)–SiO$_2$–CaO ormosil. Left: original high-resolution transmission electron microscope (TEM) image of the amorphous matrix; right: filtered high-resolution TEM image; inset: Fourier transform pattern. Distances up to 0.53 nm for (SiO$_4$)$^{4-}$ can be observed in the filtered image, indicating the Ca^{2+} presence between tetrahedra.

measured for the bioactive hybrid, suggesting that Ca is not incorporated in the nonbioactive material. Since both hybrids exhibit different kinetics of bioactive response, this behaviour can be explained in terms of both nanostructure and chemical composition.

3.9.4   Bioactive Star-Gels

In 1995 DuPont developed the *star gels* materials.[242–244] Star gels are a type of organic–inorganic hybrids that present a singular structure of an organic core surrounded by flexible arms, which are terminated in alcoxysilane groups (Figure 3.24). At the macroscopic level, star gels exhibit behaviour between conventional glasses and highly crosslinked rubbers in terms of mechanical properties. Currently, star gels are still one of the most interesting subjects in the field of hybrid materials due to their mechanical properties.[245]

Very recently, the synthesis of *bioactive star gels*, *i.e.* star gels capable of integrating with bone tissue, has been developed. Like many other class II hybrid materials, bioactive star gels are obtained by hydrolysis and condensation of alkoxysilanes containing precursors. In fact, star gels are formulated as single-component molecular precursors with flexibility built in at the molecular level. The starting materials comprise an organic core with multiple flexible arms that terminate in network-forming trialkoxysilane

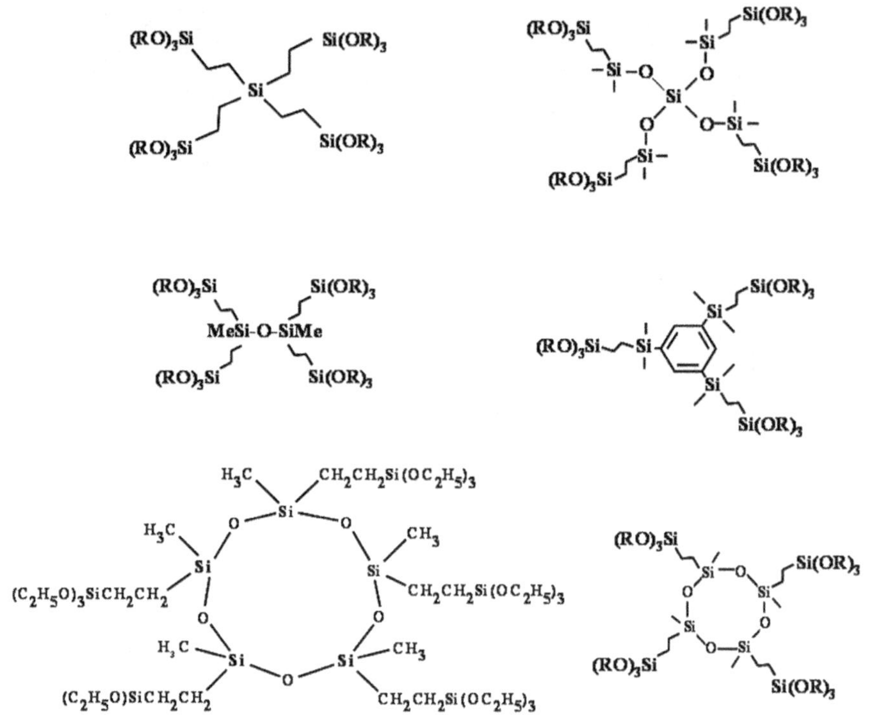

Figure 3.24   Star gel precursors.

groups. The core can be a single silicon atom, linear disiloxane segment, or ring system as seen in Figure 3.24.

The development of bioactive star gels is still in process. Only the precursors A and B in Figure 3.24 have been used so far, for the design of bioactive implants.[246] The basis of the bioactivity of star gels consists of incorporating Ca^{2+} cations into the inorganic component of the hybrid structure, thus exhibiting similar properties to conventional SiO_2–CaO sol–gel glasses, but having the flexibility supplied by the organic chains.

Not all the Ca^{2+}-containing star gels are bioactive. The relative amount of network formers (alkoxysilanes) and network modifiers (Ca^{2+} cations) determine the bioactive behaviour of star gels. More specifically, the Si/Ca ratio provides a good approximation to predict whether a star gel will be bioactive or not. All those compositions with Si/Ca ratios >9 are not bioactive, due to the high stability of these star gels at physiological pH. The chemical composition and structure of the precursors must be known, since the number of Si atoms per unit formula must be determined. All the Si atoms must be taken into account, and not only those with hydrolyzable groups as –Si–O–R. Figure 3.25 is an example of the surface evolution for a star gel obtained from a precursor with a Si/Ca ratio of 5. This figure shows the scanning electron

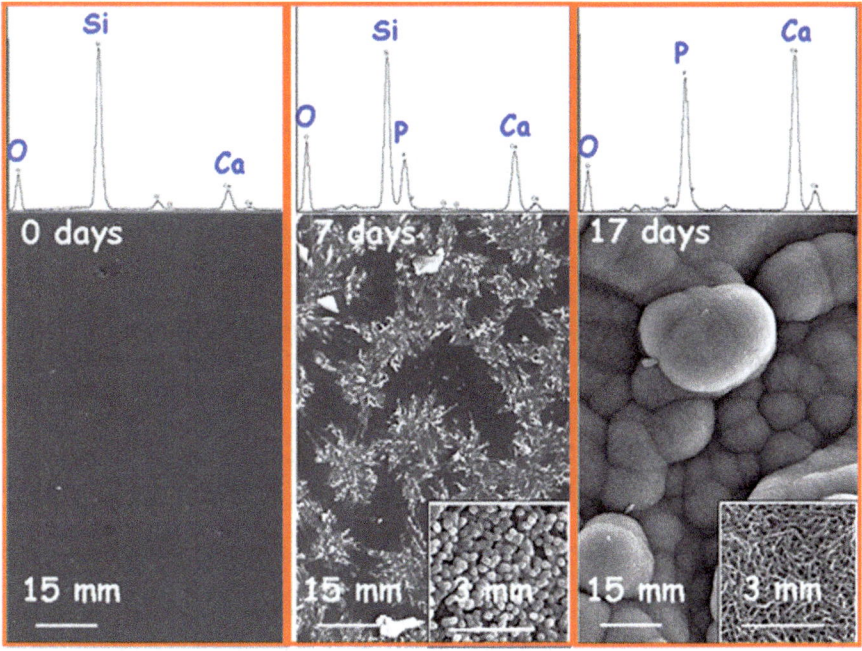

Figure 3.25 Scanning electron micrographs and energy dispersive X-ray spectra of a Ca-containing hybrid material before and after being soaked in simulated body fluid.

micrographs for this hybrid material before and after soaking in SBF for 7 and 17 days. Before soaking, the micrograph shows a smooth surface characteristic of a nonporous and homogeneous gel. EDX spectroscopy confirms the presence of Si and Ca as the only components of the inorganic phase. After 7 days in SBF, a new phase partially covers the star gel surface. This phase is formed by rounded submicron particles composed of Ca and P, as EDX spectroscopy indicates. After 17 days in SBF, the monolith surface is fully covered by a layer constituted of spherical particles, which are formed by numerous needle-shaped crystallites (characteristic of the apatite phase growth over bioactive materials surface). At this point, the EDX spectrum indicates that the surface is fully covered by a CaP with a Ca/P ratio of 1.6, *i.e.* that corresponding to a calcium-deficient apatite.

Bioactive star gels can be excellent candidates for bone tissue regeneration since they fulfil the following features: (1) easily obtained as monoliths of different shapes in order to fit to any kind of medium or large bone defect; (2) structurally homogeneous to predict their biological and mechanical response when implanted; (3) able to develop an apatite-like phase in contact with physiological fluids, *i.e.* must be bioactive; and (4) mechanical properties significantly better than those exhibited by conventional bioactive glasses.

References

1. R. L. Reis, *Curr. Opin. Solid State Mater. Sci.*, 2003, **7**, 263.
2. M. Vallet-Regí and D. Arcos, Nanostructured Hybrid Materials for Bone Implants Fabrication, in *Bioinorganic Hybrid Nanomaterials*, ed. E. Ruiz-Hitzky, K. Ariga and Y. M. Lvov, Wiley-VCH Verlag GmbH & Co. KGaA, Weinheim, 2007.
3. S. V. Dorozhkin and M. Epple, *Angew. Chem., Int. Ed.*, 2002, **41**, 3130.
4. T. Kokubo, H. Kushitani, S. Sakka, T. Kitsugi and T. Yamamuro, *J. Biomed. Mater. Res.*, 1990, **24**, 721.
5. W. E. Brown, N. Eidelman and B. Tomazic, *Adv. Dent. Res.*, 1987, **1**, 306.
6. W. E. Brown and M. U. Nylen, *J. Dental Res.*, 1964, **43**, 751.
7. L. L. Hench, R. J. Splinter, T. K. Greenly and W. C. Allen, *J. Biomed. Mater. Res.*, 1971, **2**, 117.
8. T. Kokubo, K. Hata, T. Nakamura and T. Yamamuro, in *Bioceramics*, ed. W. Bonfield, G. W. Hastings and K. E. Tunner, Butterworth-Heinemann, UK, 1991, vol. 4, p. 113.
9. M. Tanahashi, T. Kokubo and T. Matsuda, *J. Biomed. Mater. Res.*, 1996, **31**, 243.
10. H. Ohgushi and A. I. Caplan, *J. Biomed. Mater. Res.*, 1999, **48**, 913.
11. S. Leeuwenburgh, P. Layrolle, F. Barrere, J. de Bruijn, J. Schoonman, C. A. van Blitterswijk and K. de Groot, *J. Biomed. Mater. Res.*, 2001, **56**, 208.
12. L. L. Hench, *J. Am. Ceram. Soc.*, 1991, **74**, 1487.
13. S. Fujibayashi, M. Neo, J. M. Kim, T. Kokubo and T. Nakamura, *Biomaterials*, 2003, **24**, 1349.
14. T. Kokubo, H. Kushitani, C. Ohtsuki, S. Sakka and T. Yamamuro, *J. Mater. Sci.: Mater. Med.*, 1992, **3**, 79.
15. C. Du, P. Klasens, R. E. Haan, J. Bezemer, F. Z. Cui, K. de Groot and P. Layrolle, *J. Biomed. Mater. Res.*, 2002, **59**, 535.
16. A. M. Radder, H. Leenders and C. A. van Blitterswijk, *J. Biomed. Mater. Res.*, 1994, **28**, 141.
17. A. M. Radder, J. E. Davies, J. Leeners and C. A. van Blitterswijk, *J. Biomed. Mater. Res.*, 1994, **28**, 269.
18. M. Okumura, C. A. Blitterswijk, H. K. Koerten, D. Bakker, K. De Groot and H. Ohgushi, in *Advances in biomaterials*, ed. P. L. Doherty, *et al.*, Elsevier, Amsterdam, 1992, vol. 10, pp. 343–347.
19. C. A. van Blitterswijk, J. van den Brink, H. Leenders and D. Bakker, *Cells Mater.*, 1993, **5**, 55.
20. A. M. Radder and C. A. van Blitterswijk, *J. Mater. Sci.: Mater. Med.*, 1994, **5**, 320.
21. A. M. Radder, J. E. Davies, R. N. S. Sodhi, S. A. T. van der Meer, J. G. C. Wolke and C. A. van Blitterswijk, *Cells Mater.*, 1995, **5**, 320.
22. G. J. Meijer, A. van Dooren, M. L. Gaillard, R. Dalmeijer, C. De Putter and C. A. van Blitterswijk, *Int. J. Oral Maxillofac. Surg.*, 1996, **25**, 210.
23. C. Du, G. J. Meijer, C. Van de Valk, R. E. Haan, J. M. Bezemer, S. C. Hesseling, F. Z. Cui, K. De Groot and P. Layrolle, *Biomaterials*, 2002, **23**, 4649.

24. K. de Groot, R. G. T. Geesink, C. P. A. T. Klein and P. Serekian, *J. Biomed. Mater. Res.*, 1987, **21**, 1375.
25. W. L. Jaffe and D. F. Scott, *J. Bone Jt. Surg.*, 1996, **78A**, 1918.
26. P. Li, *J. Biomed. Mater. Res.*, 2003, **66A**, 79.
27. Y. F. Chou, I. Wulur, J. C. Y. Duna and B. J. Wu, in *Handbook of nanostructured biomaterials and their applications in nanobiotechnology*, ed. H. S. Nalwa, American Scientific Publishers, Stevenson Ranch, 2005, vol. 2, pp. 197–222.
28. T. Kokubo, H. M. Kim and M. Kawashita, *Biomaterials*, 2003, **245**, 2161.
29. H. M. Kim, *Curr. Opin. Solid State Mater. Sci.*, 2003, 7, 289.
30. S. V. Dorozhkin, *J. Mater. Sci.*, 2007, **42**, 1061.
31. S. Radin and P. Ducheyne, *J. Biomed. Mater. Res.*, 1996, **30**, 273.
32. M. S. A. Johnsson, E. Paschalis and G. H. Nancollas, in *Bone-biomaterial interface*, ed. J. E. Davies, University of Toronto Press, Toronto, 1991, pp. 62–75.
33. R. I. Martin and P. W. Brown, *Mater. Med.*, 1994, **5**, 96.
34. T. Kokubo, H. Kushitani, S. Sakka, T. Kitsugi, S. Kotani, K. Oura and T. Yamamuro, Apatite formation on bioactive ceramics in body environment, in *Bioceramics*, ed. H. Oonishi, H. Aoki and K. Sawai, Ishiyaku Euro America, Inc., Tokyo, 1989, vol. 1, pp. 157–162.
35. H. M. Kim, F. Miyaji, T. Kokubo and T. Nakamura, *J. Ceram. Soc. Jpn.*, 1997, **105**, 111.
36. T. Miyazaki, H. M. Kim, F. Miyaji, T. Kokubo, H. Kato and T. Nakamura, *J. Biomed. Mater. Res.*, 2000, **50**, 35.
37. Y. Abe, T. Kokubo and T. Yamamuro, *J. Mater. Sci.: Mater. Med.*, 1990, **1**, 233.
38. M. Tanahashi, T. Yao, T. Kokubo, M. Minoda, T. Miyamoto, T. Nakamura and T. Yamamuro, *J. Am. Ceram. Soc.*, 1994, **77**, 2805.
39. A. Oyane, M. Minoda, T. Miyamoto, K. Nakanishi, M. Kawashita, T. Kokubo and T. Nakamura, Apatite formation on ethylene–vinyl alcohol copolymer modified with silane coupling agent and calcium silicate, in *Bioceramics*, ed. S. Giannini and A. Moroni, Trans Tech Publications, Zurich, 2000, vol. 13, pp. 713–716.
40. H. M. Kim, K. Kishimoto, F. Miyaji, T. Kokubo, T. Yao, Y. Suetsugu, J. Tanaka and T. Nakamura, *J. Biomed. Mater. Res.*, 1999, **46**, 228.
41. H. M. Kim, K. Kishimoto, F. Miyaji, T. Kokubo, T. Yao, Y. Suetsugu, J. Tanaka and T. Nakamura, *J. Mater. Sci.: Mater. Med.*, 2000, **11**, 421.
42. W. F. Newman and M. W. Newman, *The chemical dynamics of bone mineral*, The University of Chicago Press, Chicago, 1967, p. 18.
43. A. Oyane, H. M. Kim, T. Furuya, T. Kokubo and T. Miyazaki, *J. Biomed. Mater. Res.*, 2003, **65A**, 188.
44. H. Takadama, M. Hashimoto, M. Mizuno and T. Kokubo, *Phosphorus Res. Bull.*, 2004, **17**, 119.
45. T. Hanaba, K. Asami and K. Asaoka, *J. Biomed. Mater. Res.*, 1998, **40**, 530.
46. C. A. Homsy, *J. Biomed. Mater. Res.*, 1970, **4**, 341.
47. K. Hyakuna, T. Yamamuro, Y. Kotoura, M. Oka, T. Nakamura, T. Kitsugi, T. Kokubo and H. Kushitani, *J. Biomed. Mater. Res.*, 1990, **24**, 471.

48. A. C. Lewis, M. R. Kilburn, I. Papageorgiou, G. C. Allen and C. P. Case, *J. Biomed. Mater. Res.*, 2005, **73A**, 456.
49. Y. Gao, W. Weng, K. Cheng, P. Du, G. Shen, G. Han, B. Guan and W. Yan, *J. Biomed. Mater. Res.*, 2006, **79A**, 193.
50. F. Miyaji, H. M. Kim, S. Handa, T. Kokubo and T. Nakamura, *Biomaterials*, 1999, **20**, 913.
51. F. Barrere, C. A. van Blitterswijk, K. de Groot and P. Layrolle, *Biomaterials*, 2002, **23**, 2211.
52. A. C. Tas and S. B. Bhaduri, *J. Mater. Res.*, 2004, **19**, 2742.
53. F. Barrere, C. A. van Blitterswijk, K. de Groot and P. Layrolle, *Biomaterials*, 2002, **23**, 1921.
54. F. Barrere, P. Layrolle, C. A. Van Blitterswijk and K. De Groot, *J. Mater. Sci.: Mater. Med.*, 2001, **12**, 529.
55. Y. F. Chou, W. Huang, J. C. Y. Dunn, T. Miller and B. M. Wu, *Biomaterials*, 2005, **26**, 285.
56. L. D. Warren, A. E. Clark and L. L. Hench, *J. Biomed. Mater. Res.*, 1989, **23**, 201.
57. J. H. Hanks and R. E. Wallace, *Proc. Soc. Exp. Biol. Med.*, 1949, **71**, 196.
58. Y. Shibata, H. Takashima, H. Yamamoto and T. Miyazaki, *Int. J. Oral Maxillofac. Implants*, 2004, **19**, 177.
59. P. A. P. Marques, A. P. Serro, B. J. Saramago, A. C. Fernandes, M. C. Magalhaes and R. N. Correia, *Biomaterials*, 2003, **24**, 451.
60. M. Vallet-Regí, A. J. Salinas and D. Arcos, *J. Mater. Sci.: Mater. Med.*, 2006, **17**, 1011.
61. I. Izquierdo-Barba, A. J. Salinas and M. Vallet-Regí, *J. Biomed. Mater. Res.*, 2000, **51**, 191.
62. J. Hlavac, D. Rohanova and A. Helebrant, *Ceram. Silicate*, 1994, **38**, 119.
63. S. Falaize, S. Radin and P. Ducheyne, *J. Am. Ceram. Soc.*, 1999, **82**, 969.
64. A. J. Salinas, M. Vallet-Regí and I. Izquierdo-Barba, *J. Sol-Gel Sci. Technol.*, 2001, **21**, 13.
65. G. H. Nancollas and W. Wu, *J. Cryst. Growth*, 2000, **211**, 137.
66. P. Koutsoukos, Z. Amjad, M. B. Tomson and G. H. Nancollas, *J. Am. Chem. Soc.*, 1980, **102**, 1553.
67. M. B. Tomson and G. H. Nancollas, *Science*, 1978, **200**, 1059.
68. R. Kniep and S. Bush, *Angew. Chem., Int. Ed.*, 1996, **35**, 2624.
69. S. Busch, H. dolhaine, A. Dúchense, S. Heinz, O. Hochrein, F. Laeri, O. Podebrad, U. Vietze, T. Weiland and R. Kniep, *Eur. J. Inorg. Chem.*, 1999, 1643.
70. S. Busch, U. Schwarz and R. Kniep, *Chem. Mater.*, 2001, **13**, 3260.
71. H. Tlatlik, P. Simon, A. Kawska, D. Zahn and R. Kniep, *Angew. Chem., Int. Ed.*, 2006, **45**, 1905.
72. P. Simon, D. Zahn, H. Lichte and R. Kniep, *Angew. Chem., Int. Ed.*, 2006, **45**, 1911.
73. D. Zaffe, *Micron*, 2005, **36**, 583.

74. E. L. Burger and V. Patel, *Orthopedics*, 2007, **30**, 939.
75. W. Suchanek and M. Yoshimura, *J. Mater. Res.*, 1998, **13**, 94.
76. M. Neo, T. Nakamura, T. Yamamuro, C. Ohtsuki and T. Kokubo, in *Bone-bonding biomaterials*, ed. P. Ducheyne, T. Kokubo and C. A. van Blitterswijk, Reed Healthcare Communications, Leiderdorp, Netherlands, 1993, pp. 111–120.
77. P. Ducheyne, J. Beight, J. Cuckler, B. Evans and S. Radin, *Biomaterials*, 1990, **11**, 531.
78. P. Ducheyne and J. M. Cuckler, *Clin. Orthop. Relat. Res.*, 1992, **276**, 102.
79. J. D. de Bruijn, Y. P. Novell and C. A. van Blitterswijk, *Biomaterials*, 1994, **15**, 543.
80. S. H. Maxian, J. P. Zawadski and M. G. Duna, *J. Biomed. Mater. Res.*, 1993, **27**, 111.
81. M. Jarcho, J. F. Kay, K. I. Gumaer, R. N. Doremus and H. P. Drobeck, *J. Bioeng.*, 1977, **1**, 79.
82. B. M. Tracy and R. H. Doremus, *J. Biomed. Mater. Res.*, 1984, **18**, 719.
83. G. Daculsi, R. Z. LeGeros, E. Nery, K. Lynch and B. Kerebel, *J. Biomed. Mater. Res.*, 1988, **23**, 257.
84. S. Langstaff, M. Sayer, T. Smith, S. Pugh, S. Hesp and W. Thompson, *Biomaterials*, 2001, **22**, 135.
85. K. Kurashina, H. Kurita, M. Hirano, A. Kotani, C. P. Klein and D. de Groot, *Biomaterials*, 1997, **18**, 539.
86. S. Takagi, L. C. Chow and K. Ishikawa, *Biomaterials*, 1998, **19**, 1593.
87. P. Ducheyne and Q. Qiu, *Biomaterials*, 1999, **20**, 2287.
88. R. Z. Le Geros, J. R. Parsons, G. Daculsi, F. Driessens, D. Lee, S. T. Liu, S. Metsger, D. Peterson and M. Walker, in *Bioceramics: material characteristics versus in vivo bebhavior*, ed. P. Ducheine and J. Lemons, N. Y. Acad. Sci., 1988, vol. 523, pp. 268–271.
89. T. Fujui and M. Ogino, *J. Biomed. Mater. Res.*, 1984, **18**, 845.
90. L. L. Hench, Bioactive ceramics, in *Bioceramics: Material characteristics versus in vivo behavior*, ed. P. Ducheyne and J. Lemons, N.Y. Acad. Sci., 1988, vol. 54, p. 523.
91. P. Ducheyne, S. Radin and L. King, *J. Biomed. Mater. Res.*, 1993, **27**, 25.
92. A. S. Posner, *Clin. Orthop.*, 1985, **200**, 87.
93. W. van Raemdonck, P. Ducheyne and P. de Meester, in *Metal and ceramic biomaterials*, ed. P. Ducheyne and W. Hasting, CRC Press, Boca Raton, 1984, p. 149.
94. M. Jarcho, *Clin. Orthop.*, 1981, **157**, 259.
95. J. C. Elliott, *Structure and Chemistry of the Apatites and other Calcium Orthophosphates*, Elsevier, Amsterdam, 1994.
96. R. Z. LeGeros, *Calcium phosphates in oral biology and medicine*, Basel, karger, 1991.
97. Y. Leng, J. Chen and S. Qu, *Biomaterials*, 2003, **24**, 2125.
98. S. R. Radin and P. Ducheyne, *J. Biomed. Mater. Res.*, 1993, **27**, 35.
99. R. H. Doremus, *J. Mater. Sci.*, 1992, **27**, 285.
100. K. A. Gross and C. C. Berndt, *J. Biomed. Mater. Res.*, 1998, **39**, 580.

101. T. Kobayashi, S. Shingaki, T. Nakajima and K. Hanada, *J. Long-Term Eff. Med. Implants*, 1993, **3**, 283.
102. W. Bonfield, M. D. Grynpas, A. E. Tuly, J. Bowman and J. Abram, *Biomaterials*, 1981, **2**, 185.
103. A. Sari, R. Yavuzer, S. Ayhan, S. Tuncer, O. Latifoglu, K. Atabay and M. C. Celebi, *J. Craniofac. Surg.*, 2003, **14**, 919.
104. M. C. Kruyt, W. J. A. Dhert, C. Oner, C. A. van Blitterswijk, A. J. Verbout and J. D. de Bruijn, *J. Biomed. Mater. Res.*, 2004, **69B**(2), 113.
105. S. F. Hulbert, L. L. Hench, D. Forbers and L. S. Bowman, *Ceram. Int.*, 1982, **8**, 121.
106. C. Ergun, T. J. Webster, R. Bizios and R. H. Doremus, *J. Biomed. Mater. Res.*, 2002, **59**, 305.
107. R. A. Young and P. E. Mackie, *Mater. Res. Bull.*, 1980, **15**, 17.
108. R. M. Wilson, J. C. Elliott and S. E. P. Dowker, *Am. Miner.*, 1999, **84**, 1406.
109. E. A. P. De Maeyer, R. M. H. Verbeeck and D. E. Naessens, *Inorg. Chem.*, 1993, **32**, 5709.
110. R. M. H. Verbeeck, E. A. P. De Maeyer and F. C. M. Driessens, *Inorg. Chem.*, 1995, **34**, 2084.
111. E. M. Carlisle, *Science*, 1970, **167**, 179.
112. E. M. Carlisle, *Calcif. Tissue Int.*, 1981, **33**, 27.
113. L. L. Hench and G. P. LaTorre, in *Bioceramics*, ed. T. Yamamuro, T. Kokubo and T. Nakamura, Kobunshi Kankokai, Inc., Kyoto, 1993, vol. 5, pp. 67–74.
114. C. Ohtsuki, T. Kokubo and T. Yamamuro, *J. Non-Cryst. Solids*, 1992, **143**, 84.
115. D. Arcos, D. C. Greenspan and M. Vallet-Regí, *Chem. Mater.*, 2002, **14**, 1515.
116. D. Arcos, D. C. Greenspan and M. Vallet-Regí, *J. Biomed. Mater. Res.*, 2003, **65A**, 344.
117. I. R. Gibson, J. Huang, S. M. Best and W. Bonfield, in *Bioceramics*, ed. H. Ohgushi, G. W. Hastings and T. Yoshikawa, World Scientific, Singapore, 1999, vol. 12, pp. 191–194.
118. K. A. Hing, S. Saeed, B. Annaz, T. Buckland and P. A. Revell, Transactions 7th World Biomaterials Congress, Australian Society for Biomaterials, Brunswick Lower, Vic., 2004, p. 108.
119. I. R. Gibson, S. M. Best and W. Bonfield, *J. Biomed. Mater. Res.*, 1999, **44**, 422.
120. S. M. Best, W. Bonfield, I. R. Gibson, L. J. Jha and J. D. Santos, International Patent Appl. No. PCT/GB97/02325, 1996.
121. I. R. Gibson, S. M. Best and W. Bonfield, *J. Am. Ceram. Soc.*, 2002, **85**, 2771.
122. N. Rashid, I. Harding and K. A. Hing, Transactions 7th World Biomaterials Congress, Australian Society for Biomaterials, Brunswick Lower, Vic., 2004, p. 106.

123. S. R. Kim, J. H. Lee, Y. T. Kim, D. H. Riu, S. J. Jung, Y. J. Lee, S. C. Chung and Y. H. Kim, *Biomaterials*, 2003, **24**, 1389.
124. P. A. Marques, M. C. F. Magalhaes, R. N. Correia and M. Vallet-Regí, *Key Eng. Mater.*, 2001, **192–195**, 247.
125. A. J. Ruys, *J. Aust. Ceram. Soc.*, 1993, **29**, 71.
126. S. R. Kim, D. H. Riu, Y. J. Lee and Y. H. Kim, *Key Eng. Mater.*, 2002, **218–220**, 85.
127. D. Arcos, J. Rodriguez-Carvajal and M. Vallet-Regí, *Chem. Mater.*, 2004, **16**, 2300.
128. M. Vallet-Regí and D. Arcos, *J. Mater. Chem.*, 2005, **15**, 1509.
129. D. Arcos, J. Rodriguez-Carvajal and M. Vallet-Regí, *Solid State Sci.*, 2004, **6**, 987.
130. D. Arcos, J. Rodriguez-Carvajal and M. Vallet-Regí, *Physica B*, 2004, **350**, e607.
131. D. Arcos, J. Rodriguez-Carvajal and M. Vallet-Regí, *Chem. Mater.*, 2005, **17**, 57.
132. J. R. Hupp and S. J. McKenna, *J. Oral Maxillofac. Surg.*, 1988, **46**, 533.
133. M. El Deeb and M. Roszkowski, *J. Oral Maxillofac. Surg.*, 1988, **46**, 33.
134. B. V. Rejda, J. G. J. Peelen and K. de Groot, *J. Bioeng.*, 1977, **1**, 93.
135. M. Vallet-Regí, *J. Chem. Soc., Dalton Trans.*, 2001, 97.
136. R. Ellinger, E. B. Nery and K. L. Lynch, *Int. J. Periodont. Tertor. Dent.*, 1986, **3**, 23.
137. A. Takeishi, H. Hayashi, H. Kamatsubara, A. Yokoyama, M. Kohri, T. Kawasaki, K. Micki and T. Kohgo, *J. Dent. Res.*, 1989, **68**, 680.
138. G. Daculsi, N. Passuti, S. Martin, C. Deudon, R. Z. LeGeros and S. Rather, *J. Biomed. Mater. Res.*, 1990, **24**, 379.
139. C. Schopper, F. Ziya-Ghazvini, W. Goriwoda, D. Moser, F. Wanschitz, E. Spassova, G. Lagogiannis, A. Auterith and R. Ewers, *J. Biomed. Mater. Res.*, 2005, **74B**, 458.
140. M. Trecant, J. Delecrin, J. Royer, E. Goyenvalle and G. Daculsi, *Clin. Mater.*, 1994, **18**, 233.
141. A. Sendemir-Urkmez and R. D. Jamison, *J. Biomed. Mater. Res.*, 2007, **81A**, 624.
142. C. R. Yang, Y. J. Wang, X. F. Chen and N. R. Zhao, *Mater. Lett.*, 2005, **59**, 3635.
143. S. Sánchez-Salcedo, I. Izquierdo-Barba, D. Arcos and M. Vallet-Regí, *Tissue Eng.*, 2006, **12**, 279.
144. M. I. Alam, I. Asahina, K. Ohmmaiuda and S. Enomoto, *J. Biomed. Mater. Res.*, 2000, **54**, 129.
145. O. Gauthier, J. Guicheux, G. R. Grimandi, A. Faivre-Cahuvet and G. Daculsi, *J. Biomed. Mater. Res.*, 1998, **40**, 606.
146. O. Bermúdez, M. G. Boltong, F. C. M. Driessens and J. A. Planell, *J. Mater. Sci.: Mater. Med.*, 1994, **5**, 160.
147. H. Yamamoto, S. Niwa, M. Hori, T. Hattori, K. Sawai, S. Aoki, M. Hirano and H. Takeuchi, *Biomaterials*, 1998, **19**, 1587.

148. M. Sayer, A. Stratilatov, J. Reid, L. Calderin, M. Stott, X. Yin, M. McK-enzie, J. N. Smith, J. A. Hendry and S. D. Langstaff, *Biomaterials*, 2003, **24**, 369.

149. S. Langstaff, M. Sayer, T. Smith, S. Pugh, S. Hesp and W. Thompsom, *Biomaterials*, 1999, **20**, 1727.

150. A. Pietak and M. Sayer, *Biomaterials*, 2005, **24**, 3819.

151. R. Z. LeGeros, G. Daculsi, E. Nery, K. Lynch and B. Kerebel. Transactions of the Third World Biomaterials Congress, 1988, pp. 2B1–35.

152. M. Kohri, K. Miki, D. E. Waite, H. Nakajima and T. Okabe, *Biomaterials*, 1993, **14**, 299.

153. L. L. Hench, *J. Mater. Sci.: Mater. Med.*, 2006, **17**, 967.

154. L. L. Hench, A. E. Clark and H. F. Schaake, *Int. J. Non-Cryst. Solids*, 1972, **8–10**, 837.

155. L. L. Hench and A. Paschall, *J. Biomed. Mater. Res. Symp.*, 1973, **4**, 25.

156. L. L. Hench, *Curr. Opin. Solid State Mater. Sci.*, 1997, **2**, 604.

157. M. Ogino, F. Ohuchi and L. L. Hench, *J. Biomed. Mater. Res.*, 1980, **14**, 55.

158. U. Gross, R. Kinne, H. J. Schmitz and V. Strunz, in *CRC Critical Reviews in Biocompatibility*, ed. D. L. Williams, CRC Press, Boca Raton, Florida, 1988, vol. 4, issue 2, p. 155.

159. L. L. Hench, in *Bioceramics: Materials Characteristics Versus In Vivo Behavior*, ed. J. P. Ducheyne and J. Lemmons, Annuals of the New York Academy of Sciences, 1988, vol. 523, p. 54.

160. R. Hill, *J. Mater. Sci. Lett.*, 1996, **15**, 1122.

161. K. E. Wallace, R. G. Hill, J. T. Pembroke, C. J. Brown and P. V. Hatton, *J. Mater. Sci.: Mater. Med.*, 1999, **10**, 697.

162. O. H. Anderson, K. H. Karlsson and K. Kangasmiemi, *J. Non-Cryst. Solids*, 1990, **119**, 290.

163. D. C. Greenspan and L. L. Hench, *J. Biomed. Mater. Res.*, 1976, **10**, 503.

164. Y. Ebisawa, F. Miyaji, T. Kokubo, K. Ohura and T. Nakamura, *Biomaterials*, 1997, **18**, 1277.

165. M. Brink, *J. Biomed. Mater. Res.*, 1997, **36**, 109.

166. M. Brink, T. Turunen, R-P. Happonen and A. Yli-Urpo, *J. Biomed. Mater. Res.*, 1997, **37**, 114.

167. R. Li, A. E. Clark and L. L. Hench, *J. Appl. Biomater.*, 1991, **2**, 231.

168. M. Catauro, G. Laudisio, A. Costantini, R. Fresa and F. Branda, *J. Sol-Gel Sci. Technol.*, 1997, **10**, 231.

169. I. Izquierdo-Barba, A. J. Salinas and M. Vallet-Regí, *J. Biomed. Mater. Res.*, 1999, **47**, 243.

170. A. Martínez, I. Izquierdo-Barba and M. Vallet-Regí, *Chem. Mater.*, 2000, **12**, 3080.

171. M. M. Pereira, A. E. Clark and L. L. Hench, *J. Am. Ceram. Soc.*, 1995, **78**, 2463.

172. M. M. Pereira and L. L. Hench, *J. Sol-Gel Sci.*, 1996, 7, 59.

173. T. Peltola, M. Jokinen, H. Rahiala, E. Levanen, J. B. Rosenholm, I. Kangasniemi and A. Yli-Urpo, *J. Biomed. Mater. Res.*, 1999, **44**, 12.

174. M. Vallet-Regí, D. Arcos and J. Pérez-Pariente, *J. Biomed. Mater. Res.*, 2000, **51**, 23.

175. M. Vallet-Regí and A. Rámila, *Chem. Mater.*, 2000, **12**, 961.
176. D. C. Greenspan, J. P. Zhong and G. P. LaTorre, in *Bioceramics*, ed. Turku, O. H. Anderson and A. Yli-Urpo, Butterworth- einemann Ltd., Oxford, 1994, vol. 7, p. 55.
177. M. Jokinen, H. Rahiala, J. B. Rosenholm, T. Peltola and I. Kangasniemi, *J. Sol-Gel Sci. Technol.*, 1998, **12**, 159.
178. D. Arcos, C. V. Ragel and M. Vallet-Regí, *Biomaterials*, 2001, **22**, 701.
179. M. Laczka, K. Cholewa and A. Laczka-Osyczka, *J. Alloys Compd.*, 1997, **248**, 42.
180. M. Vallet-Regí, A. M. Romero, C. V. Ragel and R. Z. LeGeros, *J. Biomed. Mater. Res.*, 1999, **44**, 416.
181. M. M. Pereira, A. E. Clark and L. L. Hench, *J. Biomed. Mater. Res.*, 1994, **28**, 693.
182. J. Pérez-Pariente, F. Balas, J. Román, A. J. Salinas and M. Vallet-Regí, *J. Biomed. Mat. Res.*, 1999, **47**, 170.
183. K. Ohura, T. Nakamura, T. Kokubo, Y. Ebisawa, Y. Kotoura and M. Oka, *J. Biomed. Mater. Res.*, 1991, **25**, 357.
184. Y. Ebisawa, T. Kokubo, K. Ohura and T. Yamamuro, *J. Mater. Sci.: Mater. Med.*, 1990, **1**, 239.
185. L. L. Hench and O. Andersson, in bioactive Glasses, *An Introduction to Bioceramics*, ed. L. L. Hench and J. Wilson, World Scintific Publishing, Singapore, 1993, p. 41.
186. C. Ohtsuki, T. Kokubo, K. Takatsuka and T. Yamamuro, *Nippon Seramikkusu Kyokai Gakujutsu Ronbunshi*, 1991, **99**, 1.
187. L. L. Hench, in *Ceramics: Towards the 21st Century*, ed. N. Soga and S. Kato, Ceramic Society of Japan, Tokyo, 1991, p. 519.
188. Ö. H. Andersson and K. H. Karlsson, *J. Non-Cryst. Solids*, 1991, **129**, 145.
189. W. D. Kingery, H. K. Bowen and D. R. Bowen, in *Introduction to Ceramics*, Wiley, New York, 2nd edn, 1960, p. 328.
190. M. Vallet-Regí, C. V. Ragel and A. J. Salinas, *Eur. J. Inorg. Chem.*, 2003, 1029.
191. D. C. Greenspan, J. P. Zhong and G. P. LaTorre, *Bioceramics*, 1995, **8**, 477.
192. D. Arcos, J. Peña and M. Vallet-Regí, *Key Eng. Mater.*, 2004, **254–256**, 27.
193. F. G. Araujo, G. P. Latorre and L. L. Hench, *J. Non-Cryst. Solids*, 1995, **185**, 41.
194. R. Li, A. E. Clark and L. L. Hench, in *Chemical Processing of Adv. Mater.*, ed. L. L. Hench and J. K. West, John Wiley and Sons, New York, 1992, p. 627.
195. F. Balas, D. Arcos, J. Pérez-Pariente and M. Vallet-Regí, *J. Mater. Res.*, 2001, **16**, 1345.
196. M. M. Pereira and L. L. Hench, *J. Sol-Gel Sci. Technol.*, 1996, **7**, 231.
197. P. Li, C. Ohtuki, T. Kokubo, K. Nakanishi, N. Soja, T. Nakamura and T. Yamamuro, *J. Am. Ceram. Soc.*, 1992, **75**, 2094.
198. K. H. Karlsson, K. Froberg and T. Ringbom, *J. Non-Cryst. Solids*, 1989, **112**, 69.
199. P. E. Wang and T. K. Chaki, *J. Mater. Sci.: Mater. Med.*, 1995, **6**, 94.
200. M. Vallet-Regí, I. Izquierdo-Barba and A. J. Salinas, *J. Biomed. Mater. Res.*, 1999, **46**, 560.

201. A. J. Salinas, A. I. Martín and M. Vallet-Regí, *J. Biomed. Mater. Res.*, 2002, **61**, 524.
202. M. Vallet-Regí, A. J. Salinas, J. Ramírez-Castellanos and J. M. González-Calbet, *Chem. Mater.*, 2005, **17**, 1874.
203. R. Z. LeGeros, *Prog. Cryst. Growth Charact.*, 1981, **4**, 1.
204. R. Z. LeGeros, J. P. LeGeros, O. R. Trantz and E. Klein, *Dev. Appl. Spectrosc.*, 1970, **7B**, 13.
205. P. Horcajada, A. Rámila, K. Boulahya, J. González-Calbet and M. Vallet-Regí, *Solid State Sci.*, 2004, **6**, 1295.
206. M. Vallet-Regí, I. Izquierdo-Barba, A. Rámila, J. Pérez-Pariente, F. Babonneau and J. M. González-Calbet, *Solid State Sci.*, 2005, **7**, 233.
207. M. Vallet-Regí, L. Ruiz-González, I. Izquierdo-Barba and J. M. González-Calbet, *J. Mater. Chem.*, 2006, **16**, 23.
208. I. Izquierdo-Barba, L. Ruiz-González, J. C. Doadrio, J. M. González-Calbet and M. Vallet-Regí, *Solid State Sci.*, 2005, **7**, 983.
209. J. M. Gomez-Vega, M. Iyoshi, K. M. Kim, A. Hozumi, H. Sugimura and O. Takai, *Thin Solids Films*, 2001, **398–399**, 615.
210. X. X. Yan, C. Z. Yu, X. F. Zhou, J. W. Tang and D. Y. Zhao, *Angew. Chem., Int. Ed.*, 2004, **43**, 5980.
211. C. J. Brinker, Y. F. Lu, A. Sellinger and H. Y. Fan, *Adv. Mater.*, 1999, **11**, 579.
212. A. López-Noriega, D. Arcos, I. Izquierdo-Barba, Y. Sakamoto, O. Terasaki and M. Vallet-Regí, *Chem. Mater.*, 2006, **18**, 3137.
213. A. García, M. Cicuéndez, I. Izquierdo-Barba, D. Arcos and M. Vallet-Regí, *Chem. Mater.*, 2009, **21**, 5474.
214. H. S. Yun, S. E. Kim and Y. T. Hyeon, *Mater. Lett.*, 2007, **61**, 4569.
215. J. Sun, Y. S. Li, L. Li, W. R. Zhao, L. Li, J. H. Gao, M. L. Ruan and J. L. Shi, *J. Non-Cryst. Solids*, 2008, **354**, 3799.
216. X. Yan, G. Wei, L. Zhao, J. Yi, H. Deng, L. Wang, G. Lu and C. Yu, *Microporous Mesoporous Mater.*, 2010, **132**, 282.
217. I. Izquierdo-Barba, D. Arcos, Y. Sakamoto, O. Terasaki, A. López-Noriega and M. Vallet-Regí, *Chem. Mater.*, 2008, **20**, 3191.
218. H. Schmidt, *J. Non-Cryst. Solids*, 1985, **73**, 681.
219. H. H. Huang, B. Orler and G. L. Wilkes, *Polym. Bull.*, 1985, **14**, 557.
220. J. D. Mackenzie, Y. J. Chung and Y. Hu, *J. Non-Cryst. Solids*, 1992, **271**, 147–148.
221. J. D. Mackenzie, *J. Sol-Gel Sci. Technol.*, 1994, **2**, 81.
222. S. Motakef, T. Suratwala, R. L. Poncone, J. M. Boulton, G. Teowee and D. R. Uhlmann, *J. Non-Cryst. Solids*, 1994, **178**, 37.
223. K. Tsuru, C. Ohtsuki, A. Osaka, T. Iwamoto and J. D. Mackenzie, *J. Mater. Sci.: Mater. Med.*, 1997, **8**, 157.
224. Q. Chen, F. Miyaji, T. Kokubo and T. Nakamura, *Biomaterials*, 1999, **20**, 1127.
225. C. Sanchez, B. Lebeau, F. Chaput and J. P. Boilot, *Adv. Mater.*, 2003, **15**, 1969.
226. C. Sanchez, B. Julián, P. Belleville and M. Popall, *J. Mater. Chem.*, 2005, **15**, 3559.

227. H. Schmidt, A. Kaiser, H. Patzelt and H. Sholze, *J. Phys.*, 1982, **12**, 275.
228. J. Livage, M. Henry and C. Sanchez, *Prog. Solid State Chem.*, 1988, **18**, 259.
229. C. Sanchez and F. Ribot, *New J. Chem.*, 1994, **18**, 1007.
230. A. I. Martín, A. J. Salinas and M. Vallet-Regí, *J. Eur. Ceram. Soc.*, 2005, **25**, 3533.
231. A. J. Salinas, J. M. Merino, N. Hijón, A. I. Martín and M. Vallet-Regí, *Key Eng. Mater.*, 2004, **254–256**, 481.
232. H. Schmidt and B. Seiferling, *Mater. Res. Soc. Symp. Proc.*, 1986, **73**, 739.
233. Y. Wei, D. Jin, "A new class of organic-inorganic hybrid dental materials" in *Abstracts of papers of the American Chemical Society* 214, 145–POLY, Part 2, Sep 7 1997.
234. J. M. Yang, C. S. Lu, Y. G. Hsu and C. H. Shih, *J. Biomed. Mater. Res.*, 1997, **38**, 143.
235. S. Rhee and J. Choi, *J. Am. Ceram. Soc.*, 2002, **85**, 1318.
236. S. Yamamoto, T. Miyamoto, T. Kokubo and T. Nakamura, *Polym. Bull.*, 1998, **40**, 243.
237. Y. Hu and J. D. Mackenzie, *J. Mater. Sci.*, 1992, **27**, 4415.
238. Y. J. Chung, S. Ting and J. D. Mackenzie, in *Better Ceramics Through Chemistry IV*, ed. B. J. J. Zelinski, C. J. Brinker, D. E. Clark and D. R. Ulrich, Materials Research Society, Pittsburg, PA, 1990, vol. 180, p. 981.
239. N. Miyata, K. Fuke, Q. Chen, M. Kawashita, T. Kokubo and T. Nakamura, *J. Ceram. Soc. Jpn.*, 2003, **111**, 555.
240. Q. Chen, N. Miyata, T. Kokubo and T. Nakamura, *J. Mater. Sci.: Mater. Med.*, 2001, **12**, 515.
241. M. Kamitakahara, M. Kawashita, N. Miyata, T. Kokubo and T. Nakamura, *J. Mater. Sci.: Mater. Med.*, 2002, **13**, 1015.
242. M. J. Michalczyk and K. G. Sharp, US Patent 5,378,790, 1995.
243. K. G. Sharp and M. J. Michalczyk, *J. Sol-Gel Sci. Technol.*, 1997, **8**, 541.
244. K. G. Sharp, *Adv. Mater.*, 1998, **10**, 1243.
245. K. G. Sharp, *J. Mater. Chem.*, 2005, **15**, 3812.
246. M. Manzano, D. Arcos, M. Rodríguez Delgado, E. Ruiz, F. J. Gil and M. Vallet-Regí, *Chem. Mater.*, 2006, **18**, 5696.

Medical Applications of Bioactive Nanoceramics

4.1 Introduction

In recent years, the development of nanotechnologies has acquired great scientific interest as they are a bridge between bulk materials and atomic or molecular structures. The properties of materials change as their size approaches the nanoscale and as the percentage of atoms at the surface of a material becomes significant. For bulk materials larger than one micrometre the percentage of atoms at the surface is minuscule relative to the total number of atoms of the material. The interesting and sometimes unexpected properties of nanoparticles are partly due to the aspects of the surface of the material dominating the properties *in lieu* of the bulk properties. The inherent nanoceramic properties allow the tackling of traditional problems of the bioceramics field, such as mechanical performance, bone-regeneration kinetics, and biocompatibility, *etc.*, as well as new challenges such as the optimisation of scaffolds for bone-tissue engineering and the design of nanodrug-delivery systems aimed to work within the bone tissue.

The contribution of nanoceramics to the field of biomaterials is also mainly justified by their surface features.[1] It must be highlighted that the final aim is the optimum tissue–implant interaction, which is a surface event. The large surface area of nanoceramics supply new reactivity features. This fact involves new expectations for events such as bioactivity, bioresorption, foreign-body responses, *etc.* Secondly, nanoceramics give the chance to tailor at the nanometric scale the interactions between the material and the osteoblast adhesion proteins, with the purpose of

RSC Nanoscience & Nanotechnology No. 39
Nanoceramics in Clinical Use: From Materials to Applications, 2nd Edition
By María Vallet-Regí and Daniel Arcos Navarrete
Published by the Royal Society of Chemistry, www.rsc.org

optimising scaffolds for bone-tissue engineering. The surfaces of nanoceramics are suitable to be easily functionalised and can incorporate biologically active molecules.[2] Since nanomaterials exhibit a maximum surface/volume ratio they are excellent candidates as vehicles for *drug delivery* applications.

One of the classical drawbacks of bioceramics for orthopaedic applications is the mechanical behaviour when implanted in high-load body locations. The brittleness of these compounds has limited the clinical applications to the filling of small bone defects within load-bearing sites. Among bioceramics having biomimetic properties, only some glass-ceramics have evidenced a good performance in spine and hip surgery of patients with extensive lesions or bone defects, for instance apatite–wollasonite (A–W) glass-ceramic,[3] due to its excellent mechanical strengths and capacity of binding to living bone. Glass-ceramics are defined as polycrystalline solids prepared by the controlled crystallisation of glasses. For instance, in the case of A–W glass-ceramic, apatite (38%) and wollastonite (34%) are homogenously dispersed in a glassy matrix (MgO 16.6, CaO_2 4.2, and SiO_2 59.2 wt%), taking the shape of a rice grain of 50–100 nm in size. Probably, this fact has inspired the use of nanoparticles to obtain highly resistant *ceramic/polymer nanocomposites*, which is currently one of the main topics in the nano-structured biomaterials field.

In addition to powder and blocks (both dense and porous) HA (hydroxyapatite) has been prepared for a long time at the micrometre scale as a coating. During recent years, significant research efforts have been devoted to nanostructure processing of HA coatings in order to obtain high surface area and ultrafine structure, which are properties essential for cell–substrate interaction upon implantation. The potential of nanoapatites as implant coatings has generated a considerable interest due to their superior biocompatibility, osteoconductivity, bioactivity, and noninflammatory nature.[4] One of the significant properties attributed to nanomaterials, namely high surface area reactivity, can be exploited to improve the interfaces between cells and implants. In addition, nanocrystallised characteristics have proven to be of superior biological efficiency. For example, compared to conventionally crystallised HA, nanocrystallised HA promotes osteoblast adhesion, differentiation and proliferation, osteointegration, and deposition of calcium-containing mineral on its surface, thereby enhancing the formation of new bone within a short period.[5]

Finally, the synthesis of hydroxyapatite nanopowders is also considered to improve the sintering processes of conventional ceramic bone implants. Sintering and densification of any ceramic depends on powder properties such as particle size, distribution, and morphology and it is believed that nanostructured calcium phosphate ceramics can improve the sintering kinetics due to a higher surface area, and subsequently improve mechanical properties significantly.[6] Figure 4.1 shows some of the clinical and potential applications of nanoceramics in the field of bone grafting.

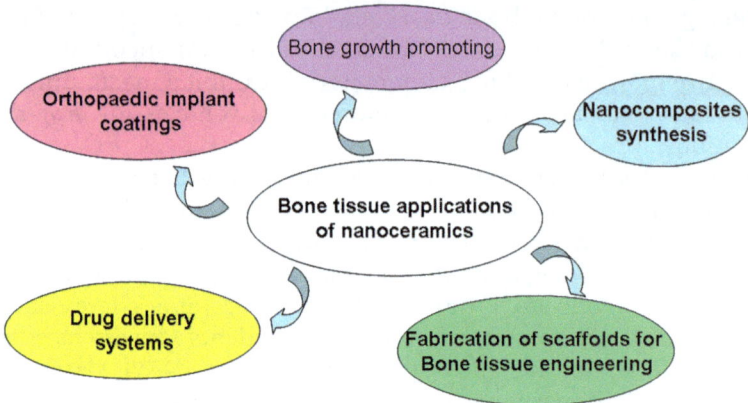

Figure 4.1   Current clinical applications of nanoceramics for bone tissue repair.

4.2   Nanoceramics for Bone-Tissue Regeneration

One of the most important aims of the nanoceramics with biomimetic properties is to provide new and effective therapies for those pathologies requiring bone regeneration. Among these diseases, osteoporosis must be highlighted due to the present and future high incidence within the world population. Osteoporosis affects 75 million people in Europe, USA and Japan[7] and 30–50% of women and 15–30% of men will suffer a fracture related to osteoporosis in their lifetime.[8] Pharmacological prevention and treatment of osteoporosis is the best strategy to date, although these therapies have some drawbacks related to bone formation in areas different from the osteoporosis sites. This fact is assumed as a consequence of drug intake through systemic administration (oral and parenteral). In the case of fractures that cannot self-heal, orthopaedic devices such as fixation devices or total hip prostheses are required. These devices have a limited average lifetime of around 15 years and it is speculated that this situation is due to lack of biomimetism exhibited by the implant surface at the nanometric scale.

Nanometrical calcium-phosphate-based biomaterials are very promising materials for both delivering drugs (see Section 4.5) and for increasing bone mass. Through the implantation of calcium phosphates with properties mimicking the natural bone mineral, it is expected that properties such as bioactivity, dissolution range, resorption, *etc.* will be close to those of natural bones. Whereas previous emphasis was upon control of the stoichiometry of these biomaterials, the aim of this past decade has been focused on controlling size and morphology.[9] Actually, although bioceramics like HA and alumina with grain sizes >100 nm are used for orthopaedic and dental implants because of their biocompatibility, these materials sometimes exhibit insufficient apposition of bone, leading to implant failure.[10-16]

Nowadays, experimental results show that ceramics, metals, polymers, and composites with nanometre grain sizes stimulate osteoblast activity leading to more bone growth.[16] Long-term functions such as cell proliferation, synthesis of alkaline phosphatase and concentration of calcium in the extracellular matrix (ECM) are enhanced when osteoblasts are seeded on nanoceramics.[17] Since there are chemical differences between osteoporotic and healthy bone,[18] calcium-phosphate-based nanoparticles can be formulated to selectively attach to areas of osteoporotic bone. The key point on osteoblast selectivity is the protein adhesion at the first stages after implantation. Cells do not directly attach onto the material's surface, but on the proteins previously linked to the implant. Therefore, the first physical–chemical arguments to explain the better performance of nanoceramics must be found in the events that occur between the nanoceramics and the serum proteins.

4.2.1 Bone Cell Adhesion on Nanoceramics: The Role of the Proteins on the Specific Cell-Material Attachment

Proteins play a fundamental role in bone cell adhesion. In fact, in the absence of serum proteins, cell attachment to a substrate is dramatically decreased, whereas a culture medium containing 10% of these proteins highly enhances the adhesion onto ceramics with grain sizes <100 nm.[19,20]

It is difficult to draw conclusions regarding optimal osteoblast adhesion as a function of the type of ceramic, since different grain sizes of several bioceramics such as alumina, titania, and HA have been tested, and enhanced osteoblast adhesion is observed on the three of them when exhibiting grain sizes <100 nm.[21] Therefore, cellular responses to nanophase ceramics are independent of surface chemistry, at least among the biocompatible ceramics mentioned above.

Nano-sized grains provide higher surface roughness in the range of tens of nanometres, which appears to be a critical characteristic that determines the biocompatibility of the nanoceramic. Moreover, the nanostructure provides a higher number of grain boundaries as well as an increased surface wettability, which is also associated with enhanced protein adsorption and cell adhesion. However, the enhanced biocompatibility of nanoceramics exhibits a much more interesting selective mechanism. When considering several protein anchorage-depending cells, for instance osteoblast, fibroblast and endothelial cells, it is possible to correlate the adsorbed protein type and concentration with the observed cell adhesion on the materials tested. Figure 4.2 summarises this mechanism, where *vitronectine* is mainly adsorbed on nanoceramics, whereas *laminin* is preferentially adsorbed on conventional ceramics. Although the mechanism is not well established, fibroblast and endothelial cells attach preferentially onto conventional ceramics, whereas osteoblasts are preferentially adhered onto nanoceramics.

The fact that nanophase ceramics adsorb greater concentrations of vitronectin while conventional ceramics adsorb greater concentrations of laminin explains the subsequent enhanced adhesion of osteoblasts and endothelial cells on nanophase ceramics and conventional ceramics, respectively. Adhesion to substrate surfaces is imperative for subsequent functions

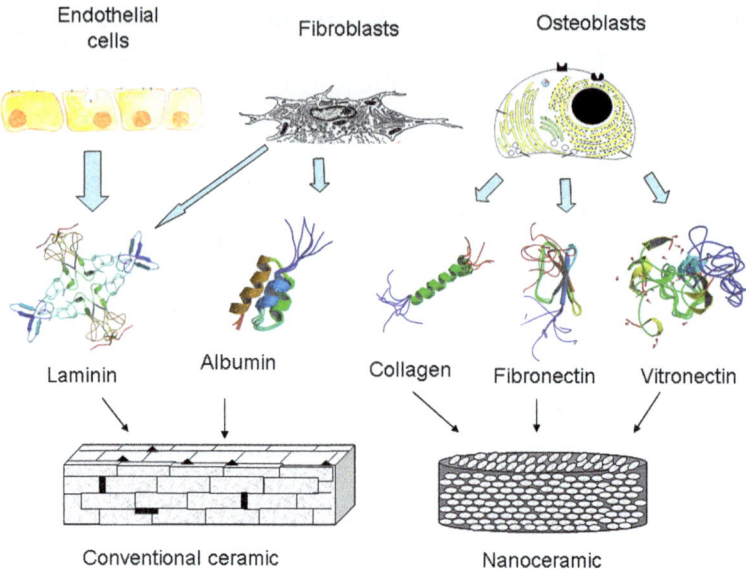

Figure 4.2 Schematic mechanism that explains ceramic–protein–cell attachment specificity.

of anchorage-dependent cells. Now the question is why vitronectine is mainly immobilised on nanoceramics, whereas laminin is more likely to attach to conventional ceramics. Webster *et al.*[21] explain this in terms of the inherent defect sizes of each kind of bioceramic. In addition to enhanced surface wettability, the roughness dictated by grain and pore size of nanophase ceramics influences interactions (such as adsorption and/or configuration/bioactivity) of determined serum proteins and thus affects subsequent cell adhesion. In this sense, vitronectin, which is a linear protein 15 nm in length[22] may preferentially have adsorbed to the small pores present in nanophase ceramics, while laminin (cruciform configuration, 70 nm both in length and width) would be preferentially adsorbed into the large pores present in conventional ceramics.

The specificity of nanoceramics with respect to the type of cell has been also observed with bone marrow mesenchymal stem cells and osteosarcoma cells.[23] When both cultures are exposed to HA nanoparticles 20–80 nm in diameter, greater cell viability and proliferation of mesenchymal stem cells were observed on the nano-HA, especially in the case of the smallest nanoparticles. In contrast, the growth of osteosarcoma cells was inhibited by the nano-HA and the smallest particles exhibited the higher inhibitory effect.

Another example of the groundbreaking possibilities of nano-HA is the behaviour of the periodontal ligament cells in contact with this nanoceramic. In previous chapters we explained the role that HA plays in the restoration of human hard tissue as well as in techniques that aim to regenerate periodontal tissues. Hydroxyapatite has osteoconductive effects but is non-bioresorbable and its use for periodontal tissue regeneration is not always effective. In fact, according to some studies, new periodontal regeneration is not always found if hydroxyapatite is used in the treatment of periodontal bone loss.[24] The key can be found in the poor response of the periodontal ligament cells to materials which have only osteoconductive but no osteoinductive effect. To summarise, when hydroxyapatite is used in periodontal osseous destruction, new periodontal regeneration is rarely found. This question has been tackled by implanting well-dispersed nano-HA powders. Appropriate particle dispersion can be achieved by using a sol–gel process in the presence of citric acid.[25] Citric acid acts as chelating reagent during the sol–gel process and prevents the agglomeration of hydroxyapatite. Nano-HA promotes the proliferation of periodontal ligament cells (PDLCs) as well as alkaline phosphatase (ALP) activity. ALP plays a key role in the formation and calcification of hard tissues, and its expression and enzyme activity are frequently used as markers of osteoblastic cells. The high expression of ALP activity in nano-HA indicates that nano-HA has the ability to induce osteogenic differentiation of PDLCs. This points out that nano-HA may be a suitable grafting material for periodontal tissue regeneration.

4.2.2 Bioinspired Nanoapatites: Supramolecular Chemistry as a Tool for Better Bioceramics

Synthetic nano-HA with high levels of structural and chemical similarities to bone has been successfully synthesised during the past 20 years and details of this advance has been collected in several reviews.[26–28] However, the ability to prepare these compounds mimicking the morphological and organised complexity analogous to their biological counterparts has not yet been attained. We have seen that bone is a composite consisting of HA nanorods embedded in a collagen matrix. In this sense, it is thought that HA nanorods are desirable as building blocks for the long-range assembly of macroscopic biomaterials with hierarchical order, aimed to improve the implant's biocompatibility.[29] Organic matrix-mediated biomineralisation is a process that principally involves the use of organic macromolecular assemblies to control various key aspects of inorganic deposition from supersaturated biological solutions. In particular, the organic matrix plays an important role in delineating the structure and chemistry of the mineralisation environment, providing site-specific nucleation centres, regulating crystal growth and morphological expression, and facilitating the construction of higher-order assemblies.[30]

Significant progress has been made, for example, in crystal morphology using water-soluble organic additives such as polyaspartic acid,[31,32] poly(acrylic acid),[33] and monosaccharides.[34] Similarly, ionic,[35–37] nonionic,[38] and blockcopolymer[39] surfactants have been used to produce calcium phosphates with specific morphologies. In addition, self-assembled organic supramolecular structures have been employed as templates for the controlled deposition of calcium phosphate. This is the case with nano-HA synthesised within liposomes[40] or templated nano-HA synthesis within a collagen matrix leading to nano-HA/collagen composites.[41] All these possibilities are based on the fact that organised organic surfaces can control the nucleation of inorganic materials by geometric, electrostatic, and stereochemical complementarities between the incipient nuclei and the functionalised substrates.[42–46]

By properly choosing organic additives that might have specific molecular complementarity with the inorganic component, the growth of inorganic nanocrystals can be rationally directed to yield products with desirable morphologies and/or hierarchical structures. Wang *et al.*[47] have prepared hydroxyapatite nanorods with tunable sizes, aspect ratios, and surface properties by properly tuning the interfaces between surfactants and the central atoms of HA based on the liquid–solid–solution strategy. This method is based on phase transfer and separation process across the liquid, solid, and solution interfaces. By properly tuning the chemical reactions at the interfaces, an extensive group of nanocrystals with tuneable sizes and hydrophobic surfaces has been prepared, demonstrating the effectiveness of controlling the chemical process occurring at the interfaces.

The preparation of HA nanorods in the presence of a cationic surfactant, cetyl trimethyl ammonium bromide (CTAB), has contributed to explain the formation mechanism of specific morphologies.[48] It is widely known that CTAB acts as a template,[49] with the template action resulting in the epitaxial growth of the product. Through the charge and stereochemistry features, molecule recognition occurs at the inorganic/organic interface.[34,50] The surfactant also binds to certain faces of a crystal or to certain ions, so these ions are also incorporated to the existing nuclei at a steady rate and the final shape and size of HA particles can be well controlled.[51]

In the case of hydroxyapatite growth, the behaviour of CTAB is also considered to correlate with the charge and stereochemistry properties. In an aqueous system, CTAB ionises completely and results in a cation with a tetrahedral structure, which can be well incorporated to the phosphate anion by the charge and structure complementarity. A probable mechanism for the templating process is that CTA^+–PO_4^{3-} mixtures form rod-like micelles, which contain many PO_4^{3-} groups on the surface, and when Ca^{2+} is added into the solution, $Ca_9(PO_4)_6$ clusters[52] are preferentially formed on the rod-shaped micellar surface due to conformation compatibility between identical hexagonal shape of the micelles and $Ca_9(PO_4)_6$ clusters.

The presence of a surfactant not only allows preparation of HA nanorods, but also can lead to self-organisation to form ordered island-like bulk crystal complex structures through oriented attachment.[53] The conventional hydrothermal crystallisation process is a transformation process where amorphous fine nanoparticles act as the precursor. The formation of tiny crystalline nuclei in a supersaturated medium occurs at first and this is then followed by crystal growth. The large particles will grow at the expense of

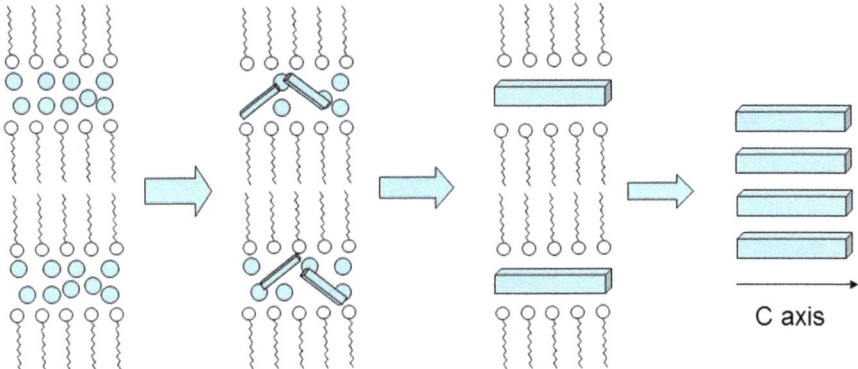

Figure 4.3 Formation process of oriented attachment hydroxyapatite nanorods assisted by dodecylamine.

the small ones due to the higher solubility of the small particles than the large particles. In the early stage, the examination of intermediate products shows the coexistence of short rods, irregular nanoparticles and longer nanorods.

However, in the presence of the self-assembled surfactant, the hydrothermal crystallisation process is limited in the controlled ordered space of the water/surfactant interface. Initially, it is similar to the conventional hydrothermal crystallisation process. The formation of tiny crystalline nuclei in a supersaturated medium occurs at first and is then followed by crystal growth. But the difference is that intermediate products show the coexistence of the short rods and irregular nanoparticles formed only in the ordered limited space of water/surfactant interface. At the same time, while the small particles grow to long nanorods, the long nanorods are self-organised as building blocks through oriented attachment by sharing a common crystallographic orientation of HA crystal and form island-like bulk crystals. This formation process of oriented attachment of nanorods is schematically illustrated in Figure 4.3.

4.3 Nanocomposites for Bone-Grafting Applications

In Chapter 2, we described how HA is widely used for bone repair and tissue engineering due to its biocompatibility, osteoconductivity, and osteoinductivity. Through osteoconduction mechanisms, HA can form chemical bonds with living tissue. However, its poor biomechanical properties (brittle, low tensile strength, high elastic modulus, low fatigue strength, and low flexibility), when compared with natural hard tissues limit its applications to components of small, unloaded, or low-loaded implants. One strategy to overcome this difficulty is to combine the bioactive ceramics with a ductile

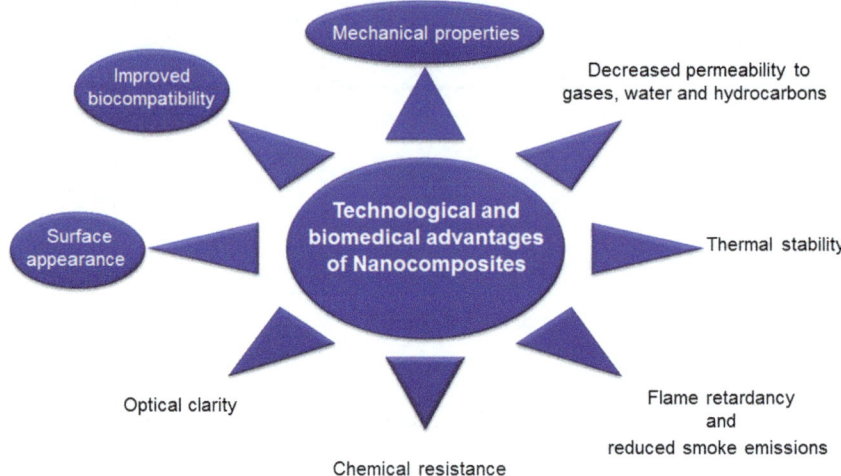

Figure 4.4   Advantages of nanocomposites over conventional composites. Those
features related to the biomaterials field are highlighted.

material, such as a polymer, to produce composites. In recent years, the
development of nanotechnology has shifted *composite* synthesis towards
nanocomposite fabrication.

Nanocomposites are materials that are created by introducing nanoparticles
(often referred to as *filler*) into a macroscopic sample material (often referred
to as *matrix*). The main characteristic of nanocomposites is that the filler has
at least one dimension in the range 1–100 nm. Currently, these materials
constitute an important topic in the field of nanotechnology. Nanomaterial
additives can provide very important advantages in comparison to both their
conventional filler counterparts and base polymer. Figure 4.4 collects some
of the most important advantages, highlighting the properties with outstanding importance for bone-grafting applications: mechanical properties
and biocompatibility and surface features improvement.

Several bioceramics have been used for the fabrication of nanocomposites. Among them we can highlight:

- alumina (Al_2O_3);
- zirconia (ZrO_2); and
- hydroxyapatite $(Ca_{10}(PO_4)_6(OH)_2)$.

Alumina and zirconia belong to the first generation of bioceramics, char-
acterised by an almost inert response after implantation and acceptable
mechanical properties. Alumina has been used as a bearing couple in total
hip replacement since the 1970s. As an artificial femoral head, alumina has
demonstrated even better mechanical behaviour than metals, since its pol-
ished surface exhibits excellent wear resistance and produces less debris.[54]
The osteoblast viability has been studied in the presence of nano-sized

alumina and titania particles, and better cell proliferation has been observed independently of the chemical composition.[55] Several inorganic–inorganic nanocomposites such as alumina–zirconia and alumina–titania have been fabricated by employing techniques such as transformation-assisted consolidation and plasma spraying.[56] These combinations have resulted in nanocomposites with better fracture toughness and mechanical strength.

Zirconia exhibits chemical stability together with a good mechanical performance. For these reasons it has been used as hard-tissue repairing biomaterial. However, zirconia presents a similar drawback to alumina, *i.e. ageing*. Degradation of zirconia is attributed to the transition of the tetragonal to the monoclinic phase, followed by the occurrence of cracks from the surface to the inner bulk. Currently, yttria-stabilised zirconia (YSZ) is the preferred material for making ball heads.[57] Hydroxyapatite/YSZ nanocomposites have been obtained with approximately 99% of the theoretical density.[58] These nanocomposites show improved mechanical properties (flexural strength and fracture toughness), which can be explained in terms of a uniform distribution of YSZ particles in a nano-HA matrix that hinders HA grain growth during thermal treatment.

Although nano-sized alumina, zirconia and titania can provide excellent mechanical properties as biomaterial components, none of them exhibit the biomimetic characteristics of nano-HA. Since this text is mainly devoted to nanoceramics with biomimetic properties, special attention will be paid to composites formed by nano-HA as the inorganic filler.

4.3.1 Nano-HA-Based Composites

Although nano-HA is an excellent artificial bone graft substitute, its inherent low strength and fracture toughness have limited its use in certain orthopaedic applications. The fracture toughness of HA does not exceed 1.0 MPa $m^{1/2}$ (compare to human bone: 2–12 MPa $m^{1/2}$). To summarise, HA behaves as a typical brittle ceramic material[59] and HA-derived nanocomposites are an excellent alternative to overcome this problem. Compared with either pure polymers or conventional polymer composites, nanocomposites generally also exhibit an outstanding improvement in their mechanical properties.

From the point of view of biological behaviour, nanocomposites promote an enhanced osteoblasts function, as has been reported by Webster *et al.*[60] Besides the conventional biopolymer composites studied,[61–70] a number of investigations have focused on determining the mineralisation, biocompatibility, and mechanical properties of the nanocomposites based on various biopolymers. These groups of biocomposites mainly cover nano-HA/polylactide and its copolymers,[71–74] nano-HA/chitosan,[75] nano-HA/collagen,[76–81] nano-HA/collagen/poly(L)lactic acid (PLA),[82] nano-HA/gelatin,[83–85] and the polycaprolactone semi-interpenetrating nanocomposites.[86]

In most cases, the improvement of mechanical properties and biological behaviour are the two main contributions provided by the apatite-derived nanocomposites.

4.3.2 Mechanical Properties of HA-Derived Nanocomposites

The incorporation of HA nanoparticles within a polymeric matrix leads to an increase of their mechanical parameters, mainly those related to the *dynamic mechanical properties* or *viscoelastic behaviour*. Dynamic mechanical analysis is a technique used to study and characterise the viscoelastic nature of some materials, especially polymers. Two methods are currently used. One is the decay of *free oscillations* and the other is *forced oscillation*. Free oscillation techniques involve applying a force to a sample and allowing it to oscillate after the force is removed. Forced oscillations involve the continued application of a force to the sample. An oscillating force is applied to a sample of material and the resulting displacement of the sample is measured. Since the sample deforms under the load, the stiffness of the sample can be determined, and the sample *storage* and *loss modulus* can be calculated. The storage and loss modulus in viscoelastic solids measure the stored energy (representing the elastic portion) and the energy dissipated as heat (representing the viscous portion), respectively. The tensile storage (E') and loss module (E'') are as follows:

$$E' = \frac{\sigma_0}{\varepsilon_0}\cos\delta \qquad (4.1)$$

and

$$E'' = \frac{\sigma_0}{\varepsilon_0}\sin\delta \qquad (4.2)$$

In these equations, σ (stress) and ε (strain) are defined as

$$\sigma = \sigma_0 \sin(t\omega + \delta) \qquad (4.3)$$

and

$$\varepsilon = \varepsilon_0 \sin(t\omega) \qquad (4.4)$$

where

   ω is period of strain oscillation,
   t is time, and
   δ is phase lag between stress and strain.

From eqn (4.1) and (4.2), $\tan\delta$ can be calculated, *i.e.* the ratio (E''/E'), useful for determining the occurrence of molecular mobility transition, such as the glass transition temperature.

   The main dynamic mechanical effect of the nano-HA incorporation is the increase of the storage modulus with respect to the polymer and, of course, to the ceramic apatites. This means that nanocomposites exhibit higher elastic behaviour than their separated precursors. The storage modulus of nanocomposites commonly increases with increased nano-HA content, indicating that hydroxyapatite has a strong reinforcing effect on the elastic properties of the polymer matrix. Since the storage modulus reveals the capability of a material to store mechanical energy and resist deformation,[66] it can be stated that the higher the storage modulus, the more resistant the material is. The *loss modulus*, representing the ability to dissipate energy, also increases upon raising the nano-HA content.

4.3.3   Nanoceramic Filler and Polymer Matrix Anchorage

A common problem with HA–polymer composites is the weak binding strength between the HA filler and the polymer matrix, since they cannot form strong bonds during the mixing process. Often, the mechanical strength of the composite is compromised due to the phase separation of the HA filler from the polymer matrix. A clear example can be found in nano-HA/collagen nanocomposites. These compounds are among the most studied systems because of their similarities with the natural bone. Actually, bone is an inorganic–organic composite material consisting mainly of collagen and HA, and its properties depend intimately on its nanoscale structures, which are dictated specifically by the collagen template.[87,88] Collagen is the major component of ECMs, such as tendons, ligaments, skin, and scar tissues in vertebrates.[89] However, the biocomposites of collagen and hydroxyapatite alone do not have adequate mechanical properties for various biomedical applications, due to the weak filler–matrix interactions.

In order to improve the durability and mechanical properties of nano-HA/collagen composites, the use of polymeric binders has been proposed. Among these binders poly(vinyl alcohol) (PVA) has shown very good performance in terms of improving the filler/matrix binding.[90] PVA hydrogels exhibit biocompatibility as well as a high elastic modulus even at relatively high water concentrations, and have been employed in several biomedical applications including drug delivery, contact lenses, artificial organs, wound healing, and cartilage, *etc.*[91,92] PVA has also been proposed as a promising biomaterial to replace diseased or damaged articular cartilage, but it has limited durability and does not adhere well to tissue.[93] However, the role that PVA can play as a binder between nano-HA and collagen fibres can be very interesting. The polar nature of PVA facilitates strong adhesion between the HA and collagen. In this sense, nano-HA links to PVA through hydrogen bonding and by the formation of the [OH–]-Ca^{2+}-[–OH] linkage, whereas the carbonyl groups of the collagen would be the active sites to bind to the nanocomposite components. The final result is an increase of the dynamic mechanical parameters, especially those related to the elastic properties (storage modulus) rather than the viscosity portion. Finally, the mechanical properties can be upgraded by cryogenic treatments, as has been already used in PVA hydrogel for heart-valve implant applications.[94,95] This effect is explained in terms of PVA crystallisation, which introduces strong interactions between the different domains of the hydrogel.

Although the binder incorporation improves the mechanical performance of nanocomposites, the drawback of weak bonding of HA with polymers is still present, since they cannot form strong bonds during the mixing process. Another alternative to overcome this problem is coating the nano-HA with a polymer film. This coating must have functional groups able to form strong bonds with the polymer matrix. The polymer coating must be degradable so that the bioactivity of the nano-HA is not shielded.

For this purpose, Nichols *et al.*[96] proposed radio-frequency plasma polymerisation technology to activate nano-HA powder surfaces by creating a degradable film with functional groups (*e.g.* nano-HA–COOH) at nanoscale thickness. With this technique, the final biodegradability properties can be controlled through the experimental parameters such as radio-frequency power or gas pressure. For instance, under high power and low pressure conditions, the conversion of carboxyl groups into hydrocarbons, esters, or ketone/aldehydes is favourable, along with significant increases in cross-linking components, leading to nondegradable coatings. In contrast, at low radio-frequency power, the degree of cross-linking is minimalised and the COOH retention on the coatings is high. Therefore, by using low plasma power in creating degradable coatings, fragmentation can be kept to a minimum and the functional groups can also be preserved from overpolymerisation.[97,98] The presence of these functional groups also provides active centres to improve the linkage between the nano-HA and the polymer matrix. As a consequence, the mechanical strength of the nano-HA–polymer scaffold is significantly improved with ultrathin degradable coatings when compared with uncoated control and nondegradable nanocoated groups.

As can be easily deduced, nanocomposite mechanical properties and biocompatibility degree are also strongly dependent on the polymer used. The alkaline nature of HA often leads to a local pH increase of the environment. Moreover, the higher wettability and solubility of nanoparticles can result in higher pH increases that are harmful for the surrounding tissue. This problem can be partially overcome with the use of polymers with weak acid character. For instance, nanocomposites of nano-HA/polylactic acid and derived copolymers have provided very good results from the point of view of the biological response.

It is not only the dynamic mechanical properties of biomaterials that are enhanced by the incorporation of nanoparticles. Parameters such as bending modulus have also been tested with different nanocomposites.[99] The bending module of nanocomposite samples of either PLA or polymethyl methacrylate (PMMA) with 30, 40, and 50 wt% of nanophase (<100 nm) alumina, hydroxyapatite, or titania loadings were significantly greater than those of relevant composite formulations with conventional, coarser grained ceramics. The nanocomposite bending modules were one to two orders of magnitude larger than those of the homogeneous, respective polymer. Figure 4.5 clearly shows the mechanical improvement for three series of nanocomposites.

As can be seen in Figure 4.5, all of the nanoceramic/polymer composites exhibit increased bending moduli that are significantly greater than those of the corresponding conventional ceramic/polymer composites. It must be highlighted that the bending moduli values for those nanocomposites with 40% by weight of nanoceramic content are in the range 1.0–3.5 GPa, *i.e.* in the range of 1–20 GPa exhibited by human bone.[100] This increase in the

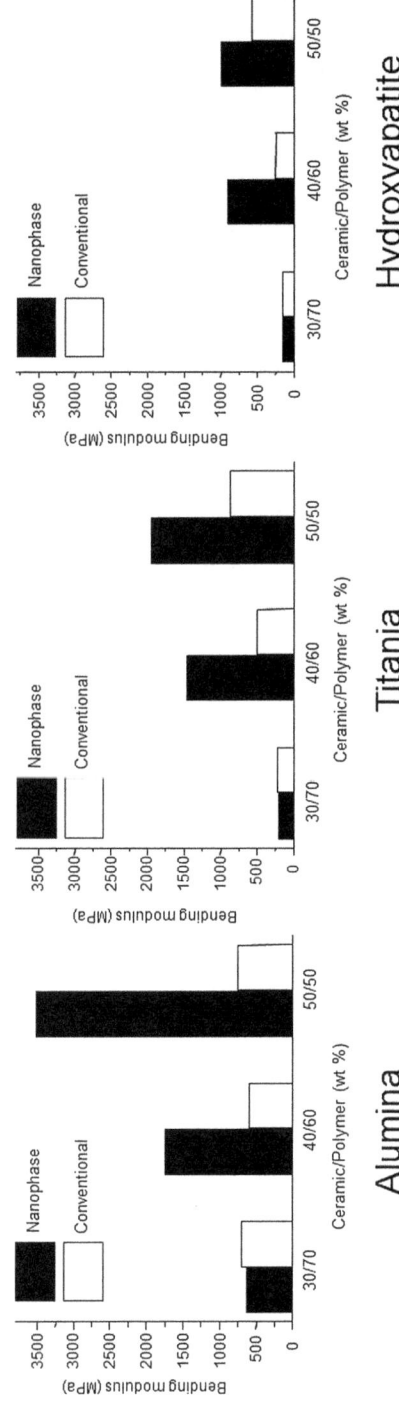

Figure 4.5 Bending moduli of ceramic/poly(L)lactic acid substrates.

strength of the nanophase ceramic/polymer composites, as compared to the conventional ceramic/polymer composites, may be attributed to the fact that nanoparticles are better dispersed in the polymer matrix and the total inter-facial area between the filler and matrix is higher for nanoscale fillers. Conse-quently, the nanoceramic powders allow enhanced interactions between the filler and the matrix compared to that of the conventional ceramic powders.

4.3.4 Significance of the Nanoparticle Dispersion Homogeneity

The importance of the nanocomposite synthesis strategy aiming to obtain a homogeneous nanoparticle dispersion has become a priority line of research in this topic. Actually, most of the different procedures for nanocomposites fabrication are aimed at avoiding this classical experimental problem, which is inherent to nanoparticles handling, *i.e.* the agglomeration of nanoparti-cles. In order to overcome it, different strategies are available. For instance, ultrasonication stirring has been proven to be an effective strategy to avoid the agglomeration of particles in the polymer,[101,102] and good nanoparti-cle dispersion can be obtained when ultrasonication is combined with a solution-casting method. After drying, nanocomposite films are obtained and subsequently shaped into the required geometry by hot pressing. In this way, Chen *et al.*[103] prepared nanocomposites based on bioresorbable polymer-poly(3-hydroxybutyrate-co-3-hydroxyvalerate) (PHBHV) by the incorporation of nano-HA using a solution-casting method.

Kim[104] has proposed the use of an amphiphilic surfactant such as oleic acid to obtain nano-HA/poly(ε)caprolactone (PCL) with the HA nanoparticles uniformly dispersed in the matrix. Oleic acid, which belongs to the fatty acid family and is generally noncytotoxic at low levels, mediates the interaction between the hydrophilic HA and hydrophobic PCL. With the mediation of oleic acid, the HA nanoparticles are distributed uniformly within the PCL matrix at the nanoscale.

4.3.5 Biocompatibility Behaviour of HA-Derived Nanocomposites

Nanoceramics in general, and nano-HA in particular are incorporated into polymeric matrices to improve the cell–material interaction. Terms that explain how nano-HA improves cell adhesion, proliferation and differenti-ation are analogous to those described in Section 3.1.3 for biomimetically grown nanoapatites. *In vivo* dissolution, adsorption of large amounts of serum proteins, increase of the surface roughness, and ion dissolution sig-nalling cells toward differentiation are upgraded features of nanocomposites when compared with conventional composites.

In 1998, Webster *et al.*[105] reported on the improved osteoblast adhesion on spherical nanosized alumina with grain size <60 nm. In this work, a first precedent of the adhesion osteoblast selectivity was provided. Actually, whereas osteoblasts exhibit a better adhesion on nanosized ceramics, fibroblasts undergo an attachment decrease with respect to that observed for conventional ceramics. This fact has been subsequently corroborated on several nanoceramic/polymer nanocomposites[99] and is a very important advantage from the point of view of implant osteo-integration. Control of fibroblast function in apposition to orthopaedic/dental implants is desirable because fibroblasts have been implicated in the clinical failure of bone prostheses. Fibrous encapsulation and callus formation are the most frequently cited causes of incomplete osteointegration of orthopaedic and dental implants *in vivo*.[106-108] For these reasons, materials that have the desired cytocompatibility, *i.e.* that selectively enhance osteoblast adhesion and subsequent functions of these cells, while at the same time minimising functions of competitive cells (such as fibroblasts), are very attractive.

When synthesising nanocomposites it must always be considered that cells do not directly attach to the material's surface, but to the adsorbed adhesion mediator proteins (fibronectin, vitronectin, laminin, and collagen). Therefore, the immediate consequences of the new properties supplied by the nanosized fillers will be modifications of the protein adsorption. Together with higher surface area, nanoceramics introduce great changes in three aspects:

(1) *Surface defects and boundaries*, which lead to a reactivity increase in those sites where nanoparticles are placed.
(2) *Surface charge*. In this sense, HA doped with trivalent cations such as La^{3+}, Y^{3+}, In^{3+}, or Bi^{3+} have shown better osteoblast adhesion, although these cationic substitutions involve a reduced grain size, making it difficult to differentiate between both concomitant effects.[109]
(3) *Surface morphology*. The dimensions of proteins that mediate cell adhesion and proliferation are at the nanometre level. Therefore, a surface with nanometre topography can increase the number of reaction sites compared to those materials with smooth surfaces.[21] The range of several tens of nanometres of surface roughness seems to be optimum for nanoceramics biocompatibility.

The enhancing of initial events during cell–biomaterial interactions (such as cell adhesion and concomitant morphology) clearly means an important advantage of nanoceramics incorporation into polymeric networks for dental and orthopaedic applications. However, evidence of long-term effects (cell proliferation, synthesis of alkaline phosphatase, and ECM mineralisation) is necessary before clinical use. These effects had been previously observed in ceramic surfaces modified with the immobilised peptide sequences arginine–glycine–aspartic acid–serine and lysine–arginine–serine–arginine[110,111] contained in ECM proteins such as vitronectin and collagen. Nanoceramics are able to enhance these functions without peptide immobilisation,[17] exhibiting a much more efficient biological behaviour than conventional bioceramics.

4.3.6 Nanocomposite-Based Fibres

Nanocomposite fibrous structures are highly useful for the fabrication of porous biodegradable scaffolds. In this case, the homogeneity of nanoparticles dispersion becomes critical, and to develop the ceramic–polymer composite system as a micro-to-nanoscale structure, in the forms of fibres, tubes, wires, and spheres, the problem related to agglomeration and mixing needs to be primarily overcome.

Due to the high surface area-to-volume ratio of the fibres and the high porosity on the submicrometre length scale of the obtained nonwoven mat, these materials have been proposed for biomedical applications,[112–115] including drug delivery, wound healing, and scaffolding for tissue engineering. The challenge in tissue engineering is the design of scaffolds that can mimic the structure and biological functions of the natural ECM.

Most of the work carried out to produce nanocomposite fibres has been through the electrostatic spinning technique. Electrostatic spinning or electrospinning is an interesting method for producing nonwoven fibres with diameters in the range of submicrometres to nanometres. In this process, a continuous filament is drawn from a polymer solution or melt through a spinneret by high electrostatic forces and later deposited on a grounded conductive collector,[116] as schemed in Figure 4.6. With this method electrospun fibres of nano-HA/polycaprolactone with different diameters have been obtained,[117] and used as scaffolding materials for the culture of preosteoblastic cells.[118]

Kim *et al.*[119] have also used this technique of synthesising a nano-HA/PLA biocomposite system to produce fibrous structures. One of the main problems inherent to fibre fabrication is the difficulty in generating continuous and uniform fibres with the composite solution because of the innate problems of agglomeration. The incorporation of surfactants as a mediator

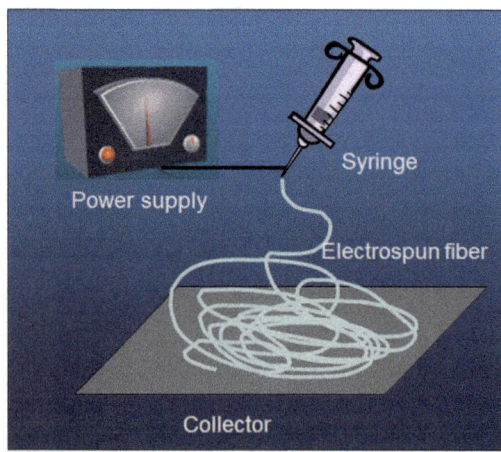

Figure 4.6 Scheme of an electrostatic spinning device.

between the hydrophilic HA and the hydrophobic PLA allows the generation of uniform fibres with diameters of 1–2 μm.

The electrostatic spinning technique has also been used to obtain nano-HA/PLA composites shaped as membranes, with applications in bone-tissue regeneration.[120] The resulting membranes exhibit better cell adhesion and proliferation than the classical PLA membranes, perhaps due to the constant pH of the environment when compared with PLA degradation. As mentioned before, PLA degradation results in a pH decrease that can be harmful for osteoblasts, depending on the culture conditions. The presence of a soluble and slightly alkaline nano-HA buffers these changes and allows a better cell proliferation on the membranes. The mechanical properties such as the tensile strength, elastic modulus, and strain to failure are highly improved, which can be explained in terms of a good HA dispersion that makes the nanofibre matrix stiffer and less plastic in deformation, as could be expected from the incorporation of a hard inorganic phase.

4.3.7   Nanocomposite-Based Microspheres

Currently, much attention is focused on the fabrication of microspheres for biomedical application, with special significance as drug- and gene-delivery systems. Microspheres are widely accepted as delivery systems because they can be ingested or injected and present a homogeneous morphology.[121–124] Various approaches have been designed to prepare microspheres, depending on the chemical features of the final product. For instance, pyrolysis of an aerosol generated by ultra-high-frequency spraying of a solution of precursors has been applied to prepare mesoporous silica microspheres encapsulating magnetic nanoparticles.[125] In the case of ceramic/polymer nanocomposites, for instance nano-HA/PLA composites, microspheres are better prepared through methods involving oil-in-water emulsions.[72,126] The keystone of this strategy is the incorporation of a hydrophilic nanoceramic within the hydrophobic matrix. During the process, the inorganic particles tend to be located in the water phase during the preparation process of the oil-in-water emulsion, thus a very small amount of inorganic particles is incorporated into the hydrophobic polymer microspheres. Qiu *et al.*[127] have proposed the functionalisation of the nano-HA surface with PLA before creating the water-in-oil emulsion. Using this strategy, these authors have prepared composite microspheres with uniform morphology and the encapsulated functionalised HA nanoparticle loading reached up to 40 wt% in the nano-HA/PLA composite microspheres.

4.3.8   Nanocomposite Scaffolds for Bone-Tissue Engineering

The regeneration of critical bone defects is one of the most challenging and difficult issues to be tackled by biomaterials science. In the case of small bone defects, for instance defects in periodontal locations, the material can be implanted as granules or mixed with blood or physiological serum to form a mouldable paste. However, the bone regeneration in a critical defect requires pieces that must be fitted to the defect size and morphology. Very often, the implantation site bears high mechanical loads, thus constraining the use of ceramic and porous materials. Finally, the implants must have pores large enough to allow bone cell colonisation and blood vessel formation.[128,129]

Ceramic 3D porous scaffolds designed for bone tissue engineering often present problems related to its brittleness and difficulty of shaping. Ceramic/ polymer composites can overcome these limitations, keeping the biocompatibility and bone regenerative properties of some bioactive ceramics.[130–137] Among the main drawbacks of ceramic/polymer nanocomposites, we can highlight the organic solvents sometimes remaining in the composites and the coating of the ceramic by the polymer, which hinders its exposure to the scaffold surface. Figure 4.7 shows a clear example of this situation. Figure 4.7a shows the surface of a bioactive glass after being exposed to a biomimetic process in simulated body fluid (SBF). The surface appears to be fully covered by an apatite phase only 1 day after being treated with this fluid at 37 °C. In contrast, Figure 4.7b shows the surface of the same bioglass making up a composite with PMMA after 1 day in SBF. The polymer skin that covers the ceramic is clearly seen and only separated apatite nuclei are observed, demonstrating an important delay of the biomimetic process.

In order to avoid these drawbacks, poly(D,L)-lactic-co-glycolic acid (PLGA)/ nano-hydroxyapatite composite scaffolds have been fabricated by the gas-forming and particulate leaching (GF/PL) method, without the use of organic solvents.[74] The GF/PL method exposed HA nanoparticles at the scaffold surface significantly more than the conventional solvent-casting and particulate leaching (SC/PL) method. The GF/PL scaffolds show interconnected porous structures without a skin layer and exhibit superior enhanced mechanical properties to those of scaffolds fabricated by the SC/PL method. The GF/PL method consists of shaping pieces of the corresponding polymer and ceramic together with NaCl. The conformed body is subsequently exposed to high-pressure CO_2 gas to saturate the polymer with the gas. Then, decreasing the gas pressure to ambient pressure creates a thermodynamic

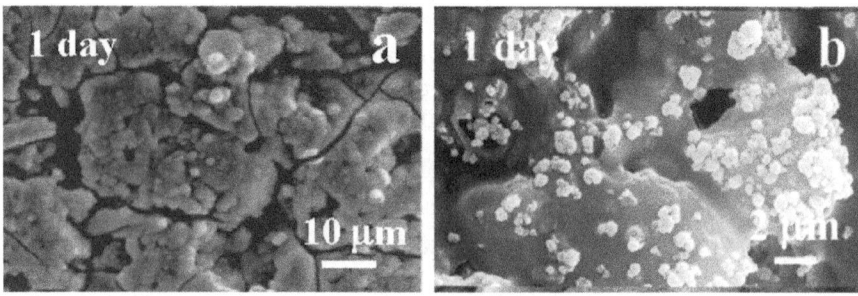

Figure 4.7   Surface of (a) a bioglass; and (b) a bioglass/polymethyl methacrylate composite after 1 day in simulated body fluid.

instability. This leads to the nucleation and growth of CO_2 pores within the polymer scaffolds. The NaCl particles are subsequently removed from the scaffolds by leaching the scaffolds in distilled water. Using this strategy, highly porous PLGA/HA composite scaffolds can be fabricated, exhibiting a higher exposure of HA at the scaffold surface and much better bone formation *in vitro* and *in vivo* than those fabricated by more conventional methods.

Sánchez-Salcedo *et al.*[138] have reported the latest processing of direct foaming methods. The method consists of the combination of the sol–gel technique with an accelerated evaporation-induced self-assembly (EISA) method including a biopolymer as a macropore foam former. In a typical synthesis, Pluronic F127 was dissolved in ethanol and water with a molar ratio of F127 : ethanol = 0.22, followed by triethyl phosphite addition and stirring for 24 h. A $Ca(NO_3)_2 \cdot 4H_2O$ dissolution in a molar ratio Ca/P of 1.67 was added to the sol and aged for 6 h at 60 °C. Sols were diluted twice with the total ethanol. After 72 h of ageing at 60 °C, the resulting mixture was transferred into an open Petri dish and placed in an oven for 48 h at 30 °C to evaporate the solvent. Afterwards, foams were placed for 1 h at 100 °C to obtain green foam-like HA. Finally, they were annealed at 550 °C for 6 h giving rise to macroporous nanocrystalline foam-like HA. After the HA foams were obtained, they were coated by soaking into different biopolymer solutions, extracted and dried by lyophilisation. HA macroporous foams coated with gelatin-glutaraldehyde biopolymers (Figure 4.8) were proposed as new potential devices for the treatment of heavy-metal intoxication by ingest and as a water purification remedy. However, both the composition and 3D interconnected architectural design of these HA/biopolymer foams suggested that they could fit all the requirements to behave as a proper bone scaffold with also applications in bone tissue engineering. Not only is it important that these characteristics would allow a cellular internalisation and adequate colonisation over entire surface as they have a high degree of porosity, but they also have a high degree of flexibility and handle-ability, which could guarantee easy surgical management.

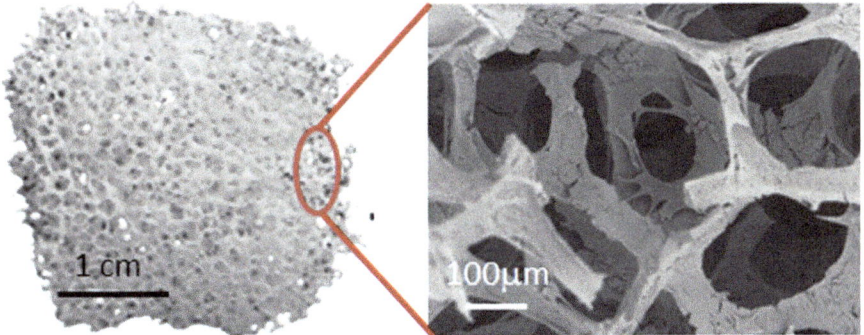

Figure 4.8   Hydroxyapatite foam as a scaffold for tissue engineering purposes.

4.4   Nanostructured Biomimetic Coatings

The integration of any implant with bone tissue depends on the chemical and physical properties of the surface. In orthopaedic surgery, metals and their alloys are the most widely used implant materials due to their good mechanical properties, although in contact with body fluids or tissues they corrode.[139] An interesting alternative for the protection of metal surfaces against corrosion is to coat the metal surface with a ceramic, which can act as an interface between the substrate and the bone, favouring the bone bonding. In this sense the calcium phosphates such as HA and β-tricalcium phosphate are common examples of such coatings.[140–142]

Nowadays, the most frequently employed technique to prepare commercial covered implants is plasma spraying.[143] However, this technique exhibits some disadvantages that cannot be easily avoided: inability to coat implants with complex shapes, differences in the chemical composition, delamination, *etc.* Other line-of-sight deposition methods such as sputtering[144] or laser ablation[145] do not solve, for instance, the coating of porous substrates. Other physical methods include magnetron sputtering, ion-beam coating, anode oxidation, and anodic spark deposition.[146–150] Chemical vapour deposition techniques have been successfully used for the preparation of calcium phosphate coatings. By using this coating strategy, thin films of calcium phosphates can be obtained, where the microstructure, crystallinity, and composition of the deposited films can be controlled by modifying the composition of the precursors solution, reactor atmosphere, and substrate temperature.[151,152]

Solution-based methods are an emerging option for the preparation of these coatings due to several features: better control of coating morphology, chemistry and structure, covering of intricate pieces, simplicity of technology, *etc.* In this section we deal with *sol–gel* and *biomimetic deposition coating* procedures, two strategies that lead to highly biocompatible nanostructured coatings with a wide range of possibilities to incorporate therapeutic agents, growth factors, adhesion proteins, peptides sequences, *etc.*, due to the low-temperature processes involved in these methods.

4.4.1   Sol–Gel-Based Nano-HA Coatings

The *sol–gel* process[153,154] is a wet-chemical technique for the fabrication of materials (typically a metal oxide) starting from a chemical solution that reacts to produce colloidal particles (*sol*). Typical precursors are metal alkoxides and metal chlorides, which undergo hydrolysis and polycondensation reactions to form a colloid, a system composed of solid particles (size ranging from 1 nm to 1 μm) dispersed in a solvent. The sol evolves then towards the formation of an inorganic network containing a liquid phase (*gel*). Formation of a metal oxide involves connecting the metal centres with oxo (M–O–M) or hydroxo (M–OH–M) bridges, therefore generating metal-oxo or metal-hydroxo polymers in solution. The *drying* process serves to remove the liquid

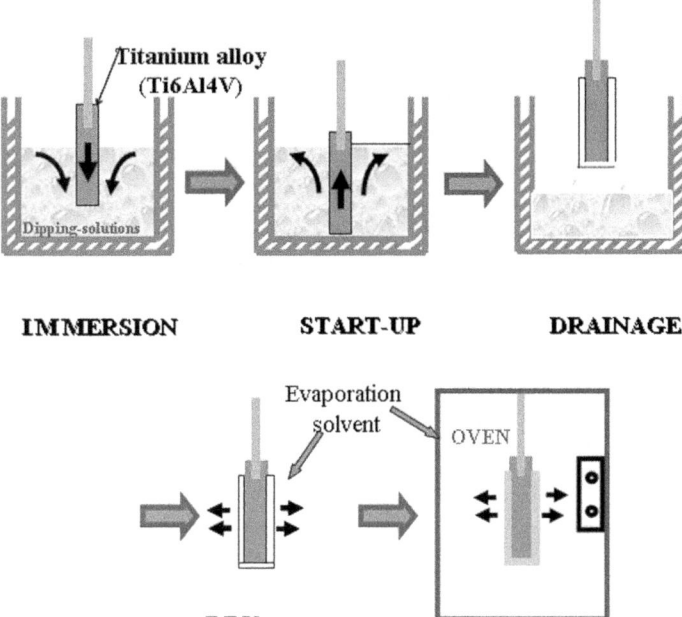

IMMERSION   **START-UP**   **DRAINAGE**

DRY

Figure 4.9 Scheme followed for the preparation of coatings by the dip-coating process.

phase from the gel, thus forming a porous material, then a thermal treatment (*firing*) may be performed in order to favour further polycondensation and enhance mechanical properties.

The precursor sol can be either deposited on a substrate to form a film (*e.g.* by dip-coating or spin-coating), cast into a suitable container with the desired shape (*e.g.* to obtain monolithic ceramics, glasses, fibres, membranes, or aerogels), or used to synthesise powders (*e.g.* microspheres, nanospheres). The sol–gel approach is interesting in that it is a cheap and low-temperature technique that allows the incorporation of drugs and osteogenic agents within the coatings. Hijón *et al.*[155] prepared bioactive nano-HA coatings deposited on Ti6Al4V by the sol–gel dipping technique (see Figure 4.9), from aqueous solutions containing triethyl phosphite and calcium nitrate, although other precursors can be also used to prepare HA coatings using the sol–gel techniques, as shown in Table 4.1.

Nano-HA coatings with particle sizes of ~75 nm and controlled roughness can be prepared by modifying the drying temperature in range of 30–60 °C. A decrease in the R value is observed insofar as the ageing temperature increases, as can be seen from the roughness profiles obtained by scanning force microscopy and shown in Figure 4.10. The R values obtained were 11, 8 and 5 nm for layers dried at 30, 40 and 60 °C, respectively. The coating thickness is ~0.2 μm by dipping cycle. To obtain HA coatings with greater thickness, the dip-coating method is repeated several times (up to 10 times).

Table 4.1 Some calcium and phosphorous precursors used in the synthesis of hydroxyapatite coatings deposited by the sol–gel technique.

Reference	Calcium precursor/solvent	Phosphorous precursor/solvent
Brendel et al.[156]	Calcium nitrate/acetone	Phenyldichlorophosphine/acetone/water
Russell et al.[157]	Calcium nitrate/2 methoxyethanol	N-butyl acid phosphate/2 methoxyethanol
Hsieh et al.[158]	Calcium nitrate/2 methoxyethanol	Triethyl phosphate/2 methoxyethanol
Goins et al.[159]	Calcium nitrate/2 methoxyethanol	Diethyl phosphite/2 methoxyethanol
You and Kim[160]	Calcium nitrate/methanol	Triethyl phosphite/methanol
Hwang and Lim[161]	Calcium nitrate/methanol	Phosphoric acid/methanol
Kojima et al.[162]	Calcium nitrate/ethanol	Triethyl phosphate/ethanol
Liu et al.[163]	Calcium nitrate/ethanol	Triethyl phosphite/ethanol/water
Gan and Pilliar[164]	Calcium nitrate/ethanol	Triethyl phosphite/water
		Ammonium dihydrogen phosphate/water
Piveteau et al.[165]	Calcium nitrate/ethanol	Phosphoric pentoxide/ethanol
Cavalli et al.[166]	Calcium nitrate/water	Diammonium hydrogen phosphate/water
Weng and Baptista[167]	Calcium glycoxide/ethyleneglycol	Phosphoric pentoxide/ethanol or butanol
Chai et al.[168]	Calcium diethoxide/ethanol/ethanediol	Triethyl phosphite/ethanol
Gross et al.[169]	Calcium diethoxide/ethanol/ethanediol	Triethyl phosphite/ethanol/ethanediol
Haddow et al.[170]	Calcium diethoxide/ethanediol	Triethyl phosphite/ethanediol
Ben-Nissan et al.[171]	Calcium diethoxide or calcium acetate/ethyleneglycol/acetic acid	Diethylhydrogenphosphonate
Tkalcec et al.[172]	Calcium 2-ethylhexanoate/ethylhexanoic acid	2-ethylhexylphosphate/ethylhexanoic acid

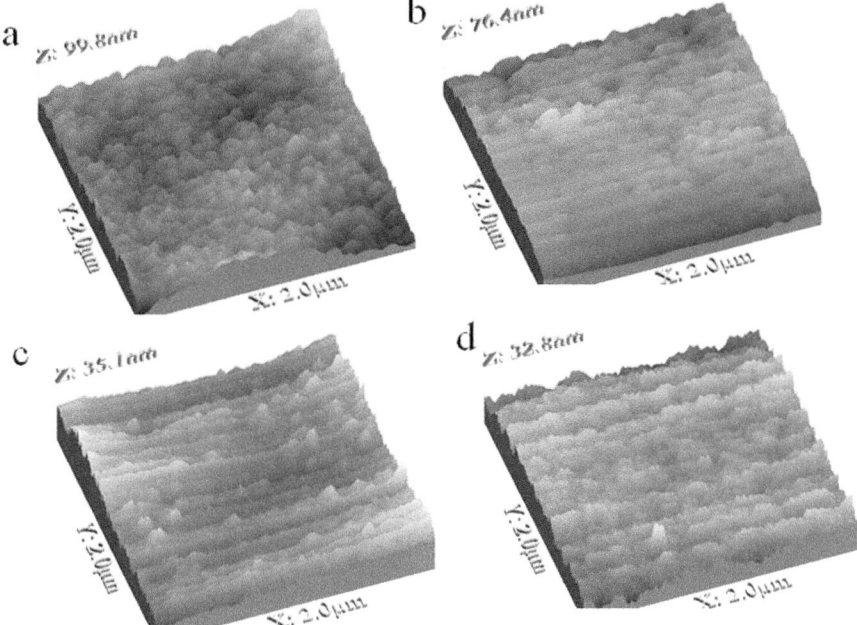

Figure 4.10 Scanning force microscopy 3D image corresponding to the nano-hydroxyapatite coating deposited from sols aged at different temperatures. (a) One layer dried at 30 °C (R = 11 nm); (b) one layer dried at 40 °C (R = 8 nm); (c) one layer dried at 60 °C (R = 5 nm); and (d) six layers dried at 60 °C (R = 4 nm).

In coatings of six or more layers, the formation of cracks on the coating are likely to occur, and a valid compromise between the coating thickness and integrity can be reached for six layers, *i.e.* five or six dipping cycles. In general, the coating roughness decreases as more layers are incorporated into the coating.

Another factor influencing final coating composition, textural properties and homogeneity is water presence in the sol. Actually, the precursors:H_2O ratio determines the hydrolysis/polycondensation kinetic and the final coating characteristics. In general, sols containing higher amounts of ethanol require longer ageing times and lead to purer HA as well as to more homogeneous coatings.[173] The adhesion of these coatings exhibit tensile strength adhesion values of ~20 Mpa,[174] comparable to those obtained for HA coatings prepared by other coating strategies, such as electrodeposition, plasma spray or pulsed laser deposition.[142,175,176]

Nano-HA coatings prepared by sol–gel dipping exhibit biomimetic behaviour when exposed to SBF at 37 °C. Contrary to other CaP-based bulk materials, the development of a new carbonated calcium-deficient apatite phase can be observed by the nucleation and growth of biomimetic crystals that are observable by scanning electron microscopy. Figure 4.11 shows this

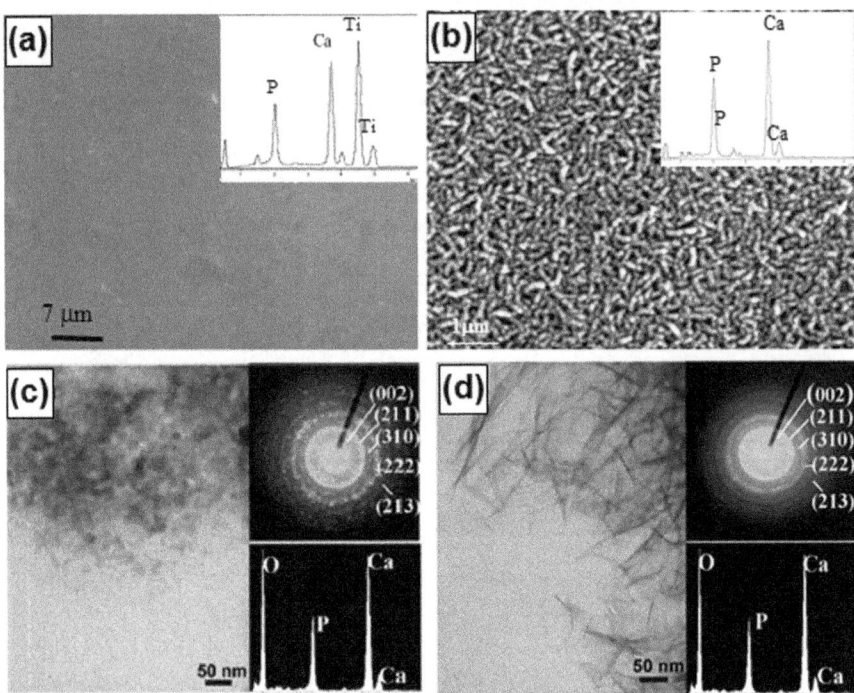

Figure 4.11 Biomimetic behaviour of apatite nanocoatings. (a) Scanning electron micrograph and energy-dispersive X-ray (EDX) spectrum of a Ti4AlV substrate coated by a nano-hydroxyapatite (HA) phase; (b) scanning electron micrograph and EDX spectrum of a Ti4AlV substrate coated by a nano-HA phase after 7 days in simulated body fluid (SBF); (c) transmission electron microscopy image, electron diffraction pattern and EDX spectrum of nano-HA sol–gel coating; (d) transmission electron microscopy image, electron diffraction pattern and EDX spectrum of biomimetic apatite growth after 7 days in SBF.

new apatite-like layer on a HA coating, which exhibits similar morphology to those biomimetic layers appearing on the surface of other highly bioactive bioceramics. Transmission electron microscopic observation (Figure 4.11c and d) demonstrates the different morphology of the nano-HA particles that constitute the coating, and the nano-HA forming the new biomimetic layer.

The crystals corresponding to the sol–gel coating (Figure 4.11c) are nanosized (<25 nm), round-shaped and have a Ca/P molar ratio of 1.7 ± 0.1, according to the energy-dispersive X-ray spectra. Those formed in the SBF solution (Figure 4.11d) are larger and show a needle-like shape; the Ca/P molar ratio of this kind of crystal was found to be 1.4 ± 0.1, similar to the ratio observed in biological apatites[177] and other apatites formed in SBF.[178] In the same way, both electron diffraction patterns show diffraction rings that can be indexed to the interplanar spacings of an apatite phase. In addition, the electron diffraction diagram in Figure 4.11c shows diffraction maxima that are indicative of the higher crystallinity of sol–gel-derived HA nanocrystals when compared to those obtained in SBF.

The bioactive behaviour of the nano-HA coatings can be improved by the incorporation of silicon into the apatite structure.[179] Through the incorporation of stoichiometric amounts of tetraethyl orthosilane, $Si(OCH_2CH_3)$, within the sol, silicon-substituted hydroxyapatite coatings can be formed, according to the formula: $Ca_{10}(PO_4)_{6-x}(SiO_4)_x(OH)_{2-x}\square_x$ where x varies from 0.25 to 1 and $\square$ expresses the anionic vacancies generated. The presence of carbonated species in the sol lead to a final coating composition with the general formula $Ca_{10}(PO_4)_{6-x-y}(SiO_4)_x(CO_3)_y(OH)_{2-x+y}$ where carbonates are included in the phosphate sites, competing with the introduced silicates.

Surface sol–gel processing, a variant of the bulk sol–gel dip-coating method, can be used to fabricate ultrathin metallic oxides with nanometre-precise control.[180] The layer-by-layer process begins with the chemisorption of a hydroxyl functionalised surface in a metal alkoxide solution followed by rinsing, hydrolysis, and drying of the film. This sol–gel reaction occurs on the surface of the substrate each time the hydroxyl groups TiOOH are regenerated to form a monolayer of TiO_2, and repetition of the entire process results in multilayers of the thin oxide film. A calcination or sintering process may be applied if a denser or more crystalline oxide is desired, but this is often unnecessary.[181] The process is readily applied to any hydroxylated surface, using a metal alkoxide reactive to OH groups, and the sol–gel procedure is independent of each cycle, which allows individual layers to be nanostructured.[182] With this strategy, coated Ti6Al4V substrates can be obtained with corrosion behaviour as good as that of TiO_2, but with an increased bioactive behaviour under biomimetic conditions.

4.4.2 Nano-HA Coatings Prepared by Biomimetic Deposition

Chapter 3 describes how supersaturated solutions with ionic compositions similar to that of human plasma can be used with the aim of mimicking the mineralisation process. In that chapter, bioceramics that induce the growth of nano-apatites in contact with biomimetic solutions were considered as "nanoapatite producers" since their chemical composition and textural characteristics induced the nucleation and subsequent growth of bone-like apatites.

Supersaturated solutions can be also employed to coat complex-shaped materials, including metals. The most widely used biomimetic solution is SBF. This solution is just slightly supersaturated with respect to the precipitation of HA, and consequently the nucleation and precipitation processes on metal surfaces are quite slow. In order to accelerate deposition, high ionic strength calcium phosphate solutions and pre-treatment with highly supersaturated solutions (3×, 5×, or even 10× SBF), can be used, as explained in Chapter 3.

Biomimetic nano-HA coatings highly increase the osteoconductivity of metallic implants and supply osteogenic induction. Li[183] prepared nano-HA coatings on grit-blasted Ti6Al4V by soaking these specimens in a biomimetic solution highly concentrated in Ca^{2+} (6.0 mM) and HPO_4^{2-} (2.4 mM). The coating formation took place after 3 days at 45 °C and the differences between the coated and uncoated specimens were evident. When implanted in the distal femur of dogs, greater bone formation was generated in those surfaces lined with the apatite coating than those of the noncoated titanium surface. Human osteoblasts also exhibit clear differences when cultured in coated and noncoated Ti6Al4V substrates. The human osteoblasts cultured on coated substrates develop a more 3D morphology as well as a higher number of anchorage elements than those on noncoated surfaces.

One of the most attractive features of biomimetic coatings is that biologically active molecules, such as drugs, osteogenic agents, growth factors, *etc.* can be coprecipitated with the apatite crystals onto metal implants,[184,185] which can be subsequently released during the coating degradation, acting as a drug delivery system. The retardant effect of serum albumin on the biomimetic nano-HA formation is well known and was explained in Chapter 3. However, under high Ca^{2+}, PO_4^{3-}, or HPO_4^{2-} concentration conditions the effect of serum albumin is reflected inchanges in the morphology of the crystallites, but not as an inhibitory effect. The nano-HA crystallites decreases in size, assume a marked curvature, and become more densely packed as a function of serum albumin concentration in the solution.[186]

Coprecipitation of active agents with biomimetic nano-coatings also provides a very important advantage with regard to the kinetic release. Although the use of apatite nanoparticles as drug delivery systems will be considered in next section, it is important to highlight the effect of the coprecipitation methods compared with adsorption onto preformed coatings. Most of the therapeutic agents adsorbed on preformed coatings are released in a single fast-burst effect. In contrast, therapeutics incorporated by coprecipitation are gradually released over several days, enhancing their potential as controlled drug delivery carriers.

Biomimetic coprecipitation methods allow the nucleation and growth of a variety of calcium phosphates. For instance, carbonate hydroxyapatites (CHA) or octacalcium phosphate (OCP) can be also precipitated as biomimetic coatings. CHA and OCP have different pH stabilities and solubilities and show different resorption under the action of osteoclasts.[187] Barrère *et al.*[188] have demonstrated that OCP biomimetic coatings exhibit higher osteogenic action when implanted in both intramuscular and bone locations. The presence of the Ca/P coating during an appropriate time period (*i.e.* coating stability) and the architecture of the implant seem to be very important conditions. Very fast coating dissolution and flat or dense surfaces do not induce an osteogenic response when implanted in intramuscular locations. Anyway, bone induced in muscular implantation is degraded with time. When Hedrocel™ cylinders (porous tantalum) are biomimetically coated with OCP and CHA and placed into bone tissue, a direct bonding between the implant and the host bone occurs. However, only OCP-coated cylinders exhibit bone ingrowth in the centre of the implant, although this new bone is not necessarily in contact with the host bone. This suggests that OCP coatings exhibit a higher osteogenic behaviour than CHA coatings. This difference in the osteogenic behaviour could be explained by the lower CHA coating in the bone environment. In this location, the resorption of the coating mainly depends on the osteoclastic activity. In this sense, osteoclastic activity is higher on biomimetic CHA, as previously demonstrated.[184] Moreover, the rougher surface provided by the larger and sharp vertical OCP crystals seems to provide a more appropriate microstructure to influence bone formation.

4.5 Nanoapatites for Diagnosis and Drug/Gene-Delivery Systems

4.5.1 Biomimetic Nanoapatites as Biological Probes

Biomedical probes posses interesting diagnostic properties as intracellular optical sensors.[189-191] Especially relevant are those luminescent probes that exhibit a fluorescent signal as response, due to the high sensibility showed by fluorescence spectroscopy. A wide variety of fluorescent organic molecules are currently used as biological probes, which enable molecules in cells to be visualised by fluorescence. Although this method is sensitive, degradation of the organic molecule under irradiation leads to a rapid fall in fluorescence intensity.

In this sense, the incorporation of quantum dots[192] into the field of diagnosis and therapy has meant a significant advance. Quantum dots are inorganic fluorophores that exhibit size-tunable emission (*i.e.* there is a predictable relationship between the size of the quantum dot and its emission wavelength), strong light absorbance, bright fluorescence, narrow symmetric emission bands, and high photostability. The problem is that quantum dot cores are usually composed of elements from groups II and VI, such as CdSe, or groups III and V, such as InP, while the shell is typically a high bandgap material such as ZnS.[193] Since cadmium and selenium can be highly toxic, the search of more compatible compounds, such as biomimetic apatites with luminescent properties, is a priority research line in the development of nanodiagnosis. In addition to the high biocompatibility, calcium phosphate nanoparticles might undergo long-term dissolution inside the cells due to the lower Ca^{2+} concentration in the intracellular compartment.[194,195]

Doat and co-workers[196,197] have synthesised and studied biomimetic calcium-deficient apatite nanocrystallites doped with trivalent europium (Eu^{3+}). The composition and crystallite sizes of such an apatite enable it to interact with living cells and therefore to be exploited as a biological probe. These apatites were synthesised at 37 °C by coprecipitating a mixture of Ca^{2+} and Eu^{3+} ions by phosphate ions in a water–ethanol medium. The nanoparticles are internalised by human epithelial cells and their luminescence stability allows their observation by confocal microscopy.

The required size-tunable properties of these probes dictates a size range of 2–6 nm, exhibiting dimensional similarity with biological macromolecules, *e.g.* nucleic acids and proteins. Most of the proposed synthetic nanoapatites routes commonly yield to slightly aggregated bioapatite nanoparticles and individualisation of the primary crystallites has to be achieved for a better spectral and spatial resolution in biological applications. In addition, in order to minimise the influence of the luminescent nanocrystals on the biological mechanisms, decreasing of the size of the individualised nanoparticles in the range of small proteins or oligonucleotides is desirable. Lebugle *et al.*[198] have described the preparation of individualised monocrystalline colloidal apatitic calcium phosphate nanoparticles stabilised at neutral pH and using amino-ethyl phosphate $(NH_3^+-CH_2CH_2-PO_4H_2)$. This strategy has been successfully applied to the synthesis of various doped calcium phosphate nanoparticles. Doping with luminescent centres such as Eu^{3+}, Tb^{3+}, *etc.* yields a range of calcium phosphate nanophosphors suitable for biological labelling. Finally, colloidal stability in neutral pH must be achieved. For this aim, the use of functional amino surface groups, which offers the further possibility of bio-conjugation, is a suitable strategy to achieve the appropriate stability.

4.5.2   Biomimetic Nanoapatites for Drug and Gene Delivery

Drug delivery systems can be described as formulations that control the rate and period of drug delivery (*i.e.* time-release dosage) and target specific areas of the body. Currently, local drug delivery is an evolving strategy that responds to the development of new active molecules and potential treatments, such as gene therapy. Actually, research efforts in the pharmaceutical field are leading to the evolution of new therapeutic agents and also to the enhancement of the mechanisms to administer them.[199-202]

The field of nanotechnology in recent years has motivated researchers to develop nano-structured materials for biomedical applications. In a similar way to silica-based mesoporous materials,[203-207] biomimetic calcium phosphate nanoparticles for drug delivery have experienced an outstanding advance in recent years. These nanosystems are especially promising for those pathological situations associated with bone surgery, such as bone tumour extirpation, infection risk, acute inflammatory response, *etc.* Therefore, it can be stated that local drug release in bone tissue is one of the most promising therapies in orthopaedic surgery. Oral administration commonly requires very high, and sometimes, low effective dosages to reach high enough drug concentrations in the poorly irrigated bone tissue. Antibiotics, growth factors, chemotherapeutic agents, antioestrogens and anti-inflammatory drugs are good candidates for the most common bone-related therapies.

4.5.2.1   Biomimetic Nanoapatites for Bone Tumour Treatment

One of the most promising therapeutic actions of drug-loaded biomimetic nanoapatites is in the treatment of bone cancer. For instance, osteosarcomas and Ewing's sarcoma are malignant bone tumours most commonly occurring in the growing bones of children. No evolution of the survival rates has been recorded for two decades in response to current treatment, associating often toxic and badly tolerated cures of chemotherapy with low therapeutic response.[208] Thus treatment for these bone cancers commonly involves surgery such as limb amputation or limb-sparing surgery, and consequently, the high loss of bone tissue is one of the main drawbacks of a bone tumour extirpation. A second problem is that malignant cells can remain around the tumour site, leading to the recurrence of the tumour with fatal consequences. In this sense, the use of biomimetic calcium phosphate grafts combined with local specific cancer treatments is an excellent alternative to restore bone defects such as those occurring after tumour extirpation.

Cis-diamminedichloroplatinum (cisplatin) is one of the most active anti-cancer agents in the treatment of osteosarcoma, but must be used in limited short-term, high-dose treatments because of nephrotoxicity and ototoxicity. The minimisation of the systemic toxicity of chemotherapeutic drugs including cisplatin has been demonstrated in local intratumoural treatments with comparable anti-tumour efficacy to that of a systemic dose.[209–217] Cisplatin can be easily adsorbed onto hydroxyapatite nanoparticles.[218] The chemical and physical characteristics of the apatite crystals, including the chemical composition, structure, porosity, particle size and surface area, as well as the ionic composition of the equilibrating solution (pH, ionic strength, and concentration of ion constituents), all play an important role in both the binding and release of the specific chemical components from calcium phosphates.[219–223]

Nanoapatites bind higher amounts of cisplatin than well-crystallized apatites. This linkage is achieved through an endothermic process, since the cisplatin immobilised onto the apatite surface is three times greater when the adsorption process is accomplished at 37 °C than when it is performed at 24 °C.[215] Regarding the drug release, nanoapatites deliver cisplatin more slowly that the more crystalline apatites, although in both cases the presence of chloride ions in the surrounding fluid is needed for cisplatin release. The greater activity of the poorly crystalline apatites in the cisplatin adsorption process, compared to the well-crystallised hydroxyapatite, can be attributed to the presence of more surface defects (nonapatitic environments) which create active binding sites. Moreover, the morphology and size of the crystals, other surface irregularities, and lower crystallinity also account for the higher reactivity of these materials. The cytotoxicity tests with these apatite/cisplatin systems demonstrate that these formulations exhibit cytotoxic effects on K8 cells with a dose-dependent decrease in the cell viability.

The adsorption and release of Pt-derived antitumoural drugs can be tuned by controlling the nanoparticle morphology of the biomimetic nano-HA. Actually, there exist several synthesis routes that allow tailoring of the shape and surface composition of the nanoparticles. For instance, Palazzo *et al.*[224] have proposed the preparation of biomimetic nano-HA with both needle-shaped and plate-shaped morphologies and different physical–chemical properties. For instance, *needle-shaped* nanocrystals can be prepared from an aqueous suspension of $Ca(OH)_2$ by slow addition of H_3PO_4,[225,226] obtaining needle-shaped nanocrystals having a granular dimension ~100 ± 20 nm. Besides, *plate-shaped* nanocrystals can be precipitated from an aqueous solution of $(NH_4)_3PO_4$ by slow addition of an aqueous solution of $Ca(CH_3COO)_2$ keeping the pH at a constant value of 10 by the addition of $(NH_4)OH$ solution. The reaction mixture is stirred at 37 °C for 72 h and then the deposited inorganic phase is isolated by filtration of the solution, repeatedly washed with water, and freeze-dried. In this way, *plate-shaped* nano-HA are obtained, having granular dimensions of 25 ± 5 nm.[227] Although the bulk Ca/P ratios are similar in both kinds of nano-HA (1.65–1.62), the surface Ca/P ratio is lower for the *needle-shaped* nanocrystals (1.30) compared with the HA particles (1.45), suggesting that the former is more surface-deficient in calcium ions.

Taking into account the specific properties of each drug (negative, positive, or neutral charge) and the nano-HA features, specific nano-HA/drug conjugates can be tailored for specific clinical situations. The adsorption and desorption kinetics are dependent on the specific properties of the drugs and the morphology of the HA nanoparticles. In addition to cisplatin, di(ethylenediamineplatinum)medronate (DPM) has been also incorporated to biomimetic nano-HA (see Figure 4.12). DPM belongs to the family of platinum(II) compounds with aminophosphonic acids and have also been proposed as a means for targeting cytotoxic moieties to the bone surface.[228,229] These compounds exhibit antimetastatic activity, reduce bone tumour volume and are less nephrotoxic than cisplatin.[230–232]

Adsorption of the platinum complexes occurs with retention of the nitrogen ligands, but the chloride ligands of cisplatin are displaced. Consequently, the positively charged aquated cisplatin is strongly adsorbed on the negatively charged nano-HA surface, while the neutral DPM complex shows lower affinity towards the negatively charged nanoceramic. The adsorption of the two platinum complexes is driven by electrostatic attractions. Consequently,

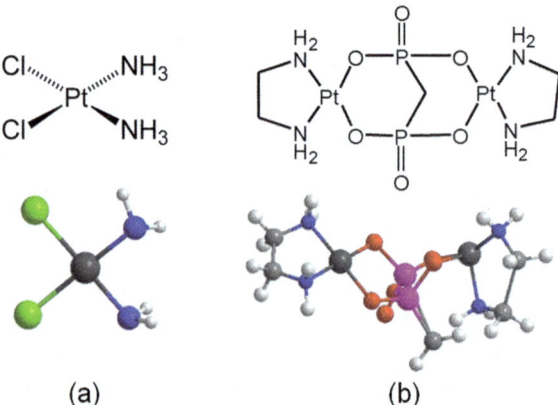

(a)   (b)

Figure 4.12   Molecular structure of (a) *cis*-diamminedichloridoplatinum (II) (cisplatin) and (b) di(ethylenediamineplatinum)medronate (DPM).

adsorption of positively charged hydrolysis species of cisplatin is more favoured on the phosphate-rich needle-shaped nano-HA surface, while the neutral DPM complex shows lower affinity for needle- or plate-shaped negatively charged apatitic surfaces.

This kind of short-range electrostatic interaction dominates the kinetic release and drug desorption is faster for neutral DPM than for charged aquated cisplatin. The release of DPM takes place through complete cleavage of the platinum–medronate bond and the release is greater when adsorbed to needle-shaped rather than for plate-shaped nano-HA. These processes are modulated to some extent by the surface composition, demonstrating that biomimetic nano-HA can be tailored for specific therapeutic applications.

4.5.2.2 Nanoapatites as Antibiotic Delivery Systems

Nano-HA is used to improve the performance of polymeric antibiotic delivery systems through the formation of nanocomposites, which exhibit an enhanced drug loading capacity as well as better release performance. For instance, nano-HA can be incorporated into polylactide-based systems,[233] increasing the potential of PLA for biomedical applications in general and for drug delivery in particular. With this aim, several biodegradable polymers have been combined with nano-HA. For instance, Wang *et al.*[234] have prepared polyhydroxybutyrate-co-hydroxyvalerate (PHBV)-nano-HA microparticles for gentamicin release. The idea is to fabricate a long-term drug release system by preparing drug-loaded HA nanoparticles, followed by the encapsulation of the nanoparticles with a biodegradable polymer, such as PHBV. A classical problem of polymeric microspheres for drug delivery is that microspheres are generally prepared by double or single emulsion solvent evaporation methods. The outer phase, which usually was the aqueous phase, induced hydrophilic drugs like gentamicin to move out of the polymer phase, resulting in a comparatively lower encapsulation efficiency of gentamicin.[235,236] In the case of nanocomposites, a kind of hybrid structure consisting of HA nanoparticles and polymers can be fabricated to increase the encapsulation efficiency. The comparatively higher encapsulation efficiency is due to the high bond affinity and hydrophilicity of nano-HA particles. When hydrophilic drugs such as gentamicin are distributed over the HA phase the amount of drug moving toward the aqueous phase is reduced. To summarise, it is like increasing the system affinity of the drugs, providing higher encapsulation efficiency. The control on the release rate is also enhanced, since the interaction of gentamicin with nano-HA avoids the initial burst effect and allows a sustained release for more than 10 weeks.

Calcium-deficient hydroxyapatite nanoparticles (nano-CDHA) have also been used to regulate the kinetic release of chitosan-derived microspheres.[237] Carboxymethyl-hexanoyl chitosan constitutes a hydrophilic matrix with a burst release profile in a highly swollen state. Incorporation of nano-CDHA has been demonstrated to regulate the release of ibuprofen, insofar as the nanoparticles amount increases in the composite, due to the inorganic nanofiller acting as a cross-link agent and diffusion barrier. In contrast, when nano-CDHA is incorporated with an *O*-hexanoyl chitosan matrix, which is a hydrophobic compound, the ibuprofen release is accelerated. This can be explained in terms of higher polymer degradation due to the hydrophilic character of nano-CDHA, facilitating water accessibility and thus enhancing the drug diffusion. The amount of nano-CDHA incorporated into chitosan matrices is not the only factor that alters the extent of filler–polymer interactions. The drug release behaviour of nano-CDHA/chitosan is strongly dependent upon the synthesis method.[238] For instance, the diffusion exponent of the CDHA/chitosan membranes is lower for that synthesised using *ex situ* processes, *i.e.* for those nanocomposites where the CDHA nanofiller was synthesised first and then added into the chitosan solution. When these nanocomposites are prepared as membranes, the permeability is lower when the nano-CDHA is synthesised in the presence of chitosan, which can be explained in terms of a better dispersion of the CDHA nanoparticles, resulting in more efficient physical barrier.

The use of silver has recently become one of the preferred methods to impede microbial proliferation on biomaterials and medical devices. Silver and silver-based compounds are highly antimicrobial by virtue of their antiseptic properties to as many as 16 kinds of bacteria, including *Escherichia coli* and *Staphylococcus aureus*.[239] Silver-loaded HA powder has shown antibacterial effects, both in nutrient-rich and -poor environments.[240] From a crystallochemical point of view, the substitution of Ag^+ (1.28 Å) ions takes place for Ca^{2+} (0.99 Å) preferentially in the Ca(1) site of HA, and this leads to an increase in the lattice parameters linearly with the amount of silver added in the range of atomic ratio Ag/(Ag + Ca) between 0 and 0.055.[241] The proposed general formula is $Ca_{10-x}Ag_x(PO_4)_6(OH)_{2-x}\square_x$, with vacancies at the hydroxyl site due to a charge imbalance caused by Ag^+ for Ca^{2+} ions, although PO_4^{3-} for HPO_4^{2-} is also a likely substitution to compensate the charge imbalance. Rameshbabu *et al.*[242] have synthesised nanosized silver-substituted HA using microwaves and incorporating $Ag(NO)_3$ to the reaction media in the required stoichiometric amounts. The microwave synthesis is a fast, simple, and efficient method to prepare nanosized materials,[243] with narrow particle size distribution due to fast homogenous nucleation.[244] Ag-substituted nano-HA prepared with this method are nanosized with needle-like morphology, with width ranging from 15 to 20 nm and length ~60–70 nm. A substitution degree of $x = 0.05$ in the general formula $Ca_{10-x}Ag_x(PO_4)_6(OH)_{2-x}\square_x$ is enough to completely inhibit the growth of *E. coli* and *S. aureus* after 24 and 48 h with 10^5 cells per mL, while exhibiting excellent osteoblast adherence and spreading.

4.5.2.3 *Nanoapatites for Nonviral Gene-Delivery Systems*

Gene therapy is becoming a rapidly growing therapeutic strategy, which consists of *transfecting* a modified gene into the genome to replace a disease-causing gene.[245,246] The incorporation of bared DNA would be the simplest method of transfection. However, in these cases gene expression is very low due to the low endosomal escape, nuclease degradation, and inefficient nuclear uptake.[247,248] Consequently, other methods must be applied and for these purpose a *vector* must be used to deliver the therapeutic gene to the patient's target cells. Efficient gene transfection is achieved when the gene delivery vector facilitates:

- physical and chemical stability to the DNA in the extracellular space;
- cellular uptake;
- DNA escape from the endosomal network;
- cytosolic transport; and
- nuclear localisation of the DNA for transcription.[249,250]

Currently, the most common vectors are viruses that have been genetically altered to carry normal human DNA. The vector infects the cell of the patient and unloads its genetic material containing the therapeutic human gene into the target cell, restoring it to a normal state.

The development of non-viral vectors is currently catching the attention of many researchers. Non-viral methods present certain advantages over viral ones, like simple large-scale production and low host immunogenicity, among others. The major limitations of non-viral gene transfer are that it must be tailored to overcome the intracellular barriers to DNA delivery, including the cellular and nuclear membranes.[251] However, recent advances in vector technology have yielded molecules and techniques with transfection efficiencies similar to those of viruses. Among these techniques, the binding of DNA to calcium phosphates is one of the most attractive options, due to the high biocompatibility, biodegradability, easy handling, and high adsorptive capacity for plasmid DNA (pDNA).[252-254] Graham and van der Eb discovered in 1973 the capability of CaP to act as a non-viral vector.[255] The method developed by these authors comprised the formation of nanoparticles on the DNA backbone. For the preparation of the DNA/CaP complex for transfection, a calcium chloride solution is mixed with the DNA and the phosphate is subsequently added a in a buffered saline solution.

The hydrothermal technique has been applied to prepare hydroxyapatite nanoparticles for evaluation as a material for pDNA transfection.[256] The pDNA binding to a previously obtained nano-HA can be achieved through the mixture of a nanoparticle suspension with the pDNA and subsequent incubation at room temperature. The DNA–nanoparticle complexes transfect pDNA into SGC-7901 mice cells *in vitro* and transmission electron microscopic examination demonstrated their biodistribution and expression within the cytoplasm and also a little in the nuclei of the liver, kidney and brain tissue cells of mice. The ratio potential to adsorb pDNA is about 1 : 36 and the procedure can only be performed under acidic and neutral conditions.

pDNA encapsulated in calcium phosphate nanoparticles have been prepared as DNA delivery carriers and have specifically targeted these particles to liver cells after appropriate surface modification.[257,258] Although the pDNA entrapped in these nanoparticles is highly protected from enzymatic degradation, the transfection efficiency of these synthetic systems is not optimal. The problem is that prolonged ultrasonication used to be a prerequisite for re-dispersion of nanoparticles in aqueous buffers, which leads to the partial disintegration of DNA molecules, thus reducing transfection efficiency. In order to overcome this situation, these methods have been optimised by forming the calcium phosphate within the droplets of microemulsions and completely avoiding ultrasonication.[259]

Olton *et al.*[260] have dealt with the influence of synthesis parameters on transfection efficiency. Their results revealed that improved, more consistent levels of gene expression can be achieved by optimising both the stoichiometry (Ca/P ratio) of the CaP particles as well as the mode in which the precursor solutions are mixed. The optimised forms of these CaP particles were ~25–50 nm in size (when complexed with pDNA) and were efficient at both binding and condensing the genetic material. Differences in gene expression are not only due to a change in size of the naked CaP particles, but are rather due to the combined effects of pDNA binding and condensation to the particle which ultimately dictates the overall size of the pDNA–nanoCaP complex size.

Epple and co-workers have used apatite nanoparticles for gene transfection purposes by the classical precipitation method of mixing aqueous solutions of calcium nitrate and diammonium hydrogen phosphate at pH 9.[261] Immediately after mixing, the DNA solution is incorporated, resulting in nanoparticles in the rage of 10–20 nm, consisting of a calcium phosphate core and an outer shell of nucleic acid. The interactions between calcium phosphate and nucleic acid are electrostatic. In order to protect the DNA from nuclease attack, the single-shell nanoparticles can be modified through the incorporation of additional layers of calcium phosphate, leading to multi-shell nanoparticles.[262]

4.6 Bioactive Glasses for Drug Delivery

The incorporation of drugs into bioceramics to potentiate their therapeutic performance has been widely considered by many authors.[263] In the case of clinically approved bioactive glasses there exists a serious inconvenience in their use as host matrices, as melt-derived glasses are not appropriate for use as drug delivery systems. The high temperatures of the melting (~1300 °C) is noncompatible with the incorporation of drugs during the synthesis process, as can be achieved with other bioceramics, for instance calcium phosphate cements. Besides, the textural properties of melt-derived glasses are very poor, with very low surface area and porosity. In this sense, silica-based bioactive glasses required upgrading to be able to host drugs.

In the early 1990s bioactive glasses were for the first time prepared by the sol–gel process.[264] Porous bioglasses could be prepared from the hydrolysis and polymerisation of metal hydroxides, alkoxides, and/or inorganic salts. In contrast to melt-derived bioglasses, sol–gel glasses are not prepared at high processing temperatures. In addition, and due to the high surface area and porosity derived from the sol–gel process, the range of bioactive compositions is wider, also exhibiting higher bone bonding rates together with excellent degradation/resorption properties.[265] During the sol–gel process, the gelling stage occurs around room temperature. Gels, aerogels, glasses, dense oxides, *etc.*, can be made by sol–gel processing, thus facilitating the incorporation of organic and biological molecules within the network,[266] or even cells within silica matrices.[267]

One of the most interesting alternatives for bone regenerative purposes is the association of osteogenic agents with bioactive glasses.[268] On the one hand, the bioactive behaviour of many bioactive glass compositions involves not only osteoconduction and osteoproduction, but also osteoinduction processes when implanted in living tissue.[269] On the other hand, this bio-functionalisation provides added values to the implant, since the graft not only fills and repairs the defect but also acts as a drug delivery system, which locally supplies osteoregenerative agents.

Bioactive glasses can be used as carriers of osteogenic agents. Those with silica based chemical compositions can be easily functionalised, thus attaching chemical groups that control the kinetic release of the osteogenic agents. In this sense, Verne *et al.*[270] have proposed the functionalisation of bioactive glasses through covalent bonding with 3-amino-propyltriethoxysilante. Thereafter, the immobilisation of bone morphogenetic protein-2 was achieved, suggesting a promising alternative to the systemic administration of osteogenic agents. Other macromolecules, such as collagen have been linked to bioactive glasses,[66] aimed to improve the subsequent cell attachment, thus increasing the bone regeneration rate of 45S5 Bioglass porous scaffolds.

Local antibiotic release is a promising and effective procedure of delivering drugs at the implantation site. This strategy is aimed at preventing implant-associated infections by reducing the concentration of bacteria and/or impeding bacterial adherence to the implant surface. The utilisation of carriers for local antibiotic release is very important in orthopaedic surgery, because prevention and surgical precision are not enough to ensure the absence of infectious microorganisms. In fact, the occurrence of osteomyelitis makes implant removal essential for the prevention of further complications, such as loss of function and septicaemia.[271,272] Systemic antibiotic administration does not always allow for efficient concentrations, mainly because of poor blood flow in the bone tissue. This necessitates the administration of large antibiotic doses in order to obtain acceptable concentrations in the affected region.

The combination of uniaxial and isostatic pressure over a bioceramic-antibiotic mixture have been proposed, resulting in implants with good osteointegration and appropriate drug levels for several weeks.[273] A bioactive sol–gel glass was mixed with gentamicin sulphate and shaped by combining both kinds of cold-pressure systems. The implants were designed for application in orthopaedic surgery as filling materials for osseous defects and prevention or defence against osseous infections. *In vivo* tests were performed using New Zealand rabbits for 1, 4, 8 and 12 weeks to study their biological response. The bone response to the implant was of perfect osseous-integration, cortical and sponge osseous tissue growth that allowed the osseous defect restoration and partial resorption of the implant at medium length. These sol–gel glasses allow osseous tissue growth from the periphery to the implant. The newly formed bone accesses the implant and, at the same time, the glass is resorbed slowly but progressively. The local gentamicin levels detected in osseous tissue were above of the minimal inhibitory gentamicin concentration for the majority of the resident microorganisms. In addition, there was a progressive decrease of gentamicin levels in the osseous tissue with time, but they remained above the minimal inhibitory concentration until the end of the assay, which impeded the development of microorganisms sensitive to the antibiotic.

Radin *et al.*[274] showed that the sol–gel method is an excellent strategy to incorporate antibiotics into silica xerogels, as well as the subsequent delivery. More recently, Munusamy *et al.*[275] used the soft synthesis conditions of the sol–gel method for the incorporation of gentamicin within silica xerogel particles. This procedure has been tested in a mouse model for targeted intracellular delivery in *Salmonella*. The authors could obtain a delayed release of the drug with the sol–gel carrier reaching the intracellular compartment of the bacteria.

4.7   Sol–Gel Silica Glasses as Scaffolds for Bone Tissue Engineering

The manufacturing of 3D macroporous scaffolds based on bioactive sol–gel glasses has been one of the most important research lines in the field of bioceramics for bone regeneration purposes. For instance, Zhang *et al.* prepared 3D-ordered macroporous scaffolds with binary SiO_2–CaO and ternary SiO_2–CaO–P_2O_5 compositions.[276] The *in vitro* bioactivity of these materials has been studied as a function of chemical composition, micro- and macrostructure. The formation of apatite and degradation of the glass were slightly enhanced for the phosphate-containing composition. In addition, large particles formed less apatite and degraded less completely compared with small particles. Lastly, an increase in macropore size slowed down the glass degradation and apatite-formation processes, an effect related to the decreased internal surface area of the larger pore materials, although macroporous scaffolds exhibit better *in vitro* bioactive behaviour than non-3D structured glasses.[277]

Sol–gel glasses can be foamed to produce scaffolds that mimic cancellous bone macrostructure.[278] Jones and Hench have developed a method to produce 3D macroporous bioactive scaffolds, exhibiting a hierarchical structure, with interconnected macropores, enhancing the bioactive behaviour as well as the release of ionic products.[279,280] The foams are produced by including a surfactant at the sol stage and stirring vigorously. When the foam is formed, HF acid is added to quench the foam structure by fast polycondensation of the species. These macroporous glasses have been tested in osteoblast cell cultures.[281] Osteoblast proliferation has been shown to be higher in the presence of the foams, as is the collagen secretion. In addition, viable osteoblasts colonise the foams, suggesting that porous glass foams are promising materials for bone repair. The nanoporosity of these foams have been optimised, improving their compressive mechanical properties,[282] although fracture toughness and pore strength are still problems that must be overcome for future applications.

4.8 Mesoporous Bioactive Glasses for Drug Delivery

In 2001 the SiO_2-based mesoporous material MCM-41 was first proposed as a potential drug delivery system.[204] MCM-41 belongs to the M41S materials family that resulted from the incorporation of a structure directing agent, cetylammonium bromide (CTAB) to the SiO_2 synthesis *via* the sol–gel method.[283] The CTAB molecules self-organise into micelles that co-assemble with the soluble SiO_2 precursors, resulting in ordered SiO_2–CTAB structures. After CTAB removal by calcination or acid extraction, a SiO_2 mesoporous material is obtained which exhibits excellent textural properties for drug delivery purposes,[203] such as:

- ordered mesoporous network with homogeneous pore size, which allows a close control of the kinetic release of the drugs incorporated;
- high surface area and pore volume to adsorb the required drug dosage; and
- high density of silanol (Si–OH) groups at the surface, which can be functionalised for a better control of drug load and release.

The fast development of drug delivery systems based on pure SiO_2 mesoporous materials (such as MCM-41 or SBA-15) facilitated the incorporation of this function to mesoporous bioactive glasses (MBGs). Since MBGs retain the excellent surface and porosity properties, as well as the capability of functionalisation, several research groups understood the potential of MBGs as a matrix for local drug delivery in bone tissue.[284] Thus, their excellent bioactive behaviour could be added to the local delivery of antibiotic, antitumoural, antiosteoporotic and osteogenic agents, as can be seen in Table 4.2.

One of most common and complicated problems in orthopaedic surgery is implant infection. The incidence of osteomyelitis caused by *S. aureus* and *S. epidermidis* is relatively high. The problem is more complicated when these microorganisms develop a biofilm over the implant surface. In this scenario, the systemic administration of antibiotics is not efficient and the only safe solution is the withdrawal of the implant to avoid septicaemia. For this reason, controlled and local drug delivery from the implant is considered to be an excellent alternative treatment against infection. Xia and Chang proposed for the first time the incorporation of antibiotics within the porous network of MBGs.[285,286]

These authors applied a conventional drug impregnation strategy by soaking the 58S-MBGs into a highly concentrated solution of gentamicin sulphate. They observed that MBGs could incorporate three times more drug

Table 4.2 List of therapeutic agents incorporated into mesoporous bioactive glasses.

Therapeutic purpose	Drug/growth factor	Reference
Osteogenic drugs and growth factors	Dexamethasone	287
	Bone morphogenetic proteins	288
	β-fibroblast growth factor	289
Angiogenic agents	Vascular endothelial growth factor	290
	Dimethyloxallyl glycine	291
Anti-inflammatory	Ibuprofen	292
Antibiotic/antiseptic	Gentamicin	285
	Ampicillin	293
	Triclosan	294
Antitumoral	Doxorubicin	295

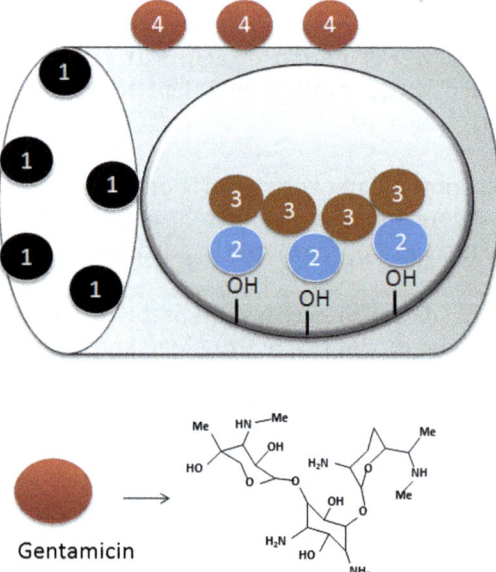

Figure 4.13 Different locations of gentamicin molecule with respect to the meso-pores in mesoporous bioactive glasses. (1) outside the window of the pore; (2) linked to the inner pore wall; (3) linked to another gentamicin molecule within the pore; (4) on the external surface.

than a conventional sol–gel glass with the same composition. Besides, the kinetics of the gentamicin delivery profile can be explained in terms of the different sites of the pores that gentamicin molecules can occupy and which are schemed in Figure 4.13:

- outside the window of the pore (location 1);
- linked to the inner pore wall by hydrogen bonding (location 2);
- within the inner part of the pore linked to another gentamicin molecule (location 3); and
- on the external surface of the MBG (location 4).

The drug-loading capability and the kinetic release not only depend upon porosity, but also on chemical composition, as Zhao *et al.* demonstrated.[296] For instance, the amount of tetracycline absorbed in MBGs is also a function of the CaO content. CaO acts as chelating agent in the pore wall. Consequently, the antibiotic release is slower as the CaO content increases in the systems SiO_2–CaO–P_2O_5, as the affinity of tetracyclines to Ca^{2+} would control release to the surrounding fluids. Several studies have been undertaken to test drug loading and release as a function of the structure-directing agent used, for instance P123 and F127. Some differences could be observed, which were explained in terms of the different textural properties for each MBG.[297]

Even considering the importance of the textural properties for the control of drug release, previous experience with pure SiO_2 mesoporous materials had demonstrated that only surface functionalisation could provide a satisfactory control on drug release.[238] Otherwise, the burst effect and the impossibility of achieving zero-order kinetics constrained the use of MBGs as control drug delivery matrices. However, the functionalisation of MBGs can result in the loss of bioactivity as the surface is modified.[298] A systematic study was undertaken into how functionalisation agents control drug release without significantly affecting the bioactivity of MBGs.[299] By means of functionalisation with thiol, amino, hydroxyl, and phenyl groups, zero-order kinetics were obtained. The kinetic constants could be controlled by tailoring the hydrogen bonding strength between the drug and the different functional groups.

4.8.1 Bioactive Mesoporous Microspheres

In addition to the mesoporous structure and chemical composition, the morphology and size of the particles is one the most important features of MBGs intended for drug delivery systems. A system formed by small-sized spheres offers much better possibilities for reproducibility and controlled release compared with irregular bulky particles with different sizes. The research team of Professor Stucky proposed the preparation of bioactive mesoporous microspheres and tested their haemostatic properties.[300] These materials have been proposed as first-aid agents in haemorrhages with massive blood loss.[301] Moreover, these haemostatic materials do not show the thermal effect commonly shown by other systems, thus avoiding the necrosis of surrounding healthy tissues. These spheres have diameters ranging between 100 nm and 1 μm and can be prepared in the system SiO_2–CaO–P_2O_5 using the surfactant P123 as a structure-directing agent.

There are several synthesis methods for the preparation of bioactive mesoporous spheres. Among them, aerosol-assisted methods and basic precipitation methods must be highlighted. Aerosol-assisted methods combine the EISA method with the formation of very small aerosol droplets containing all the precursors that subsequently will constitute the bioactive mesoporous microspheres. The ordered phase is formed within these droplets when the solvent is partially evaporated. Thereafter, the particles are pyrolised to form systems like SiO_2–CaO–P_2O_5 or similar. Some of these microspheres have been proposed for bone grafting and drug delivery systems.[294] These particles are intended for bone grafting and augmentation in small bone defects, for instance in periodontal surgery before the implantation of endosseous titanium abutments. In this sense, we can obtain a therapeutic synergy by combining the regenerative capability of MBGs and the pharmacological effect when the particles are loaded with antiseptic agents.

A second strategy to prepare mesoporous bioactive microspheres is the precipitation from diluted solutions of precursors and surfactant in alkaline medium. This method produces smaller and monodisperse spheres than those prepared by aerosol. However, the agglomeration of the particles is very likely to occur,[302] and the Ca^{2+} precipitation as calcium hydroxide often takes place under basic pH. Obtaining monodisperse nanoparticles would allow their incorporation into the blood system. The hydrodynamic stability in the blood torrent is reached when the particle size is kept between 50 and 300 nm. Larger particles are retained in the lungs and liver, whereas those with smaller sizes can cross the vascular endothelium and are non-specifically distributed all over the body. Finally, the combination of MBGs with biocompatible polymers must be highlighted. These systems keep the most of the bioactive properties while ensuring a controlled drug delivery kinetic.[303] In these composites the MBG does not form the microsphere itself but takes part as a discrete phase into the continuous polymeric phase. In these microspheres the drug can be incorporated into the mesoporous structure of MBGs, within the polymeric matrix, and in both.

Ultrafine hollow fibres have been prepared with MBGs,[304] by means of an electrospinning technique. Through the control of the water/ethanol ratio in the precursor solution and the addition of a phase-separation agent, the spinoidal decomposition of the solution is reached during the fibres' fabrication. Consequently, the formation of hollow fibres with mesoporous structures in the walls takes place. The drug loading and controlled-release capabilities are closely related to the length of the fibres, while exhibiting excellent bioactive behaviour.

Stimuli-responsive or smart systems are very interesting drug delivery devices to achieve highly controlled and specific drug release. These systems have been successfully applied to pure SiO_2 mesoporous materials.[203] Stimuli-responsive mesoporous systems consist of the design of open/close gates at the entrance of the pores, which are controlled by external stimuli. In this sense, smart devices have been tailored based on molecular gates, nanoparticles and nanomachines. In the field of MBGs, there are a few studies comprising stimuli-responsive systems. Recently, Lin *et al.* prepared a molecular gate system based on the photodimerisation of the cumarine molecule.[305] This molecular gate is activated by means of ultraviolet light, in such a way that irradiation with light of 310 nm or higher leads to cumarine dimerisation, thus closing the gate. In contrast, when the system is irradiated with ultraviolet light of wavelengths of 250 nm or lower, the monomer is regenerated, thus opening the gate and releasing the drug entrapped within the mesopores. Another stimuli-responsive drug-delivery system is that combining magnetic properties and controlled drug delivery. The incorporation of magnetic nanoparticles within mesoporous systems supplies new capabilities to MBGs. Through the action of an external magnetic field, the MBG nanoparticle can be vectored toward localised regions. Moreover, the capability of superparamagnetic nanoparticles of heating under an AC field opens the possibility of using them as thermoseeds for the treatment of tumours using hyperthermia.[306]

4.9   MBG Scaffolds for Regenerative Bone Therapies

The first attempt to shape MBGs as pieces was carried out by Stucky and co-workers[307] These authors prepared an injectable paste by mixing a SiO_2–CaO–P_2O_5 MBG with a buffered solution of ammonium phosphate. This paste settled in a similar way to calcium phosphate cements, allowing the formation of solid pieces with certain mechanical strength and high bioactive behaviour. However, these materials do not exhibit the porous architecture required for scaffolding purposes in regenerative treatments. For this purpose, interconnected macropores with sizes ~400 μm are needed. The pioneering team shaping MBGs as macroporous scaffolds was the research group of Professor Yun.[308] This group incorporated rapid prototyping methods into the preparation of MBG macroporous scaffolds with ordered macroporous architecture, as well as to the free-form fabrication of MBG implants through computer-assisted designs (Figure 4.14). These scaffolds show three different porous systems in the nanometre, micrometric and macroscopic levels. For this purpose, the following combination is required:

- a structure-directing agent to achieve an ordered mesoporous arrangement at the nanometre level;
- a polymer, such as methylcellulose, to leave pores of several micrometres after calcination; and
- a macroporous architecture tailored by rapid prototyping.

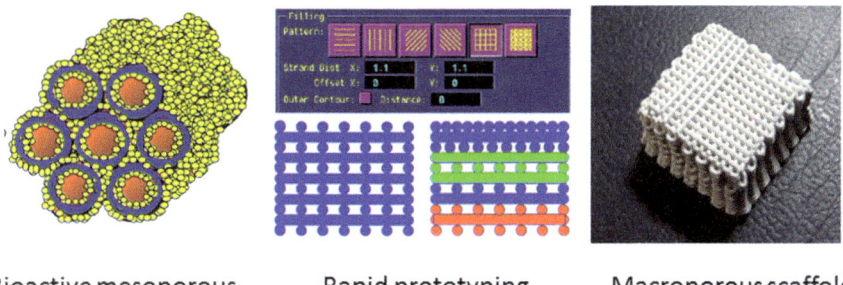

Bioactive mesoporous   Rapid prototyping   Macroporous scaffold
material (powder)   design (computer file)   (bone implant)

Figure 4.14   The three main stages in the fabrication of mesoporous bioactive glass macroporous implants for bone tissue regeneration.

The main drawback of the scaffolds so prepared is their fragility, which limits their clinical application in the orthopaedic field. Although the formation of the new apatite phase during the bioactive behaviour reinforces the mechanical strength of MBG-based scaffolds, this enhancement is not enough to satisfy the mechanical requirements in load-bearing locations.[309] The combination of MBGs with biocompatible polymers, such as polycaprolactone (PCL), have been proposed.[310] These composite scaffolds have shown excellent results in terms of mechanical improvement, while keeping their bioactive properties under *in vitro* conditions. *In vitro* biocompatibility tests using osteoblastic cells showed a better cell adhesion and osteoblast proliferation compared to those scaffolds made of only PCL. The excellent cell biocompatibility has been subsequently confirmed by the same group and other authors using different cell cultures.[311-316]

The macroporous architecture of the scaffolds can be also designed from polymers that are easily removable by calcination. This methodology has been also applied using polyurethane sponges.[317] After being removed by calcination, pieces with highly interconnected macropores are obtained, thus mimicking the morphology of the cancellous bone. Based on these kind of structures and considering the antiosteoporotic activity of strontium renalate, Wu *et al.*[318] prepared mesoporous scaffolds in the $SrO–SiO_2$ system. This composition does not have bioactive properties, but the capability to release strontium comprises antiosteoporotic action and facilitates the scaffold's degradation. In this sense, the development of macroporous scaffolds manufactured with MBGs, and combined with antiosteoporotic agents, opens very promising perspectives for the treatment of fractures in osteoporotic patients.[319]

References

1. T. Traykova, C. Aparicio, M. P. Ginebra and J. A. Planell, *Nanomedicine*, 2006, **1**, 91.
2. D. Aronov, R. Rosen, E. Z. Ron and G. Rosenman, *Process Biochem.*, 2006, **41**, 2367.
3. K. Kawanabe, H. Iida, Y. Matsusue, H. Nishimatsu, R. Kasai and T. Nakamura, *Acta Orthop. Scand.*, 1998, **69**, 237.
4. B. Ben-Nissan, *MRS Bull.*, 2004, **29**, 28.
5. Y. F. Chou, W. Huang, J. C. Y. Duna, T. A. Millar and B. M. Wu, *Biomaterials*, 2005, **26**, 285.
6. S. Bose and S. K. Saha, *Chem. Mater.*, 2003, **15**, 4464.
7. EFFO and NOF, *Osteoporosis Int.*, 1997, **7**, 1.
8. A. Randell, P. N. Sambrook PN and T. V. Nguyen, *Osteoporosis Int.*, 1995, **5**, 427.
9. L. M. Rodríguez-Lorenzo and M. Vallet-Regí, *Chem. Mater.*, 2000, **12**, 2460.
10. A. Toni, C. G. Lewis and A. Sudanese, *J. Arthroplasty*, 1994, **9**, 435.
11. L. L. Hench and J. Wilson, in *Bioceramics*, ed. R. Z. LeGeros and J. P. LeGeros, World Scientific, New York City, NY, 1998, vol. 11, p. 31.

12. R. D. Bloebaum and J. A. Dupont, *J Arthroplasty*, 1993, **8**, 195.
13. A. R. Biesbrock and M. Edgerton, *Int. J. Oral Maxillofac. Implants.*, 1995, **10**, 712.
14. E. W. Morscher, A. Hefti and U. Aebi, *J. Bone Jt. Surg.*, 1998, **80**, 267.
15. T. Ichikawa, K. Hirota and H. Kanitani, *J. Oral Implantol.*, 1996, **22**, 232.
16. T. J. Webster and J. U. Ejiofor, *Biomaterials*, 2004, **25**, 4731.
17. T. J. Webster, C. Ergun, R. H. Doremus, R. W. Siegel and R. Bizios, *Biomaterials*, 2000, **21**, 1803.
18. X. Wang, X. Shen, X. Li and C. M. Agarwal, *Bone*, 2002, **31**, 1.
19. T. J. Webster, R. W. Siegel and R. Bizios, *Nanostruct. Mater.*, 1999, **12**, 983.
20. T. J. Webster, R. W. Siegel and R. Bizios, *Soc. Biomater. Trans.*, 1999, 88.
21. T. J. Webster, C. Ergun, R. H. Doremus, R. W. Siegel and R. Bizios, *J. Biomed. Mater. Res.*, 2000, **51**, 475.
22. S. Ayad, R. Boot-Handford, M. J. Humpries, K. E. Kadler and A. Shuttleworth, *The Extracellular Matrix Facts Book*, Academic Press Inc., San Diego, 1994, p. 29.
23. Y. R. Cai, Y. K. Liu, W. Q. Yan, Q. H. Hu, J. H. Tao, M. Zhang, Z. L. Shi and R. K. Tang, *J. Mater. Chem.*, 2007, **17**, 3780.
24. A. Scabbia, *J. Clin. Periodontol.*, 2004, **31**, 348.
25. W. Sun, C. Chu, J. Wuang and H. Zhao, *J. Mater. Sci.: Mater. Med.*, 2007, **18**, 677.
26. M. Vallet-Regí and J. M. Gónzalez-Calbet, *Prog. Solid State Chem.*, 2004, **32**, 1.
27. M. Vallet-Regí, *J. Chem. Soc., Dalton Trans.*, 2001, **2**, 97.
28. S. V. Dorozhkin and M. Epple, *Angew. Chem., Int. Ed.*, 2002, **41**, 3130.
29. M. Yoshimura, H. Suda, K. Okamoto and K. Ioku, *J. Mater. Sci.*, 1994, **29**, 3399.
30. S. Mann, *Biomineralization: Principles and Concepts in Bioinorganic Materials Chemistry*, Oxford University Press, Oxford, U.K., 2001.
31. E. M. Burke, Y. Guo, L. Colon, M. Rahima, A. Veis and G. H. Nancollas, *Colloids Surf., B*, 2000, **17**, 49.
32. A. Bigi, E. Boanini, D. Walsh and S. Mann, *Angew. Chem., Int. Ed.*, 2002, **41**, 2163.
33. E. Bettoni, A. Bigi, G. Falini, S. Panzavolta and N. Roveri, *J. Mater. Chem.*, 1999, **9**, 779.
34. D. Walsh, J. L. Kingston, B. R. Heywood and S. Mann, *J. Cryst. Growth*, 1993, **133**, 1.
35. S. Sarda, M. Heughebaert and A. Lebugle, *Chem. Mater.*, 1999, **11**, 2722.
36. M. Bujan, M. Sikiric, F. Vincekovic, N. Vdovic, N. Garti and F. H. Milhofer, *Langmuir*, 2001, **17**, 6461.
37. L. Hovarth, I. Smit, M. Sikiric and F. Vincekovic, *J. Cryst. Growth*, 2000, **219**, 91.
38. L. Qi, J. Ma, H. Cheng and Z. Zhao, *J. Mater. Sci. Lett.*, 1997, **16**, 1779.
39. M. Antonietti, M. Breulmann, C. Goltner, H. Colfen, K. Wong, D. Walsh and S. Mann, *Chem.–Eur. J.*, 1998, **4**, 2493.
40. H. A. Schmidt and A. E. Ostafin, *Adv. Mater.*, 2002, **14**, 532.

41. W. Zhang, S. S. Liao and F. Z. Cui, *Chem. Mater.*, 2003, **15**, 3221.
42. S. Mann, D. D. Archibald, J. M. Didymus, T. Douglass, B. R. Heywood and F. C. Meldrum, *Science*, 1993, **261**, 1286.
43. S. Mann, *Nature*, 1993, **365**, 499.
44. D. D. Archibald and S. Mann, *Nature*, 1993, **364**, 430.
45. I. Weissbuch, F. Frolow, L. Addadi, M. Lahav and L. Leiserowitz, *J. Am. Chem. Soc.*, 1990, **112**, 7718.
46. A. Firouzi, D. Kumar, L. M. Bull, T. Besier, P. Sieger and Q. Huo, *Science*, 1995, **267**, 1138.
47. X. Wang, J. Zhuang, Q. Peng and Y. Li, *Adv. Mater.*, 2006, **18**, 2031.
48. Y. Wang, S. Zhang, K. Wei, N. Zhao, J. Chen and X. Wang, *Mater. Lett.*, 2006, **60**, 1484.
49. F. C. Meldrum, N. A. Kotov and J. H. Fendler, *J. Phys. Chem.*, 1994, **98**, 4506.
50. D. H. Gray, S. Hu, E. Juang and D. L. Gin, *Adv. Mater.*, 1997, **9**, 731.
51. L. Yan, Y. D. Li, Z. X. Deng, J. Zhuang and X. M. Sun, *Int. J. Inorg. Mater.*, 2001, **3**, 633.
52. K. Onuma and A. Ito, *Chem. Mater.*, 1998, **10**, 3346.
53. J. D. Chen, Y. J. Wang, K. Wei, S. H. Zhang and W. T. Shi, *Biomaterials*, 2007, **28**, 2275.
54. K. S. Katti, *Colloids Surf., B*, 2004, **39**, 143.
55. L. G. Gutwein and T. J. Webster, *Biomaterials*, 2004, **25**, 4175.
56. B. H. Kear, J. Coliazzi and W. E. Mayo, *Scr. Mater.*, 2001, **44**, 2065.
57. C. Piconi and G. Maccauro, *Biomaterials*, 1999, **20**, 1.
58. Y. M. Sung, Y. K. Shin and J. J. Ryu, *Nanotechnology*, 2007, **18**, 065602.
59. W. Suchanek and M. Yoshimura, *J. Mater. Res.*, 1998, **13**, 94.
60. T. J. Webster, R. W. Siegel and R. Bizios, *Biomaterials*, 1999, **20**, 1222.
61. I. B. Lonor, A. Ito, K. Onuma, N. Kanzaki and R. L. Reis, *Biomaterials*, 2003, **24**, 579.
62. A. R. Boccaccini and V. Maquet, *Compos. Sci. Technol.*, 2003, **63**, 2417.
63. U. Arnold, K. Lindenhayn and C. Perka, *Biomaterials*, 2002, **23**, 2303.
64. N. Tamai, A. Myoui, M. Hirao, T. Kaito, T. Ochi, J. Tanaka, *et al.*, *Osteoarthr. Cartilage*, 2005, **13**, 405.
65. C. Doyle, E. T. Tanner and W. Bonfield, *Biomaterials*, 1991, **12**, 841.
66. J. Ni and M. Wang, *Mater. Sci. Eng. C*, 2002, **20**, 101.
67. Y. E. Greish, J. D. Bender, S. Lakshmi, P. W. Brown, H. R. Allcock and C. T. Laurencin, *Biomaterials*, 2005, **26**, 1.
68. D. Choi, K. G. Marra and P. N. Kumta, *Mater. Res. Bull.*, 2004, **39**, 417.
69. R. A. Sousa, R. L. Reis, A. M. Cunha and M. J. Bevis, *Compos. Sci. Technol.*, 2003, **63**, 389.
70. A. P. Marques and R. L. Reis, *Mater. Sci. Eng. C*, 2005, **25**, 215.
71. X. M. Deng, J. Y. Hao and C. S. Wang, *Biomaterials*, 2001, **22**, 2867.
72. Z. K. Hong, P. B. Zhang, C. L. He, X. Y. Qiu, A. X. Liu, L. Chen, *et al.*, *Biomaterials*, 2005, **26**, 6296.
73. J. H. Lee, T. G. Park, H. S. Park, D. S. Lee, Y. K. Lee, S. C. Yoon, *et al.*, *Biomaterials*, 2003, **24**, 2773.

74. S. S. Kim, M. S. Park, Q. Jeon, C. Y. Choi and B. S. Kim, *Biomaterials*, 2006, **27**, 1399.

75. Q. L. Hu, B. Q. Li, M. Wang and J. C. Shen, *Biomaterials*, 2004, **25**, 779.

76. C. Du, F. Z. Cui, Q. L. Feng, X. D. Zhu and K. de Groot, *J. Biomed. Mater. Res.*, 1998, **42**, 540.

77. M. Kikuchi, S. Itoh, S. Ichinose, K. Shinomiya and J. Tanaka, *Biomaterials*, 2001, **22**, 1705.

78. M. Kikuchi, H. N. Matsumoto, T. Yamada, Y. Koyama, K. Takakuda and J. Tanaka, *Biomaterials*, 2004, **25**, 63.

79. A. K. Lynn, T. Nakamura, N. Patel, A. E. Porter, A. C. Renouf, P. R. Laity, *et al.*, *J. Biomed. Mater. Res.*, 2005, **74A**, 447.

80. M. C. Chang and J. Tanaka, *Biomaterials*, 2002, **23**, 4811.

81. M. C. Chang and J. Tanaka, *Biomaterials*, 2002, **23**, 3879.

82. S. S. Liao, F. Z. Cui and Y. Zhu, *J. Bioact. Compat. Polym.*, 2004, **19**, 117.

83. M. C. Chang, C. C. Ko and W. H. Douglas, *Biomaterials*, 2003, **24**, 2853.

84. M. C. Chang, C. C. Ko and W. H. Douglas, *Biomaterials*, 2003, **24**, 3087.

85. K. Hae-Won, K. Hyoun-Ee and V. Salih, *Biomaterials*, 2005, **26**, 5221.

86. J. Y. Hao, L. Yu, Z. Shaobing, L. Zhen and D. Xianmo, *Biomaterials*, 2003, **24**, 531.

87. T. A. Taton, *Nature*, 2001, **412**, 491.

88. J. D. Hartgerink, E. Beniash and S. I. Stupp, *Science*, 2001, **294**, 1684.

89. G. N. Ramachandran and A. H. Reddi, *Biochemistry of Collagen*, Plenum Press, New York, 1976.

90. N. Degirmenbasi, D. M. Kalyon and E. Birinci, *Colloids Surf., B*, 2006, **48**, 42.

91. N. A. Peppas and N. K. Mongia, *Eur. J. Pharm. Biopharm.*, 1997, **43**, 51.

92. T. H. Young, W. Y. Chuang, M. Y. Hsieh, L. W. Chen and J. P. Hsu, *Biomaterials*, 2002, **23**, 3495.

93. M. Kobayashi, J. Toguchida and M. Oka, *Biomaterials*, 2003, **24**, 639.

94. H. Jiang, G. Campbell, D. Boughner, W. K. Wan and M. Quantz, *Med. Eng. Phys.*, 2004, **26**, 269.

95. H. Jiang, G. Campbell and F. Xi, *Med. Eng. Phys.*, 2005, **27**, 175.

96. H. L. Nichols, N. Zhang, J. Zhang, D. Shi, S. Bhaduri and X. Wen, *J. Biomed. Mater. Res.*, 2007, **82A**, 373.

97. L. O'Toole, A. J. Beck, A. P. Ameen, F. R. Jones and R. D. Short, *J. Chem. Soc., Faraday Trans.*, 1995, **91**, 3907.

98. M. R. Alexander and T. M. Duc, *J. Mater. Chem.*, 1998, **8**, 937.

99. A. J. McManus, R. H. Doremus, R. W. Siegel and R. Bizios, *J. Biomed. Mater. Res.*, 2005, **72A**, 98.

100. T. M. Keaveny and W. C. Hayes, *Bone*, 1993, **7**, 285.

101. C. L. Wu, W. G. Weng, D. J. Wu and W. L. Yan, *Polymer*, 2003, **44**, 1781.

102. A. G. Rozhin, Y. Sakakibara, M. Tokumoto, H. Kataura and Y. Achiba, *Thin Solid Films*, 2004, **464–465**, 368.

103. D. Z. Chen, C. Y. Tang, K. C. Chan, C. P. Tsui, P. H. F. Yu, M. C. P. Leung and P. S. Uskokovic, *Compos. Sci. Technol.*, 2007, **67**, 1617.

104. H. W. Kim, *J. Biomed. Mater. Res.*, 2007, **83A**, 169.

105. T. J. Webster, R. W. Siegel and R. Bizios, in *Bioceramics*, ed. R. Z. LeGeros and J. P. LeGeros, World Scientific Publishing Co., New York, USA, 1998, vol. 11, pp. 273–276.
106. J. B. Brunski, *Clin. Mater.*, 1992, **10**, 153.
107. G. Heimke, *Osseo-integrated implants volume I: basics, materials and joint replacements*, CRC Press, Boca Raton, FL, 1990, pp. 31–80.
108. P. Griss, M. H. Hackenbroch, M. Jager, B. Preussner, T. Schafer, R. Seebauer, W. van Eimeren and W. Winkler, *Aktuelle Probl. Chir. Orthop.*, 1982, **21**, 1.
109. T. J. Webster, E. A. Massa-Schlueter, J. L. Smith and E. B. Slamovich, *Biomaterials*, 2004, **25**, 2111.
110. J. P. Bearinger, D. G. Castner and K. E. Healy, *J. Biomater. Sci., Polym. Ed.*, 1998, **7**, 652.
111. K. C. Dee, T. T. Andersen and R. Bizios, *J. Biomed. Mater. Res.*, 1998, **40**, 371.
112. E. R. Kenawy, G. L. Bowlin, K. Mansfield, J. Layman, D. G. Simpson, E. H. Sanders and G. E. Wnek, *J. Controlled Release*, 2002, **81**, 57.
113. B. M. Min, L. Jeong, Y. S. Nam, J. M. Kim and W. H. Park, *Int. J. Biol. Macromol.*, 2004, **34**, 281.
114. H. Yoshimoto, Y. M. Shin, H. Terai and J. P. Vacanti, *Biomaterials*, 2003, **24**, 2077.
115. W. J. Li, R. Tuli, C. Okafor, A. Derfoul, K. G. Danielson, D. J. Hall and R. S. Tuan, *Biomaterials*, 2005, **26**, 599.
116. A. Formhals, US 1 975 504, 1934.
117. P. Wutticharoenmongkol, N. Sanchavanikit, P. Pavasant and P. Supaphol, *Macromol. Biosci.*, 2006, **6**, 70.
118. P. Wutticharoenmongkol, P. Pavasant and P. Supaphol, *Biomacromolecules*, 2007, **8**, 2602.
119. H. W. Kim, H. H. Lee and J. C. Knowles, *J. Biomed. Mater. Res.*, 2006, **79A**, 643.
120. G. Sui, X. Yang, F. Mei, X. Hu, G. Chen, X. Deng and S. Ryu, *J. Biomed. Mater. Res.*, 2007, **82A**, 445.
121. J. K. Vasir, K. Tambwekar and S. Garg, *Int. J. Pharm.*, 2003, **255**, 13.
122. U. Edlund and A. C. Albertsson, *Adv. Polym. Sci.*, 2002, **157**, 67.
123. H. Kawaguchi, *Prog. Polym. Sci.*, 2000, **25**, 1171.
124. S. Freiberg and X. X. Zhu, *Int. J. Pharm.*, 2004, **282**, 1.
125. E. Ruiz-Hernández, A. López-Noriega, D. Arcos, I. Izquierdo-Barba, O. Terasaki and M. Vallet-Regí, *Chem. Mater.*, 2007, **19**, 3455.
126. X. Y. Qiu, Z. K. Hong, J. L. Hu, L. Chen, X. S. Chen and X. B. Ping, *Biomacromolecules*, 2005, **6**, 1193.
127. X. Qiu, Y. Han, X. Zhuang, X. Chen, Y. Li and X. Jing, *J. Nanopart. Res.*, 2007, **9**, 901.
128. D. W. Hutmacher, *Biomaterials*, 2000, **21**, 2529.
129. M. M. Stevens and J. George, *Science*, 2005, **310**, 1135.
130. D. Arcos, C. V. Ragel and M. Vallet-Regí, *Biomaterials*, 2001, **22**, 701.
131. D. Arcos, J. Peña and M. Vallet-Regí, *Chem. Mater.*, 2003, **15**, 4132.

132. C. V. Ragel and M. Vallet-Regí, *J. Biomed. Mater. Res.*, 2000, **51**, 424.
133. A. Rámila, R. P. del Real, R. Marcos, P. Horcajada and M. Vallet-Regí, *J. Sol-Gel Sci. Technol.*, 2003, **26**, 1195.
134. S. Padilla, R. P. del Real and M. Vallet-Regí, *J. Controlled Release*, 2002, **83**, 343.
135. S. N. Khorasani, S. Deb, J. C. Behiri, M. Braden and W. Bonfield, *Bioceramics*, 1992, **5**, 225.
136. Y. M. Khan, D. S. Katti and C. T. Laurencin, *J. Biomed. Mater. Res.*, 2004, **69A**, 728.
137. A. Piattelli, M. Franco, G. Ferronato, M. T. Santello, R. Martinetti and A. Scarano, *Biomaterials*, 1997, **18**, 629.
138. S. Sánchez-Salcedo, M. Vila, I. Izquierdo-Barba, M. Cicuéndez and M. Vallet-Regí, *J. Mater. Chem.*, 2010, **20**, 6956.
139. S. A. Brown, L. Farnsworth, K. Merrit and T. D. Crowe, *J. Biomed. Mater. Res.*, 1988, **22**, 321.
140. L. L. Hench and J. Wilson, *Biomater. Sci.*, 1984, **226**, 630.
141. W. Suchanec and M. Yoshimura, *J. Mater. Res.*, 1998, **13**, 94.
142. R. Z. LeGeros, *Clin. Orthop. Relat. Res.*, 2002, **395**, 81.
143. L. Sun, C. C. Berndt, K. A. Gross and A. Kucuk, *J. Biomed. Mater. Res.*, 2001, **58**, 570.
144. Y. Yang, K. H. Kim and J. L. Ong, *Biomaterials*, 2005, **26**, 327.
145. F. J. García-Sanz, M. B. Mayor, J. L. Arias, J. Pou, B. León and M. Pérez-Amor, *J. Mater. Sci.: Mater. Med.*, 1997, **8**, 861.
146. J. G. C. Wolke, J. P. C. M. van der Waerden, H. G. Schaeken and J. A. Jansen, *Biomaterials*, 2003, **24**, 2623.
147. Z. S. Luo, F. Z. Cui and W. Z. Li, *J. Biomed. Mater. Res.*, 1999, **46**, 80.
148. H. Ishizawa and M. Ogino, *J. Biomed. Mater. Res.*, 1997, **34**, 15.
149. R. Chiesa, E. Sandrini, M. Santin, G. Rondelli and A. Cigada, *J. Appl. Biomater. Biomech.*, 2003, **1**, 91.
150. A. Bigi, B. Bracci, F. Cuisinier, R. Elkaim, R. Giardino, I. Mayer, I. N. Mihailescu, G. Socol, L. Sturba and P. Torricelli, *Biomaterials*, 2005, **26**, 2381.
151. M. V. Cabañas and M. Vallet-Regí, *J. Mater. Chem.*, 2003, **13**, 1104.
152. M. Cifuentes, M. V. Cabañas and M. Vallet-Regí, *Key Eng. Mater.*, 2001, **192–195**, 135.
153. L. L. Hench and J. K. West, *Chem. Rev.*, 1990, **90**, 33.
154. C. J. Brinker and G. W. Scherer, *Sol-Gel Science: The Physics and Chemistry of Sol-Gel Processing*, Academic Press, San Diego, 1990.
155. N. Hijón, M. V. Cabañas, I. Izquierdo-Barba and M. Vallet-Regí, *Chem. Mater.*, 2004, **16**, 1451.
156. T. Brendel, A. Engel and C. Rüssel, *J. Mater. Sci.: Mater. Med.*, 1992, **3**, 175.
157. S. W. Russell, K. A. Luptak, C. T. Suchicital, T. L. Alford and V. C. Pizziconi, *J. Am. Ceram. Soc.*, 1996, **79**, 837.
158. M. Hsieh, L. Perng and T. Chin, *Mater. Chem. Phys.*, 2002, **74**, 245.
159. L. Goins, S. Holliday and A. Staniskevsky, *Mater. Res. Soc. Symp. Proc.*, 2004, **EXS-1**, H6.301-3.

160. C. You and S. Kim, *J. Sol–Gel Sci. Technol.*, 2001, **21**, 49.
161. K. Hwang and Y. Lim, *Surf. Coat. Technol.*, 1999, **115**, 172.
162. Y. Kojima, A. Shiraishi, K. Ishii, T. Yasue and Y. Arai, *Phosphorus Res. Bull.*, 1993, **3**, 79.
163. D. Liu, T. Troczynski and W. J. Tseng, *Biomaterials*, 2001, **22**, 1721.
164. L. Gan and R. Pilliar, *Biomaterials*, 2004, **25**, 5303.
165. L. D. Piveteau, M. I. Girona, L. Schlapbach, P. Barboux, J. P. Boilot and B. Gasser, *J. Mater. Sci.: Mater. Med.*, 1999, **10**, 161.
166. M. Cavalli, G. Gnappi, A. Montenero, D. Bersani, P. P. Lottici, S. Kaciulis, G. Mattogno and M. Fini, *J. Mater. Sci.*, 2001, **36**, 3253.
167. W. Weng and J. L. Baptista, *Biomaterials*, 1998, **19**, 125.
168. C. S. Chai, K. A. Gross and B. Ben-Nissan, *Biomaterials*, 1998, **19**, 2291.
169. K. A. Gross, C. S. Chai, G. S. K. Kannangara, B. Ben-Nissan and L. Hanley, *J. Mater. Sci.: Mater. Med.*, 1998, **9**, 839.
170. D. B. Haddow, P. F. James and R. Van Noort, *J. Sol–Gel Sci. Technol.*, 1998, **13**, 261.
171. B. Ben-Nissan, A. Milev and R. Vago, *Biomaterials*, 2004, **25**, 4971.
172. E. Tkalcec, M. Sauer, R. Nonninger and H. Schmidt, *J. Mater. Sci.*, 2001, **36**, 5253.
173. I. Izquierdo-Barba, N. Hijón, M. V. Cabañas and M. Vallet-Regí, *Key Eng. Mater.*, 2004, **254–256**, 363.
174. N. Hijón, M. V. Cabañas, I. Izquierdo-Barba, M. A. García and M. Vallet-Regí, *Solid State Sci.*, 2006, **8**, 685.
175. S. J. Lin, R. Z. LeGeros and J. P. LeGeros, *J. Biomed. Mater. Res.*, 2003, **66A**, 819.
176. Y. W. Gu, K. A. Khor and P. Cheang, *Biomaterials*, 2003, **24**, 1603.
177. R. Z. LeGeros, Calcium Phosphates in enamel, dentin and bone, in *Calcium Phosphates in Oral Biology in Medicine*, Monographs in Oral Science, ed. H. M. Myers, Karge, Zurich, 1991, pp. 108–129.
178. M. Vallet-Regí, C. V. Ragel and A. J. Salinas, *Eur. J. Inorg. Chem.*, 2003, 1029.
179. N. Hijón, M. V. Cabañas, J. Peña and M. Vallet-Regí, *Acta Biomater.*, 2006, **2**, 567.
180. I. Ichinose, H. Senzu and T. Kunitake, *Chem. Lett.*, 1996, **257**, 258.
181. J. He, I. Ichinose, S. Fujikawa, T. Kunitake and A. Nakao, *Chem. Mater.*, 2002, **14**, 3493.
182. K. Acharya and T. Kunitake, *Langmuir*, 2003, **19**, 2260.
183. P. Li, *J. Biomed. Mater. Res.*, 2003, **66A**, 79.
184. J. D. De Bruijn and C. A. Van Blitterswijk, in *Biomaterials in surgery*, ed. G. Walenkamp, Geory Thieme Verlag, Stuttgart, 1998, pp. 77–72.
185. C. M. Agrawal, J. Best, J. D. Heckman and B. D. Boyan, *Biomaterials*, 1995, **16**, 1255.
186. Y. Liu, P. Layrolle, J. de Bruijn, C. van Blitterswijk and K. De Groot, *J. Biomed. Mater. Res.*, 2001, **57**, 327.
187. S. Leeuwenburgh, P. Layrolle, F. Barrere, J. de Bruijn, J. Schoonman, C. A. van Blitterswijk and K. de Groot, *J. Biomed. Mater. Res.*, 2001, **56**, 208.

188. F. Barrère, C. M. van der Valk, R. A. J. Dalmeijer, G. Meijer, C. A. van Blitterswijk, K. De Groot and P. Layrolle, *J. Biomed. Mater. Res.*, 2003, **66A**, 779.

189. K. K. W. Lo, T. K. M. Lee, J. S. Y. Lau, W. L. Poon and S. H. Cheng, *Inorg. Chem.*, 2008, **47**, 200.

190. K. Hanaoka, K. Kikuchi, S. Kobayashi and T. Nagano, *J. Am. Chem. Soc.*, 2007, **129**, 13502.

191. K. K. V. Lo, W. K. Hui, C. K. Chung, K. H. K. Tsang, D. C. M. Ng, N. Y. Zhu and K. K. Cheung, *Coord. Chem. Rev.*, 2005, **249**, 1434.

192. H. M. E. Azzazy, M. M. H. Manssur and S. C. Kazmierczak, *Clin. Biochem.*, 2007, **40**, 917.

193. A. P. Alivisatos, W. Gu and C. Larabell, *Annu. Rev. Biomed. Eng.*, 2005, **7**, 55.

194. D. E. Clapham, *Cell*, 1995, **80**, 259.

195. Y. Kakizawa, S. Furukawa and K. Kataoka, *J. Controlled Release*, 2004, **97**, 345.

196. A. Doat, M. Fanjul, F. Pellé, E. Hollande and A. Lebugle, *Biomaterials*, 2003, **24**, 3365.

197. A. Doat, F. Pellé, N. Gardant and A. Lebugle, *J. Solid State Chem.*, 2004, **177**, 1179.

198. A. Lebugle, F. Pellé, C. Charvillat, I. Rousselot and J. Y. Chane-Ching, *Chem. Commun.*, 2006, 606.

199. V. P. Torchilin, *Nat. Rev.*, 2005, **4**, 145.

200. J. W. Yoo and C. H. Lee, *J. Controlled Release*, 2006, **112**, 1.

201. M. Malmeten, *Soft Matter*, 2006, **2**, 760.

202. M. Vallet-Regí, *Chem.–Eur. J.*, 2006, **12**, 5934.

203. M. Vallet-Regí, F. Balas and D. Arcos, *Angew. Chem., Int. Ed.*, 2007, **46**, 7548.

204. M. Vallet-Regí, A. Rámila, R. P. del Real and J. Pérez-Pariente, *Chem. Mater.*, 2001, **13**, 308.

205. F. Balas, M. Manzano, P. Horcajada and M. Vallet-Regí, *J. Am. Chem. Soc.*, 2006, **128**, 8116.

206. M. Vallet-Regí, *Dalton Trans.*, 2006, 5211.

207. B. Muñoz, A. Rámila, J. Pérez-Pariente, I. Díaz and M. Vallet-Regí, *Chem. Mater.*, 2003, **15**, 500.

208. F. Lamoureux, V. Trichet, C. Chipoy, F. Blanchard, F. Gouin and F. Redini, *Expert Rev. Anticancer Ther.*, 2007, **7**, 169.

209. P. K. Bajpai and H. A. Benghuzzi, *J. Biomed. Mater. Res.*, 1988, **22**, 1245.

210. E. P. Goldberg, A. R. Hadba, B. A. Almond and J. S. Marotta, *J. Pharm. Pharmacol.*, 2002, **54**, 159.

211. K. J. Harrington, F. Rowlinson-Busza and K. N. Syringos, *Clin. Cancer. Res.*, 2000, **6**, 2528.

212. A. Lebugle, A. Rodrigues, *et al.*, *Biomaterials*, 2002, **23**, 3517.

213. K. O. Lillehei, Q. Kong, S. J. Withrow and B. Kleinschmidt-DeMasters, *Neurosurgery*, 1996, 1191.

214. S. Miura, Y. Mii and Y. Miyauchi, *Jpn. J. Clin. Oncol.*, 1995, **25**, 61.

215. R. C. Straw, S. J. Withrow and E. B. Douple, *J. Orthop. Res.*, 1994, **12**, 1.
216. Y. Tahara and Y. Ishii, *J. Orthop. Sci.*, 2001, **6**, 556.
217. A. Uchida, Y. Shinto, N. Araki and K. Ono, *J. Orthop. Res.*, 1992, **10**, 440.
218. A. Barroug, L. T. Kuhn, L. C. Gerstenfeld and M. J. Glimcher, *J. Orthop. Res.*, 2004, **22**, 703.
219. A. Barroug and M. J. Glimcher, *J. Orthop. Res.*, 2002, **20**, 274.
220. A. Barroug, J. Lemaitre and P. G. Rouxhet, *Colloids Surf.*, 1989, **37**, 339.
221. A. Barroug, E. Lernous, J. Lemaitre and P. G. Rouxhet, *J. Colloid Interface Sci.*, 1998, **208**, 147.
222. J. Guicheux, G. Grimandi and M. Trecant, *J. Biomed. Mater. Res.*, 1997, **34**, 165.
223. V. C. Honnorat-Benabbou, A. Lebugle, B. Sallek and D. Lagarrigue, *J. Mater. Sci.*, 2001, **12**, 107.
224. B. Palazzo, M. Iafisco, M. Laforgia, N. Margiotta, G. Natile, C. L. Bianchi, D. Walsh, S. Mann and N. Roveri, *Adv. Funct. Mater.*, 2007, **17**, 2180.
225. S. P. A. Guaber, G. Gazzaniga, N. Roveri, L. Rimondini, B. Palazzo, M. Iafisco and P. Gualandi, EU Patent 005 146, 2006.
226. E. Landi, A. Tampieri, G. Celotti and S. Sprio, *J. Eur. Ceram. Soc.*, 2000, **20**, 2377.
227. L. Sz-Chian, C. San-Yuan, L. HsinYi and B. Jong-Shing, *Biomaterials*, 2004, **25**, 189.
228. F. Wingen and D. Schmahl, *Drug Res.*, 1985, **35**, 1565.
229. M. J. Bloemink, B. K. Keppler, H. Zahn, J. P. Dorenbos, R. J. Heetebrij and J. Reedijk, *Inorg. Chem.*, 1994, **33**, 1127.
230. T. Klenner, P. Valenzuela-Paz, B. K. Keppler, G. Angres, H. R. Scherf, F. Wingen, F. Amelung and D. Schmahl, *Cancer Treat. Rev.*, 1990, **17**, 253.
231. T. Klenner, F. Wingen, B. K. Keppler, B. Krempien and D. Schmahl, *J. Cancer Res. Clin. Oncol.*, 1990, **116**, 341.
232. T. Klenner, P. Valenzuela-Paz, F. Amelung, H. Muench, H. Zahn, B. K. Keppler and H. Blum, *Met. Complexes Cancer Chemother.*, 1993, 95.
233. S. Singh and S. S. Ray, *J. Nanosci. Nanotechnol.*, 2007, **7**, 2596.
234. Y. Wang, X. Wang, K. Wei, N. Zhao, S. Zhang and J. Chen, *Mater. Lett.*, 2007, **61**, 1017.
235. J. M. Xue and M. Shi, *J. Controlled Release*, 2004, **98**, 209.
236. S. Prior, C. Gamazo, J. M. Irache, H. P. Merkle and B. Gander, *Int. J. Pharm.*, 2000, **196**, 115.
237. T. Y. Liu, S. Y. Chen, S. C. Chen and D. M. Liu, *J. Nanosci. Nanotechnol.*, 2006, **6**, 2929.
238. T. Y. Liu, S. Y. Chen, J. H. Li and D. M. Liu, *J. Controlled Release*, 2006, **112**, 88.
239. J. A. Spadaro, T. J. Berger, S. D. Barranco, S. E. Chapin and R. O. Becker, *Antimicrob. Agents Chemother.*, 1974, **6**, 637.
240. K. Zhao, Q. Feng and G. Chen, *Tsinghua Sci. Technol.*, 1999, **4**, 1570.
241. L. Badrour, A. Sadel, M. Zahir, L. Kimakh and A. E. Hajbi, *Ann. Chim. Sci. Mater.*, 1998, **23**, 61.

242. N. Rameshbabu, T. S. S. Kumar, T. G. Prabhakar, K. V. G. K. Murty and K. P. Rao, *J. Biomed. Mater. Res.*, 2007, **80A**, 581.

243. O. Palchik, J. Zhu and A. Gedanken, *J. Mater. Chem.*, 2000, **10**, 1251.

244. B. L. Cushing, V. L. Kolesnichenko and C. J. O'Connor, *Chem. Rev.*, 2004, **104**, 3893.

245. T. Niidome and L. Huang, *Gene Ther.*, 2002, **9**, 1647.

246. H. Boulaiz, J. A. Marchal, J. Prados, C. Melguizo and A. Arenaga, *Cell. Mol. Biol.*, 2005, **51**, 3.

247. E. Orrantia and L. C. Chan, *Exp. Cell Res.*, 1990, **190**, 170.

248. C. W. Pouton, K. M. Wagstaff, D. M. Roth, G. W. Moseley and D. A. Jans, *Adv. Drug Delivery Rev.*, 2007, **59**, 698.

249. D. Luo and W. M. Saltzman, *Nat. Biotechnol.*, 2000, **18**, 33.

250. C. M. Wiethoff and C. R. Middaugh, *J. Pharm. Sci.*, 2003, **92**, 203.

251. K. M. Wagstaff and D. A. Jans, *Biochem. J.*, 2007, **406**, 185.

252. S. P. Wilson, F. Liu, R. E. Wilson and P. R. Housley, *Anal. Biochem.*, 1995, **226**, 212.

253. P. Batard, M. Jordan and F. Wurm, *Gene*, 2001, **270**, 61.

254. I. Roy, S. Mitra, A. Maitra and S. Mozumdar, *Int. J. Pharm.*, 2003, **250**, 25.

255. F. L. Graham and A. J. van der Eb, *Virology*, 1973, **52**, 456.

256. H. Zhu, B. Y. Huang, K. C. Zhou, S. P. Huang, F. Liu, Y. M. Li, Z. G. Xue and Z. G. Long, *J. Nanopart Res.*, 2004, **6**, 307.

257. G. Bhakta, R. Singh, S. Mitra, S. Mozumdar and A. N. Maitra, *Proc. Controlled Release Soc.*, 2003, **30**, 669.

258. A. N. Maitra, S. Mozumdar, S. Mitra and I. Roy. US Patent no. 6555376, 29 April 2003.

259. S. Bisht, G. Bhakta, S. Mitra and A. Maitra, *Int. J. Pharm.*, 2005, **288**, 157.

260. D. Olton, J. Li, M. E. Wilson, T. Rogers, J. Close, L. Huang, P. N. Kumta and C. Sfeir, *Biomaterials*, 2007, **28**, 1267.

261. T. Welzel, I. Radtke, W. Meyer-Zaika, *et al.*, *J. Mater. Chem.*, 2004, **14**, 2213.

262. V. V. Sokolova, I. Radtke, R. Heumann and M. Epple, *Biomaterials*, 2006, **27**, 3147.

263. D. Arcos and M. Vallet-Regí, *Acta Mater.*, 2013, **61**, 809.

264. R. Li, A. E. Clark and L. L. Hench, *J. Appl. Biomater.*, 1991, **2**, 231.

265. J. Zhong and D. C. Greenspan, *J. Biomed. Mater. Res.*, 2000, **53**, 694.

266. D. Avnir and S. Braun, *Biochemical aspects of sol-gel science and technology: a special issue of the journal o sol-gel science and technology*, Springer-Verlag, New York, 1996.

267. A. Nieto, S. Areva, T. Wilson, R. Viitala and M. Vallet-Regí, *Acta Biomater.*, 2009, **5**, 3478.

268. M. Vallet-Regí, *J. Intern. Med.*, 2010, **267**, 22.

269. L. L. Hench, *J. Mater. Sci.: Mater. Med.*, 2006, **17**, 967.

270. E. Verne, C. Vitale-Bovarone, E. Bui, C. L. Bianchi and A. R. Boccaccini, *J. Biomed. Mater. Res.*, 2009, **90A**, 984.

271. C. J. Wang, T. W. Huang, J. W. Wang and H. S. Chen, *J. Arthroplasty*, 2002, **17**, 608.

272. K. Saleh, M. Olson, S. Resig, B. Bershadsky, M. Kuskowski, T. Gioe, *et al.*, *J. Orthop. Res.*, 2002, **20**, 506.
273. L. Meseguer-Olmo, M. J. Ros-Nicolás, M. Clavel-Sainz, V. Vicente-Ortega, M. Alcaraz Baños, A. Lax-Pérez, *et al.*, *J. Biomed. Mater. Res.*, 2002, **61**, 458.
274. S. Radin, P. Ducheyne, T. Kamplain and B. H. Tan, *J. Biomed. Mater. Res.*, 2001, **57**, 31377.
275. P. Munusamy, M. N. Seleem, H. Alqublan, R. Tyler, N. Sriranganathan and G. Pickrell, *Mater. Sci. Eng. C*, 2009, **29**, 231378.
276. K. Zhang, H. W. Yan, B. C. Bell, A. Stein and L. F. Francis, *J. Biomed. Mater. Res.*, 2003, **66A**, 860.
277. H. W. Yan, K. Zhang, C. F. Blanford, L. F. Francis and A. Stein, *Chem. Mater.*, 2001, **13**, 1374.
278. J. R. Jones and L. L. Hench, *Curr. Opin. Solid State Mater. Sci.*, 2003, **7**, 301.
279. J. R. Jones and L. L. Hench, *J. Mater. Sci.*, 2003, **38**, 3783.
280. P. Sepúlveda, J. R. Jones and L. L. Hench, *J. Biomed. Mater. Res.*, 2002, **59**, 340.
281. P. Valerio, M. H. R. Guimaraes, M. M. Pereira, M. F. Leite and A. M. Goes, *J. Mater. Sci.: Mater. Med.*, 2005, **16**, 851.
282. J. R. Jones, *J. Eur. Ceram. Soc.*, 2009, **29**, 1275.
283. C. T. Kresge, M. E. Leonowicz, W. J. Roth, J. C. Vartuli and J. S. Beck, *Nature*, 1992, **359**, 710.
284. C. Wu and J. Chang, *J. Controlled Release*, 2014, **193**, 282.
285. W. Xia and J. Chang, *J. Controlled Release*, 2006, **110**, 522.
286. W. Xia and J. Chang, *J. Non-Cryst. Solids*, 2008, **354**, 1338.
287. C. Wu, R. Miron, A. Sculeaaan, S. Kaskel, T. Doert, R. Schullze and Y. Zhang, *Biomaterials*, 2011, **32**, 7068.
288. C. Dai, H. Guo, J. Lu, J. Shi, J. Wei and C. Liu, *Biomaterials*, 2011, **32**, 8506.
289. R. A. Perez, A. El-Fiqi, J. H. Park, T. H. Kim, J. H. Kim and H. W. Kim, *Acta Biomaterialia*, 2013, **10**, 520.
290. C. Wu, J. Fan, Y. Chang and Y. Xiao, *J. Biomater. Appl*, 2013, **28**, 367.
291. C. Wu, Y. Zhou, J. Chang and Y. Xiao, *Acta Biomater.*, 2013, **9**, 9159.
292. J. Lin, Y. Fan, P. P. Yang, S. S. Huang, J. H. Jiang and H. Z. Lian, *J. Phys. Chem., C*, 2009, **113**, 7826.
293. C. Wu, Y. Zhou, W. Fan, P. Han, J. Chang, J. Yuen, M. Zhang and Y. Xiao, *Biomaterials*, 2012, **33**, 2076.
294. D. Arcos, A. López-Noriega, E. Ruiz-Hernández, O. Terasaki and M. Vallet-Regí, *Chem. Mater.*, 2009, **21**, 1000.
295. C. Wu, W. Fan and J. Chang, *J. Mater. Chem. B*, 2013, **1**, 2710.
296. L. Z. Zhao, X. X. Yan, X. F. Zhou, L. Zhou, H. N. Wang, J. W. Tang and C. Z. Yu, *Microporous Mesoporous Mater.*, 2008, **109**, 210.
297. Y. F. Zhao, S. C. J. Loo, Y. Z. Chen, F. Y. C. Boey and J. Ma, *J. Biomed. Mater. Res.*, 2008, **85A**, 1032.
298. J. Sun, Y. S. Li, L. Li, W. R. Zhao, L. Li, J. H. Gao, M. L. Ruan and J. L. Shi, *J. Non-Cryst. Solids*, 2008, **354**, 3799.

299. A. López-Noriega, D. Arcos and M. Vallet-Regí, *Chem.–Eur. J.*, 2010, **16**, 10879.
300. T. A. Ostomel, Q. H. Shi, C. K. Tsung, H. J. Liang and G. D. Stucky, *Small*, 2006, **2**, 1261.
301. T. A. Ostomel, Q. Shi and G. D. Stucky, *J. Am. Chem. Soc.*, 2006, **128**, 8384.
302. S. Zhao, B. Li and D. X. Li, *Microporous Mesoporous Mater.*, 2010, **135**, 67.
303. X. Li, X. P. Wang, L. X. Zhang, H. R. Chen and J. L. Shi, *J. Biomed. Mater. Res. Appl. Biomater.*, 2009, **89B**, 148.
304. Y. L. Hong, X. S. Chen, X. B. Jing, H. S. Fan, Z. W. Gu and X. D. Zhang, *Adv. Funct. Mater.*, 2010, **20**, 1503.
305. H. M. Lin, W. K. Wang, P. A. Hsiung and S. G. Shyu, *Acta. Biomater.*, 2010, **6**, 3265.
306. X. Li, X. P. Wang, Z. Hua and J. L. Shi, *Acta Mater.*, 2008, **56**, 3260.
307. Q. H. Shi, J. F. Wang, J. P. Zhang, J. Fan and G. D. Stucky, *Adv. Mater.*, 2006, **18**, 1038.
308. H. S. Yun, S. E. Kim and Y. T. Hyeon, *Chem. Commun.*, 2007, 2139–2141.
309. D. Arcos, M. Vila, A. López-Noriega, F. Rossignol, E. Champion, F. J. Oliveira and M. Vallet-Regí, *Acta. Biomater.*, 2011, **7**, 2952.
310. H. S. Yun, S. E. Kim, Y. T. Hyun, S. J. Heo and J. W. Shin, *Chem. Mater.*, 2007, **19**, 6363.
311. H. S. Yun, S. E. Kim, Y. T. Hyun, S. J. Heo and J. W. Shin, *J. Biomed. Mater. Res., Part B*, 2008, **87B**, 374.
312. X. P. Wang, X. Li, K. Onuma, A. Ito, Y. Sogo, K. Kosuge and A. Oyane, *J. Mater. Chem.*, 2010, **20**, 6437.
313. M. Alcaide, P. Portolés, A. López-Noriega, D. Arcos, M. Vallet-Regí and M. T. Portolés, *Acta Biomater.*, 2010, **6**, 892.
314. Y. F. Zhu, C. T. Wu, Y. Ramaswamy, E. Kockrick, P. Simon, S. Kaskel and H. Zreiqat, *Microporous Mesoporous Mater.*, 2008, **112**, 494.
315. C. J. Shih, H. T. Chen, L. F. Huang, P. S. Lu, H. F. Chang and I. L. Chang, *Mater. Sci. Eng. C*, 2010, **30**, 657.
316. G. F. Wei, X. X. Yan, J. Yi, L. Z. Zhao, L. Zhou, Y. H. Wang and C. Z. Yu, *Microporous Mesoporous Mater.*, 2011, **143**, 157.
317. Y. F. Zhu and S. Kaskel, *Microporous Mesoporous Mater.*, 2009, **118**, 176.
318. C. T. Wu, W. Fan, M. Gelinsky, Y. Xiao, P. Simon, R. Schulze, T. Doert, Y. X. Kuo and G. Cuniberti, *Acta Biomater.*, 2011, **7**, 1797.
319. D. Arcos, A. R. Boccaccini, M. Bohner, A. Díez-Pérez, M. Epple, E. Gómez-Barrena, A. Herrera, J. A. Planell, L. Rodríguez-Mañas and M. Vallet-Regí, *Acta Biomater.*, 2014, **10**, 1793.

CHAPTER 5

Mesoporous Materials: From Macro to Nano

5.1   What is a Mesoporous Material?

In the early 1990s, the Japanese group headed by Kuroda[1] and the group from Mobil oil corporation headed by Kresge[2,3] pioneered the synthesis of mesoporous materials. It would be some years later that the research group of Terasaki[4-8] performed for the first time an excellent characterization of these new materials, enabling the understanding of their behavior and their projection to new and important applications, as discussed elsewhere.[9-12]

Mesoporous materials are included within the group of porous materials, and can be defined according to IUPAC as materials with pore diameter in the range of 2–50 nm. These pores may be distributed in an ordered or disordered fashion.

Porous solids form a large family of materials which can be classified in two groups: the *inorganic* group, which includes zeolites and metal phosphates; and the *organic–inorganic hybrids* group, with metal organic networks as a key example.

All porous solids exhibit a scaffold which contains the pores; in both groups mentioned above, this scaffold might be amorphous or crystalline, and the pores within may be ordered or disordered. Two kinds of surface can be clearly distinguished in these materials: the inner surface of the pores and the external surface of the scaffold, as depicted in Figure 5.1. Depending on its composition, the scaffold can be responsible for several physical properties, such as magnetic, electrical or optical phenomena. The inner surface of the pore dictates the adsorption properties of the material and is therefore critical in the field of catalysis, as well as in other important and expanding

RSC Nanoscience & Nanotechnology No. 39
Nanoceramics in Clinical Use: From Materials to Applications, 2nd Edition
By María Vallet-Regí and Daniel Arcos Navarrete
© M. Vallet-Regi and D. Arcos Navarrete 2016
Published by the Royal Society of Chemistry, www.rsc.org

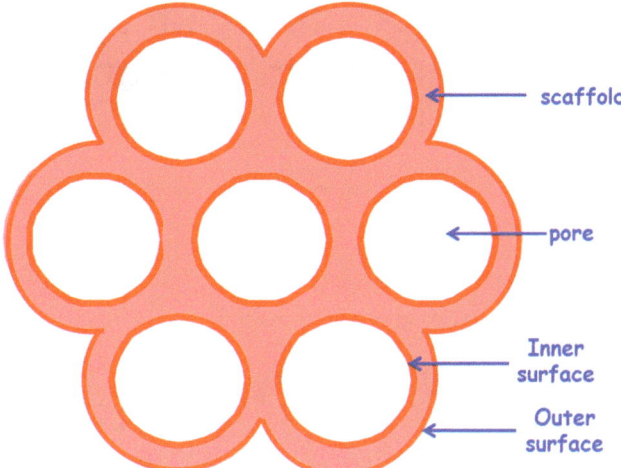

Figure 5.1 Parts of a mesoporous material. Different species can be confined inside the pores. Inner and outer surfaces can be functionalized.

applications as discussed in this chapter. The outer surface of the scaffold also plays a very important role in different properties of the material. Both surfaces can be functionalized, modifying their chemical composition and adapting it to the desired properties for a given porous material.

Pore diameter is the parameter used by IUPAC to define three groups of porous materials:

- *microporous materials*, with pore diameter values <2 nm, such as aluminum silicates, metal phosphates and hybrids;
- *mesoporous materials*, where pore diameter is within the 2–50 nm range; several aluminum silicates, metal phosphates and hybrids fall in this category, as well as metal chalcogenides; and
- *macroporous materials*, i.e. all those with pore diameter >50 nm; two examples are opals and the bone tissue of vertebrates.

In micro- and mesoporous structures, the scaffolds can be either amorphous or crystalline both in the inorganic and hybrid groups.

Figure 5.2 illustrates this IUPAC classification.

Ordered materials present a distinctive and unique feature from the structural point of view: they are ordered on the mesoscopic scale (2–50 nm) and disordered on the atomic scale. The scaffold exhibits periodically arranged pore channels within an amorphous silica matrix. This material, termed ordered mesoporous silica, displays high values of surface area and pore volume, its pore size and morphology are regular and can be tailored; besides, the mesostructured is stable and the morphology and size of its particles can also be controlled. It must be stressed that this material is non-toxic and biocompatible; moreover, the silica walls can be functionalized with diverse organic groups.[11,13–15]

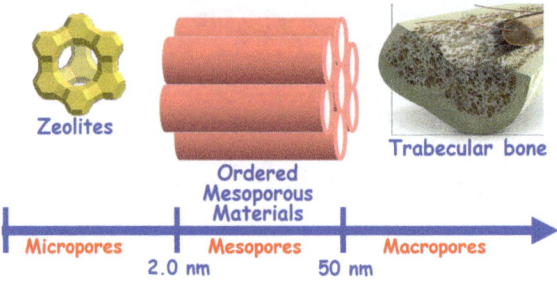

Figure 5.2   IUPAC classification of porous materials.

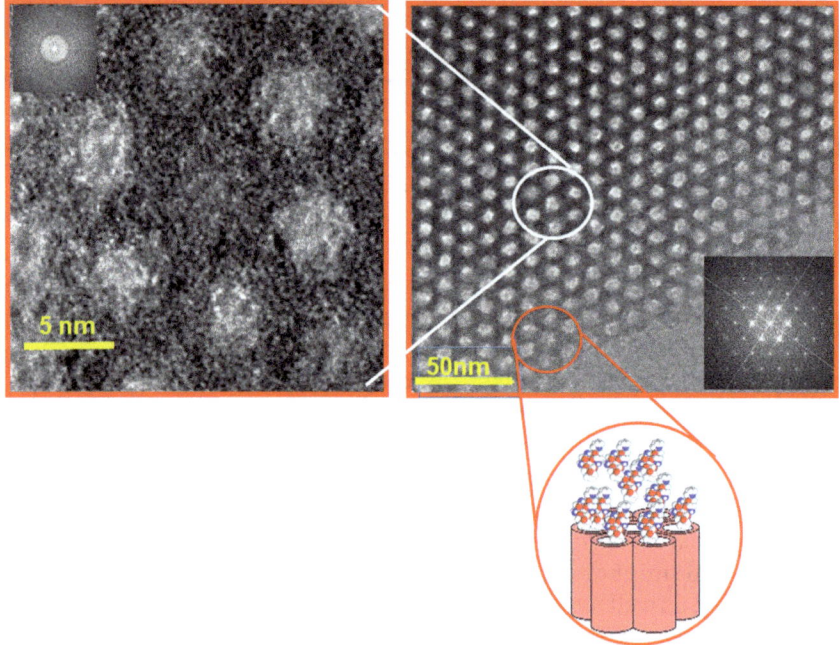

Figure 5.3   Transmission electron microscope images showing ordered mesopores.

Figure 5.3 shows the ordered mesostructure of these materials, as revealed in transmission electron (TEM) micrographs. If such ordered mesoporous silica materials are intended for use in biomedicine, their properties must be optimized according to the requirements of the desired biomedical application. In this sense, two important biomedical applications of these materials are drug delivery systems and smart nanodevices.

5.2   Discovery

The origin of mesoporous materials dates back to the 1970s, when the zeolites industry carried out a large body of experimental work in the field of chemistry in confined spaces. In 1972, Mobil Corporation developed a synthetic path to convert methanol into gasoline using the zeolite ZSM-5 (Zeolite Socony Mobil) as catalyst. This result acted as an inspiration to find zeolites with larger pore sizes, which could improve the conversion rates. The purpose of that research was to obtain cheaper gasoline from acid–base reactions taking place in the small cavities of those zeolites. However, zeolites did not achieve the desired result due to their small pore size, being microporous materials; researchers were then encouraged to design materials with larger pores.

This scenario, with the industry trying to find materials with larger pore sizes, led in the early 1990s to the first silica-based mesoporous material, produced almost simultaneously by Mobil Corporation in the USA,[2,16] and the Kuroda research team in Japan.[1,17] Mesoporous silica materials with large pores were produced combining silica with an amphiphilic surfactant (surface active agents). The surfactants were used as structure-directing agents to produce the M41S and KSW-*n* mesoporous materials, respectively.

However, this pioneering work by the American and Japanese groups was only the starting point of research in the field of mesoporous solids, which has expanded dramatically in the past 20 years. In fact, many research teams are focused in the production and evaluation at a fundamental level of the mechanisms involved in the formation of mesoporous materials. As a consequence, new reaction pathways have been found which lead to more complex nanostructures with a wide compositional range. Additionally, new synthesis routes and characterization techniques go in parallel with new potential applications, which in the early days of mesoporous materials were restricted to catalytic applications; this diversification plays a decisive role in the fast growth of research in this field.

Chronologically, the most important landmarks in the development of ordered mesoporous materials can be summarized as follows:

- the pioneering work on silica mesophases;
- expansion and variation on the inorganic framework;
- use of novel polymeric surfactants; and
- nanocasting, bringing a new dimension to the technology of ordered mesoporous materials.

Therefore, the early work in the 1990s opened up a whole new research area, which is currently very active, finding new developments with their respective commercial applications to be achieved in the near future.

5.3 Chemistry

As already mentioned, ordered mesoporous materials are unique and defined by an ordered and repetitive pore mesostructure while remaining disordered at the atomic level. Their synthesis is based on the use of surfactants, which are amphiphilic organic molecules that act as templates to direct the arrangement and subsequent condensation of the inorganic precursors.

The whole process leads to a network of periodically arranged cavities. Advances in synthesis methods of mesoporous materials have been clearly linked to the improvement of sol–gel technology, which is based on inorganic polymerization reactions that take place under mild conditions; the final product is the combination of inorganic phases with organic and/or biological systems. Thus, this sol–gel chemistry provided a mesoporous structure to traditional ceramic compositions, leading to novel and interesting properties.[18] Figure 5.4 shows the synthesis of ordered silica by molecular self-assembly or by cooperative self-assembly.

In the particular case of mesoporous silica, the development of different synthesis routes allows the preparation of materials with specific morphologies, narrow pore size distributions and high values of surface area, highlighting the adequacy of such materials for many different chemical processes, *e.g.* gas separation, controlled drug delivery, adsorption and catalysis, multiple separation techniques or production of sensing devices.[19–21]

Since the discovery of silica mesoporous materials back in the 1990s,[1,2,4] many structurally different mesoporous materials have been produced through the templating method using the self-organization properties of amphiphilic compounds under either acidic or basic conditions. In this sense, it is worth mentioning the use of block co-polymers to produce ordered mesostructure silica with pores ranging from 5 to 30 nm. The most famous material produced using those block co-polymers is SBA 15, which

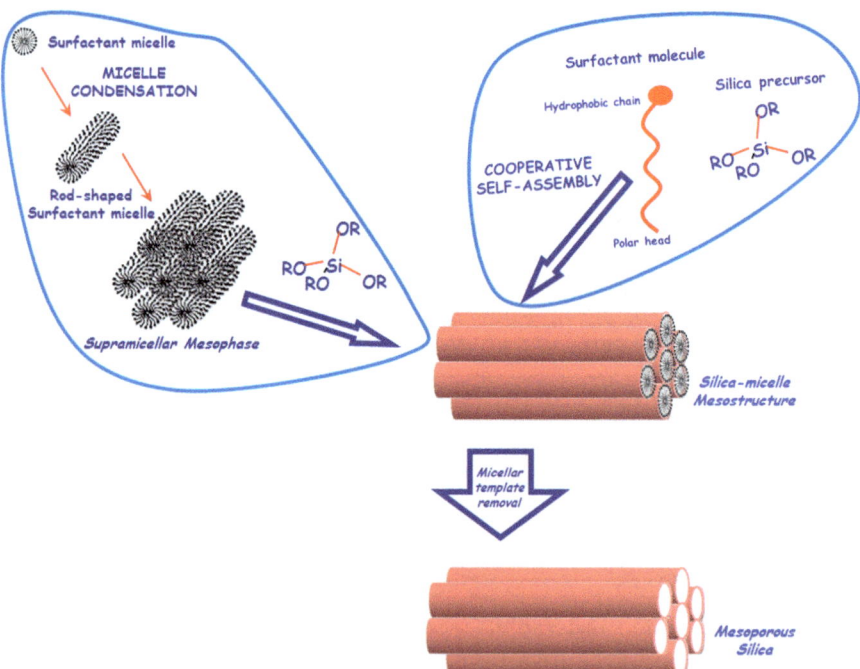

Figure 5.4 Synthesis schematics of mesoporous materials by two different routes.

was discovered in Santa Barbara, CA, USA in 1998.[22] Additionally, there are many research groups that have developed many other ordered mesoporous materials with different structures and, consequently, properties.

The combination of traditional techniques such as sol–gel with more recent advances in the field of molecular nanoscience, such as molecular self-assembly with surfactants, has greatly expanded the number of materials with controlled porosity that can be synthesized.[23]

Material functionalization using conventional techniques for surface modification exhibits certain disadvantages, which can be solved implementing *in situ* techniques where material formation and chemical functionalization of the structure are two simultaneous processes. As a result of this approach, using specific techniques, surface chemistry can be tailored as a function of the final desired application.[24]

The synthesis of mesoporous materials employs surfactants as structure-directing agents for the assembly and subsequent condensation of inorganic SiO_2 precursors. Once the surfactant is removed by calcination or extraction, these obtained materials exhibit outstanding features such as stable mesoporous structure, high surface area, large pore volume, regular and tunable nano-pore size, homogeneous pore morphology (hexagonal and cubic pore arrangements) and large amounts of silanol groups in the surface which can be chemically functionalized.

As a consequence of the synthesis by templating method, the porosity properties of ordered mesoporous materials would rely on the type of surfactant used during the liquid-crystal templating mechanism.[1,2]

In detail, the synthesis process is based on the dissolution of surfactant molecules into polar solvents to yield the so-called liquid crystals. The amount of surfactant dissolved in the solution plays a very important role in the synthetic process, and when the concentration is above the critical micellar concentration (cmc), the surfactant molecules aggregate together to form micelles. Obviously, the features of these micelles will depend on the nature of the surfactant and/or the experimental conditions, *e.g.* temperature, pH or concentration.[25]

Once formed, those micelles will in turn aggregate one another to yield supramicellar structures with determined geometries, which again will depend on the chemical nature of the surfactant and the processing conditions. The resulting mesoporous framework depends on these supramicellar geometries, such as hexagonal, cubic, laminar, *etc.*

One of the great achievements of chemistry is to design and prepare new structures with different complexities and numbers of applications. Mesoporous materials are a clear example of this. Thanks to advances in this field, new challenges continuously appear, such as, in particular, how to control the structure at different scales. Nature has always achieved this in the bone structure of vertebrates, where different porosity scales coexist, in hierarchical form, and where each pore has its own mission. New materials, with new properties, can be obtained in the artificial world, but it is necessary to control and modify the synthesis conditions of these materials. In order to achieve a precise control on the pore hierarchy of a solid material, new synthesis techniques based on weak interactions had to be developed, as well as new biomimetic methods and the combined used of organic and inorganic precursors. All this combined effort leads to the production of hierarchical solids, resembling nature, with outstanding applications in catalysis, separation, adsorption, sensors and nanomedicine.

The synthesis strategy based on templates has been extremely useful and versatile. Not only have conventional organic molecules have been used as templates; supramolecular structures, dendrimers, polymers, colloidal suspensions of nanoparticles, latex spheres and even biological materials such as butterfly wings, DNA or viruses have been used.

In synthesis by molecular self-assembly, the molecules are arranged forming superstructures, either before the solid formation or during the formation of the nanomaterial. The former case is known as the liquid-crystal mechanism, and the latter is named the co-operative mechanism: the interaction between organic molecules and inorganic precursor induces order in the system. In both cases, organic molecules are responsible for the solid structuring, and they are amphiphilic, formed by at least two domains with very different properties (hydrophilic and hydrophobic). This typical feature of surfactants induces supramolecular arrangement or self-assembly of these molecules in micelles with different geometries (Figure 5.5).

Once the supramicellar aggregates are formed, the silica precursors (usually alkoxysilanes) are added to the so-called liquid crystals and the sol–gel chemistry starts to hydrolyze and condensate those alkoxysilanes to yield highly condensed silicon oxide surrounding the supramicellar mesostructure. If a different inorganic composition is requested, such as alumina or zirconia, their respective precursors should be added instead of the alkoxysilanes.

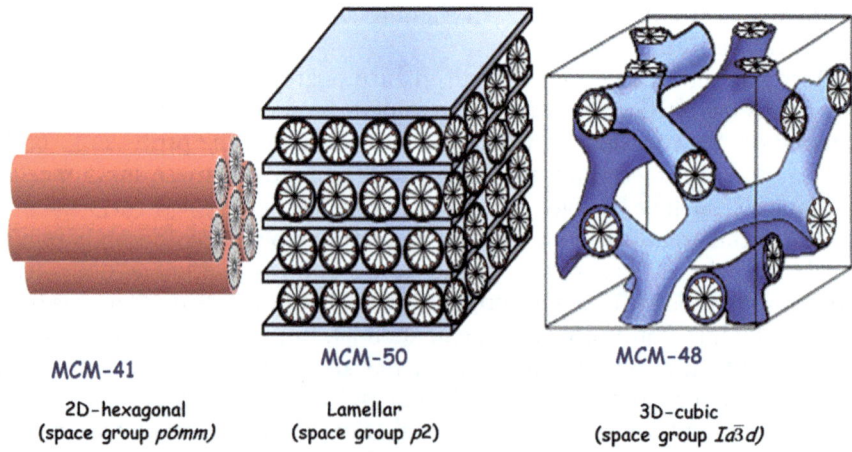

MCM-41

2D-hexagonal
(space group *p6mm*)

MCM-50

Lamellar
(space group *p2*)

MCM-48

3D-cubic
(space group $Ia\bar{3}d$)

Figure 5.5   Three mesoporous structures.

It is also possible to obtain these mesoporous materials following a different path, which is based on the co-assembly between the inorganic precursor and the structure-directing agent. To achieve this, both the precursors and surfactant should be added at the same time and the mesophase formation would take place simultaneously to hydrolysis and condensation.

In any case, the final step of the synthesis process is the removal of surfactant from the product to yield a network of cavities within the inorganic framework. This removal from the product can be achieved by calcination at high temperatures or by solvent extraction under softer conditions.

5.3.1 From Bulk to Mesoporous Nanoparticles

All the chemistry described up to this point, which has been applied in bulk form, could be easily modified to obtain mesoporous nanoparticles. Based on the same principles for the use of templates in mesoporous materials, it is possible to design synthesis strategies to obtain materials both in bulk and in nano form; again, the chemistry allows the production of materials in various forms, but preserving an ordered mesostructure in all cases.

Figure 5.6 illustrates the link between bulk materials, nanoparticles and smart materials. The know-how from each one of these fields can be of interest in the adjacent fields, hence generating a remarkable synergy which improves the design and production of new materials with tailored properties, as we shall see later.

Experience in the chemical synthesis of mesoporous materials can be used to obtain nanoparticles with similar composition. Simultaneously, experience in smart materials can be applied both to bulk and nanoparticle materials. One of the great advantages of these ordered mesoporous materials is the great versatility of the synthetic process, which allows their production in bulk, but also as microcapsules and even as nanoparticles.

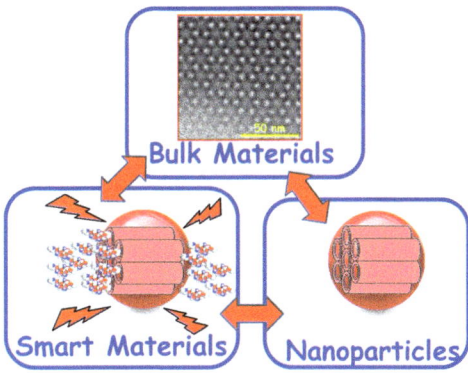

Figure 5.6 The inter-relationship between bulk materials, nanoparticles and smart materials.

5.3.2 Expansion in Compositions: Not Only Silica

As already mentioned, there are different types of structures in mesoporous silica materials and each of them exhibits distinct features, but chemistry can expand this type of mesoporous structure to a large variety of compositions. In mesoporous form, the successful synthesis of many oxides is documented: titanium, aluminum, zirconium, niobium, tin, iron, manganese and several phosphates. These phosphates, in particular, are especially interesting in the field of tissue regeneration and controlled drug release, as will be discussed later.

As this text is focused on medical applications, we shall describe in more detail those mesoporous materials which are in use (or could be used in the near future) in the clinical field, such as the *mesoporous bioactive glasses* (MBG). These materials are obtained from a silicon base with possible additions of phosphorus and calcium oxides, among others, in varying proportions; they are glasses, and therefore their composition can be that of any other glass. Their added value is the ordered mesostructure obtained during their synthesis in the presence of surfactants acting as templates and generating mesoporosity.

By definition an MBG is a nanostructured bioceramic which shares the same structural and textural properties as silica mesoporous materials (SMMs) exhibiting ordered mesoporous channels, whose walls consist of a vitreous network of composition SiO_2–CaO–P_2O_5 similar to the conventional sol–gel glasses. In this sense, at the atomic level these MBGs are very similar to conventional glasses, but have the added value of exhibiting a highly ordered mesoporous arrangement of cavities arranged periodically on lattices like artificial atoms or molecules in ordinary crystals (Figure 5.7).

This ordered mesoporous arrangement originates a surface area and pore volume significantly higher than those obtained by conventional sol–gel methods with analogous composition.[13] The research teams of Professor Zhao and Professor Vallet-Regí[26,27] synthesized these mesoporous bioactive glasses in the early 2000s. The synthesis of these materials combines the chemical sol–gel of the multicomponent SiO_2–CaO–P_2O_5 system of bioactive glasses with the supramolecular chemistry using surfactants as structure-directing agents. However, the success of the MBG synthesis does not only depend on the surfactant addition; it also requires new strategies such as the presence of a multicomponent inorganic system that includes CaO.

The said synthesis is, roughly speaking, a combination of both sol–gel and supramolecular chemistries. Once again, prior knowledge of synthesis methods enables their combination to obtain new materials with different properties (Figure 5.8). In this case, the combination of supramolecular chemistry and sol–gel chemistry leads to the synthesis of mesoporous materials with a glassy scaffold; hence, the optimum reactivity due to the disordered structure of glasses is combined with the ordered mesostructure obtained by the presence of surfactants in the synthesis process, arriving at a new material with much larger specific surface. Both features bring much higher values

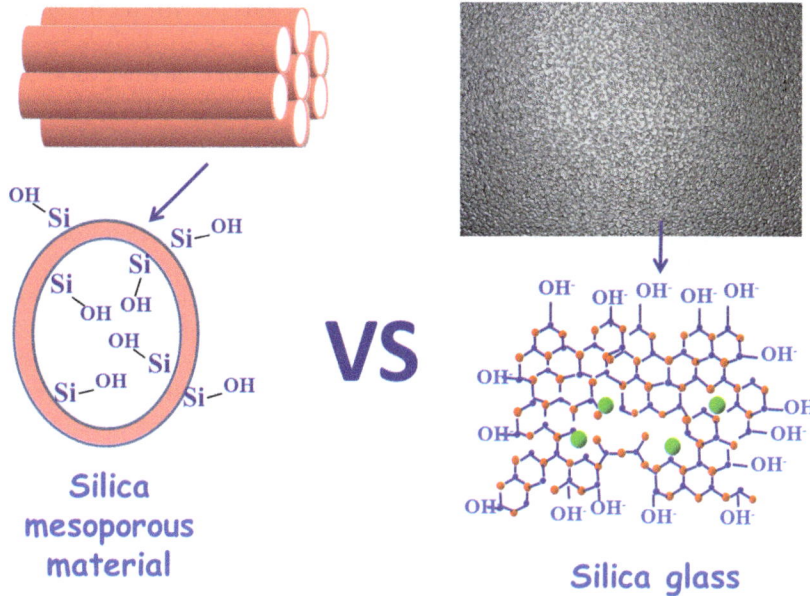

Figure 5.7   The presence of silanol groups both in silica mesoporous materials and in silica glass.

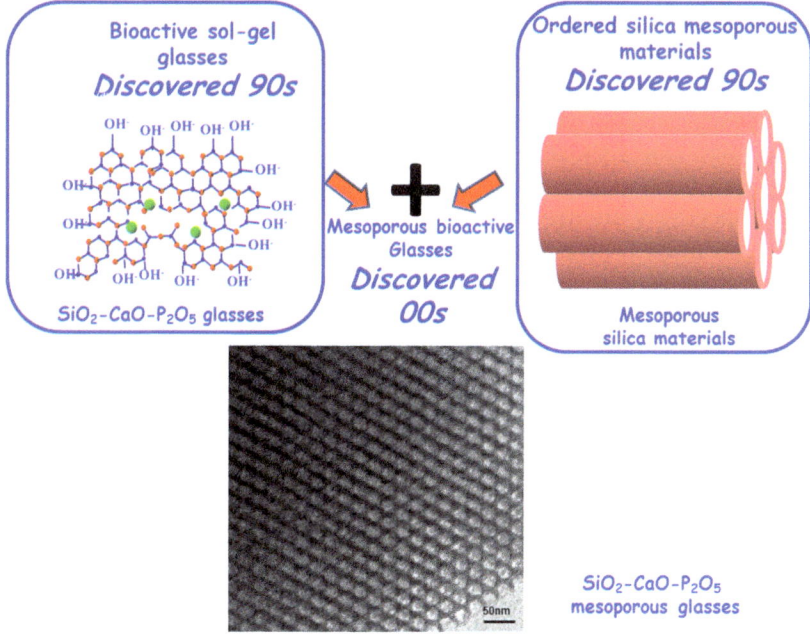

Figure 5.8   Bioactive sol–gel glasses and ordered silica mesoporous materials were discovered in the 1990s. Work on mesoporous bioactive glasses started in the 2000s.

of reaction kinetics and these new mesoporous glasses will be much more reactive.[28]

Synthesis of MBGs was successful thanks to the use of the evaporation-induced self-assembly (EISA) process.[29] The self-assembly starts with a homogeneous solution of glass precursors and surfactant prepared in ethanol/water with an initial concentration, c0 ≪ cmc. System concentration is gradually increased by the evaporation of ethanol, which drives the self-assembly of inorganic-surfactant micelles and further organization into a liquid crystalline mesophase.

The EISA process is used for the following two important reasons derived from the combination of sol–gel and supramolecular chemistries: (1) during the synthesis of SiO_2–CaO–P_2O_5 sol–gel glass, the aging and drying of the gel requires temperatures 100 °C,[30] which are incompatible with an ordered micellar phase; and (2) the classical hydrothermal preparation of SMM is not compatible with a multicomponent system, where the presence of CaO could be an obstacle for the interactions between silica and surfactant. Therefore, the EISA method uses a much diluted precursor solution by using a volatile solvent such as ethanol, which is gradually evaporated at room temperature until cmc is reached, leading to self-assembly of the micelles arranged in more robust stages in terms of structural stability.[31,32]

Through this synthesis process the textural and structural characteristics can be tailored, offering numerous advantages in the final features of these materials, including the bioactive response. As pure silica mesoporous materials (SMMs), the ordered mesoporous arrangement is mainly guided by different parameters such as surfactant nature and concentration, solvent, additives, pH, temperature, *etc.*, which determine the final structural and textural parameters. Thus, structures such as a 2D hexagonal arrangement with *p6mm* planar groups and cage-type 3D-cubic structures with *Im3m* space groups can be obtained if the surfactants Pluronic® P123 and Pluronic® F127, respectively, are used.[33,34]

In this sense, 2D hexagonal structures are formed by unidirectional open channels in both directions, which are distributed in a hexagonal arrangement. 3D cubic cage-type structures are formed by a body-centered cubic structure interconnected to form a multidirectional mesoporous network.[35]

Furthermore, it is possible to modulate the structure of MBGs from a 2D hexagonal structure to a 3D bi-continuous cubic structure with *Ia3d* space groups by using the same surfactant and varying the CaO amount in the SiO–CaO–P_2O_5 system.[27,36,37]

The possibility to tailor the pore structure and textural properties at nanometric level allows an exhaustive control over their final properties as bioactivity/reactivity and capability to load and control release of biologically active molecules. In this sense the bioactive process where a material reacts with living tissue is a phenomenon that begins at the interface between the implant surface and the bone. The amount of matter that can be disseminated and exchanged between implant and bone, determines the kinetics that take place between both surfaces. This diffusion of material is higher in

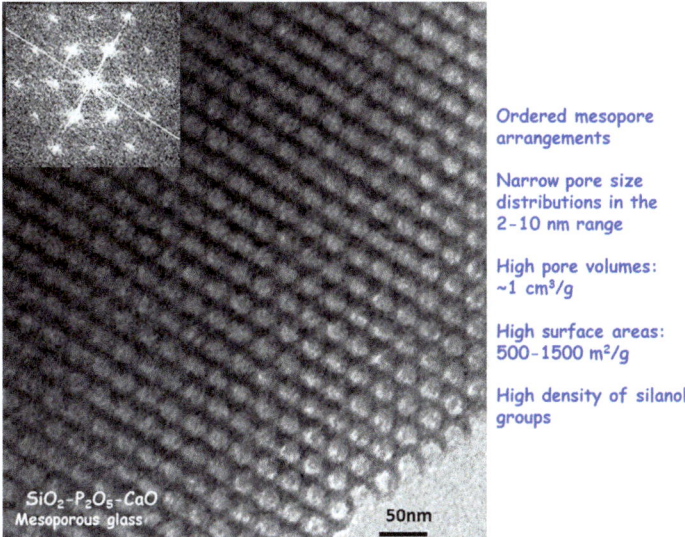

Ordered mesopore arrangements

Narrow pore size distributions in the 2-10 nm range

High pore volumes: ~1 cm³/g

High surface areas: 500-1500 m²/g

High density of silanol groups

SiO_2-P_2O_5-CaO Mesoporous glass

50nm

Figure 5.9 Transmission electron microscope image and essential features of silica mesoporous materials.

3D-cubic structures compared to 2D-hexagonal structures. This turned out to be one of the most notable differences between conventional sol–gel glasses and MBGs. In the first case, the greater amount of CaO yielded better bioactive behavior. MBGs, in contrast, offered better texture and structure parameters that determined a faster bioactive process under *in vitro* conditions.[38]

Concerning the *in vitro* bioactivity, MBGs exhibit better bioactivity than conventional glasses (melt and sol–gel glasses) due to their outstanding values of surface area and porosity.[39,40] While melt-prepared glasses exhibit bioactivity after 7 days of soaking in a simulated body fluid (SBF), sol–gel glasses exhibit bioactivity after 3 days and mesoporous bioactive glasses at only 4 hours. There is current controversy regarding which factors contribute to this great improvement in bioactivity. Although, in part, it is believed that the composition of MBGs plays a predominant role, similar to the findings on melt and sol–gel glasses, the enhanced textural properties are acknowledged as the dominating contribution to the superior *in vitro* bioactivity, which has been attributed mainly to the highly ordered arrangement of uniform-sized mesopores, as illustrated in Figure 5.9.

5.4 Structural Features

Synthesis conditions play an important role in the mesoscale order, which can show local structural variations. The use of electron crystallography has been a major breakthrough in the analysis of those structural properties and porosity characteristics. In this sense, X-ray diffraction (XRD) powder profiles of mesoporous materials normally show a few broad peaks, which

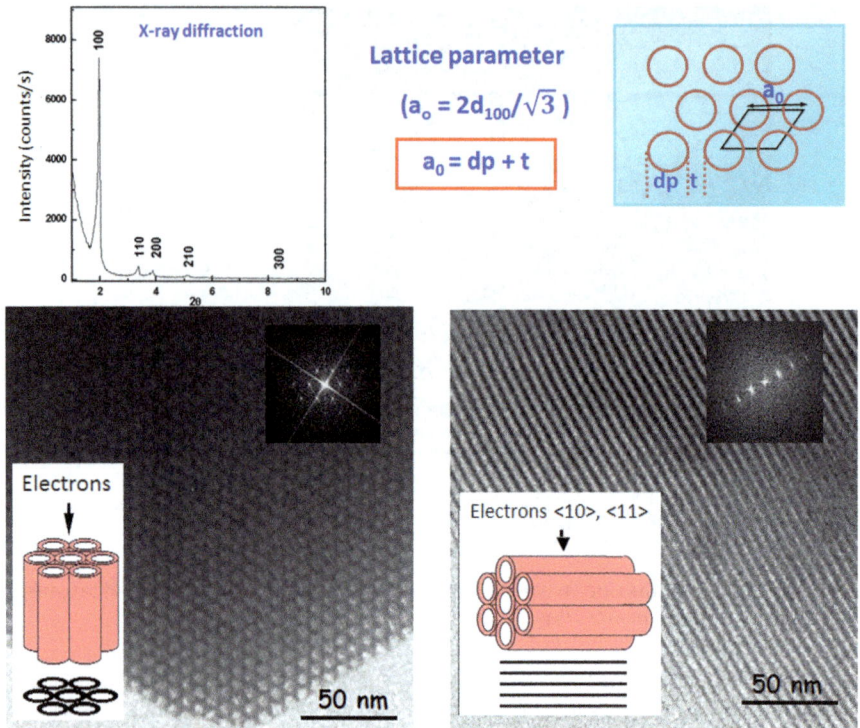

Figure 5.10   Transmission electron micrographs showing the pores (left) and pore walls (right) of a mesoporous material. Each micrograph includes a diagram (inset) for clarification purposes. The low angle X-ray diffraction scan (above, left) reveals the mesopore order.

makes structural characterization employing only XRD very difficult. Taking into account that electrons interact with materials much more strongly than X-rays, and electron scattering amplitudes are ~104 times larger than those of X-rays, it is possible to obtain the same structural information using electrons from a crystal volume 10^{-8} smaller than with X-rays. It is for this reason that crystals in the order of tens of nanometers are sufficiently large for single crystal TEM observation.

Figure 5.10 exhibits two TEM images with incidences parallel and perpendicular to the channel, as evidence of *p6mm* symmetry and a monodimensional channel system.

In 1992, Kresge *et al.* proposed the combination of electron microscopy observations with powder XRD results to solve the structure of the first mesoporous material discovered, MCM-41.[2] This work clearly evidences the importance of TEM studies in the structural characterization of mesoporous materials. However, there are many publications in this area with an unfortunate mix of speculative structures and structure solutions. In this sense, the work carried out by Terasaki and co-workers is extremely valuable because it shows the way that structural characterization should be undertaken.[4,41–44] His experience in structural characterization techniques helped him to combine high resolution TEM with electron crystallography, leading to the solution of the structure of a considerable number of mesoporous materials. Thus, the exhaustive characterization work carried out by Terasaki's research team was the key to settle the relationship between mesostructure and material properties, opening the gates to novel and exciting applications for those mesoporous materials.

Even though TEM analysis is a very powerful technique, it should be recalled that a TEM image is essentially projected structural information of a given specimen along the direction of the incident electrons. Therefore, in order to obtain images of 3D structures, it is necessary to combine images along different projections.[45,46] The combination of thousands of projected images through "filtered back-projection" allows the determination of the structure of a 3D object. This technique, called tomography, is widely employed in medical imaging and is also very useful to study certain nanostructured materials.

However, when the analyzed material is a crystalline periodic system, it is possible to employ crystallography rather than tomography. Thus, thanks to crystallography it is possible to dramatically reduce the number of images required. This way, the higher the crystal symmetry the fewer is the number of images necessary. In addition, it is also possible to reduce the S/N ratio because all the information is concentrated only on reciprocal points. The information on the periodically averaged structure can be collected over the region of high-resolution TEM micrographs where the Fourier diffractogram is obtained.

5.5 Potential Modifications to the Mesostructure

In silica mesoporous materials it is possible to modify the pore diameter, the chemical composition of both the inner and outer pore surface as well as to include different elements onto the walls that form the scaffold.

The easiest modification is to change the pore size. The only requirement is to choose surfactant micelles with a different length of carbonated chain. Different choices allow the modulation of the pore size.[47] Self-assembly techniques in the synthesis of mesoporous silica using surfactants enable the increase in pore diameter using either longer surfactants or swelling agents.

It is important to pay special attention to the qualities of the inorganic precursor. Metal alcoxides, for instance, are optimum reactants in the synthesis of many porous solids, for the following reasons: they can be obtained with high purity, their slow hydrolysis favors interactions with templates and surfactants and there is a wide range of commercially available types. It is also possible to use metal alcoxides substituted with groups of interest for the subsequent functionalization of surfaces, such as organic trialcoxides, alcoxilanes in particular. When added to the synthesis mixture, the obtained mesoporous silica can be populated on its surface with various functional groups. In this case, surfactant removal has to be performed by extraction instead of calcination to preserve the free functional groups on the silica

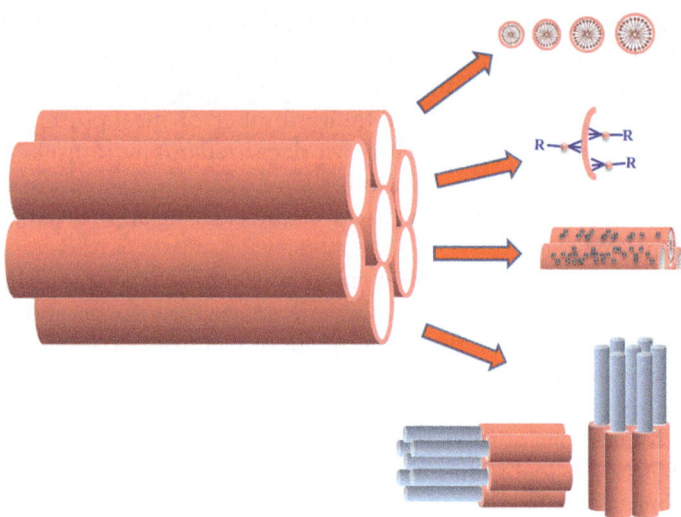

Figure 5.11 A mesoporous material and its potential modifications in terms of pore size, surface functionalization, ion entrapment and/or production of nanofibers inside the pores.

surface. An alternative route would be to functionalize the solid surface after its calcinations to remove the template. Both methods are valid. In this way, the chemical composition of both surfaces of the scaffold can be modified according to the desired purpose of the final material, achieving particular properties previously designed for a particular application.

These materials can also be used as templates in nanowire fabrication. Therefore, a material formed using micelles as templates is in turn used as template for other applications, such as the production of platinum wires, or nanowires made of other metals (Figure 5.11).

One of the reasons behind the predominant use of silicon in the synthesis of controlled porosity materials is the slow hydrolysis of silica alcoxides, enabling the desired interaction with surfactant molecules. Besides, some metal oxides become too unstable during the surfactant removal process and the structure collapses. Isomorphic partial replacement of silicon atoms in the structure with other metal atoms is a valid alternative; it allows combine of the stability and porosity of mesoporous silica with the chemical properties of the replacement phase in the structure.

5.6 Significance in the Clinical Field

As mentioned above, ordered mesoporous materials were initially developed for catalysis applications. However, many researchers quickly recognized their potential for other applications in many different research areas, such as magnetism, sensors, optical materials, photo catalysis, fuel cells, thermo electrics and even in the area of healthcare research.

In 2001, our research group proposed for the first time the use of these ordered mesoporous materials as drug delivery systems.[48] The essential idea was to load pharmaceutical agents into the mesoporous cavities to then be released as required in the body. Obviously, the lack of toxicity is a must for this type of application.

Silica mesoporous materials exhibit two outstanding properties. The first is their surface: they are formed by a silica network with silanol groups on the external surface. The second property is based on their textural properties, especially in their ordered mesoporosity. Both features make them particularly interesting as bioceramics for bone tissue regeneration and drug delivery systems. Figure 5.12 shows that the presence of silanol groups enables nanoapatite formation in a similar fashion as for bioglasses, and that the ordered mesostructure accepts the inclusion of different pharmaceutical agents. All these questions are discussed in detail in Chapter 6; let us just mention that silica-based ordered mesoporous materials have experienced a growing interest from biotechnological researchers due to their potential to host very different guest molecules. An important point within this technology is the host–guest interaction that would occur between the silanol groups located at the surface of the host matrices and the functional groups from the guest molecules. This interaction would have a strong effect on the drug adsorption and release properties of the carrier matrices.[49,50]

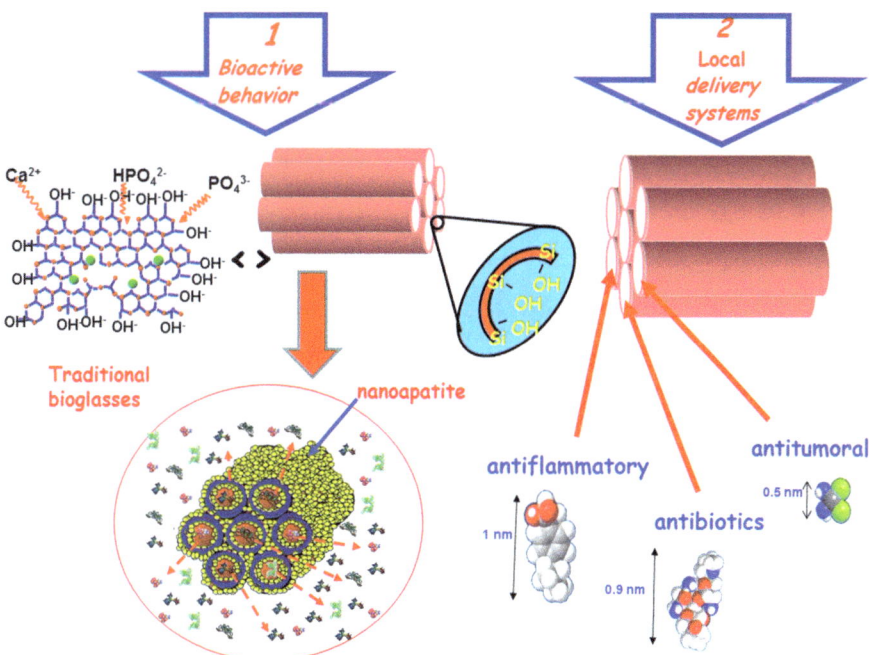

Figure 5.12 Mesoporous materials can be used in the medical field due to their bioactive behavior and their capacity as local delivery systems.

Additionally, textural and structural properties have been observed to modulate the adsorption and release characteristics of these ordered meso-porous materials.[51]

When the aim is to use these ordered mesoporous materials for drug delivery, the first and perhaps most important condition is the correct choice of the mesoporous material from all those available, depending on the molecule to be hosted.[52] In this sense, when dealing with molecules with a size of few nanometers, MCM 41 silica materials seem to be the correct choice since they are stable, with pores of ~2–3 nm, large surface area and are very easy to functionalize to modulate the adsorption and release rates.

A major breakthrough in this technology emerged from the possibility of visualizing the drug molecule inside the mesopores. Previously, it was necessary to use several indirect characterization techniques to verify the drug load into the inner part of the mesopores. The use of aberration-corrected high-resolution electron microscopy by Vallet-Regí *et al.* enabled the detection of the drug molecules confined in the pore channels of silica-based ordered mesoporous materials.[53] This work was possible using Cs correctors incorporated to a scanning TEM microscope, allowing the illumination of an individual atom with the electron beam to identify an unknown substance, and therefore ensuring that the drug molecule was inside the mesopore channel, as depicted in Figure 5.13.

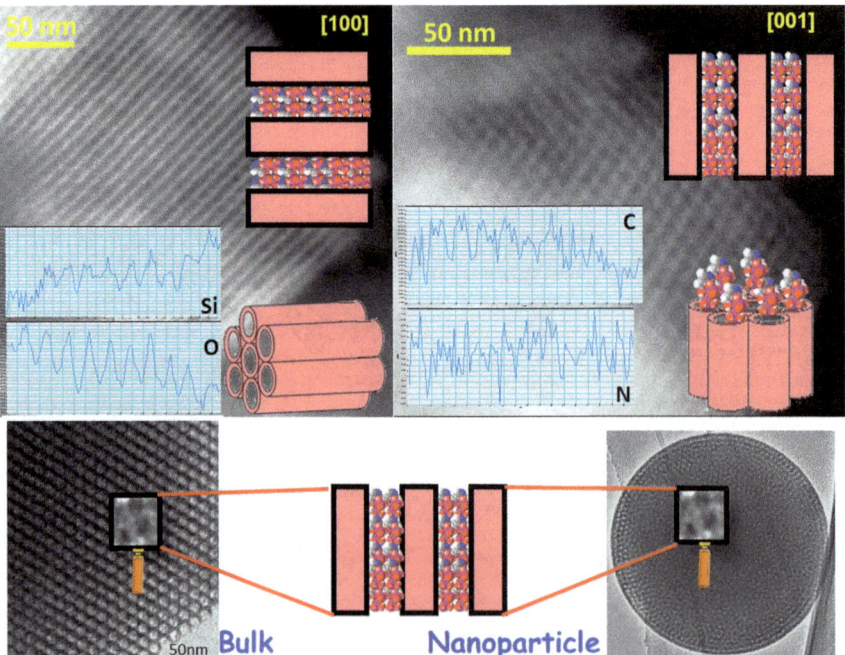

Figure 5.13   Mesoporous materials, both in bulk and in nanoparticle form, can host drugs. Electron energy loss spectroscopy helps to determine the species inside the material and also the composition of the wall.

However, to calculate the amount of adsorbed drug, it is necessary to combine different indirect methods such as porosity, specific surface area measurements, thermogravimetry and chemical analysis.

The release of the drug from the mesoporous matrix would take place through diffusion of the pharmaceutical molecule throughout the pore channels. However, when targeting a truly controlled release, it is necessary to design stimulus-responsive systems, where the drug is released under certain stimuli, such as pH, temperature, ultrasound or light.

An important feature of these silica-based ordered mesoporous materials is the possibility of organically modifying their surface through the covalent attachment of many different functional groups.[54] As a consequence of the functionalization, organic–inorganic hybrid materials are obtained, which can adsorb many different drug molecules through weak interactions.

The functionalization of the silica walls may be necessary for several reasons. In some cases, there are certain drugs with remarkable hydrophobic nature that would not penetrate into the hydrophilic mesoporous silica.[55] The functionalization with hydrophobic functional groups is a good alternative to enable the load of different hydrophobic drugs. This strategy is also employed to delay the release kinetics of certain drugs from the mesoporous channels to the aqueous release medium due to the decrease in the wettability degree of the material surface.[56]

There are other situations in which the pharmaceutical molecule can be confined into the mesoporous channels. However, higher loads and slower release kinetics can be achieved if the mesoporous silica wall is functionalized with different functional groups. Among the existing organic groups, the functionalization with amino moieties has been widely reported.[57]

Certainly, it is unlikely that the Kuroda and Kresge teams would have imagined that their discovery, devoted to catalytic applications, could be later used as an excellent tool in medicine and nanomedicine. Nowadays, magnetic nanoparticles have been encapsulated into inorganic mesoporous nanoparticles of silica for applications against cancer, designing smart nanoparticles able to simultaneous and synergically perform a double function, to deliver in a controlled way cytotoxic drugs and to give off heat (hyperthermia). The stimulus-responsive effect can be applied for on-demand release of the drug.[58]

The presence of silanol groups on the surface of silica facilitates the reaction with alcohols and organosilanes, stabilizing the suspensions in non-aqueous media, as well as providing groups for the attachment of specific ligands. Gene transport and delivery by magnetic conjugates represents a particular challenge. DNA fragments have to overcome several biological barriers when reaching the cell nucleus from the extracellular environment.[59]

The covalent bonding of dendrimer molecules to magnetic nanoparticles has been revealed as a promising strategy to perform magnetic force-assisted transfection *in vitro*. In this kind of system, the role of the magnetic component is to facilitate an intimate contact between the cell culture and gene vectors, thus providing a reduction of transfection times.

The development of novel advanced multifunctional materials for a broad range of technological applications brings renewed hope in many different fields. In particular, recent research breakthroughs in the biomedical arena have emerged as the basis for future personalized treatments and diagnostic techniques with a hitherto unsuspected selectivity.

With the amazing advances in the preparation and characterization techniques of nanotechnology products, the possibility of manufacturing devices capable of establishing an intimate interaction with the biological world has been opened. This represents a precise control over the processes of therapeutic substances release, and means an opportunity to improve the specificity of the therapeutic action, as well as to reconsider some of the promising drugs for certain diseases that were once discarded because of their low levels of tolerance.

The targeting ability of these new nanodevices should lead to tailor-made dosing regimens, with a significant reduction of severe side-effects associated with some diseases, such as cancer, and should also eventually result in a more efficient allocation of healthcare resources. Moreover, the use of these engineered products may allow the combination of the therapeutic potential and nanoscale diagnosis with low tissue invasiveness. In many of these cases, silica mesoporous nanoparticles are being used as a consequence of the discovery made in the early 1990s, even though the discoverers were probably unaware of these potential applications, since they were involved in the world of catalysts. Nonetheless, they were able to establish the basis to develop new synthetic pathways to be used in many different research areas, among them the medical field.

Chemistry currently offers many different possibilities of producing silica mesoporous materials, with porosity 2–50 nm. Although the initial synthesis was carried out in bulk materials, nowadays it is possible to produce mesoporous nanoparticles with a similar composition.[60]

As discussed earlier, in the specific field of silica mesoporous materials we can induce weak interactions between this ceramic matrix and different drugs in order to obtain efficient controlled drug delivery systems. Conversely, if we induce strong interactions between the organic matrix and biologically active molecules that enable bone tissue regeneration, we will produce efficient systems for hard tissue regeneration. Simultaneously, we can combine weak and strong interactions in the same system, which would then be suitable for both purposes: drug delivery and tissue engineering.[55]

Concerning hybrid bioceramics, the combination of both inorganic and organic constituents produces a remarkable synergy that makes them appropriate for diverse medical applications.[61,62] The weak interaction of hybrid bioceramics with drug molecules hosted in the mesoporous cavities enables the design of controlled delivery systems. In addition, mesoporous silica can be used as starting material for the manufacture of 3D scaffolds for bone tissue engineering.[50,63–66] To attain a satisfactory biological response, different osteoinductive molecules, such as peptides, hormones and growth factors should be strongly attached to the bioceramic surface, acting as signaling agents for cells to assist the bone regeneration process.[60,67,68] In addition, there are many research examples regarding hybrid bioceramics as stimuli-responsive drug delivery systems and nanosystems for cancer cell targeting and gene transfection.[11,55,59,69–79]

Any multifunctional nanoplatform suitable for nanomedicine applications must satisfy strict specifications: the biocompatibility of the nanocarrier material is a mandatory issue; the nanocarriers must provide high loading and safeguard capability of desired therapeutic or imaging agents; the nanosystem must avoid premature departure of entrapped drugs before reaching its target, while exhibiting cell-type or tissue specificity and site targeting capability, efficient cellular uptake, effective endosomal escape and tunable delivery rate to accomplish effective local concentration. Many different types of materials are being used as drug nanocarriers, as shown in Figure 5.14. Here, we focus on mesoporous silica nanospheres.

Multifunctional nanosystems that have been widely investigated for applications in nanomedicine are either organic or inorganic materials. Among the organic nanoparticles we can find dendrimers,[80] liposomes,[81,82] polymers[83,84] and virus-like particles.[85]

Inorganic nanosystems have been also widely explored, including gold nanoparticles,[86] semiconductor nanocrystals,[87] superparamagnetic nanoparticles[88] and silicon-[89] and silica-based nanoparticles.[90,91] Currently, inorganic

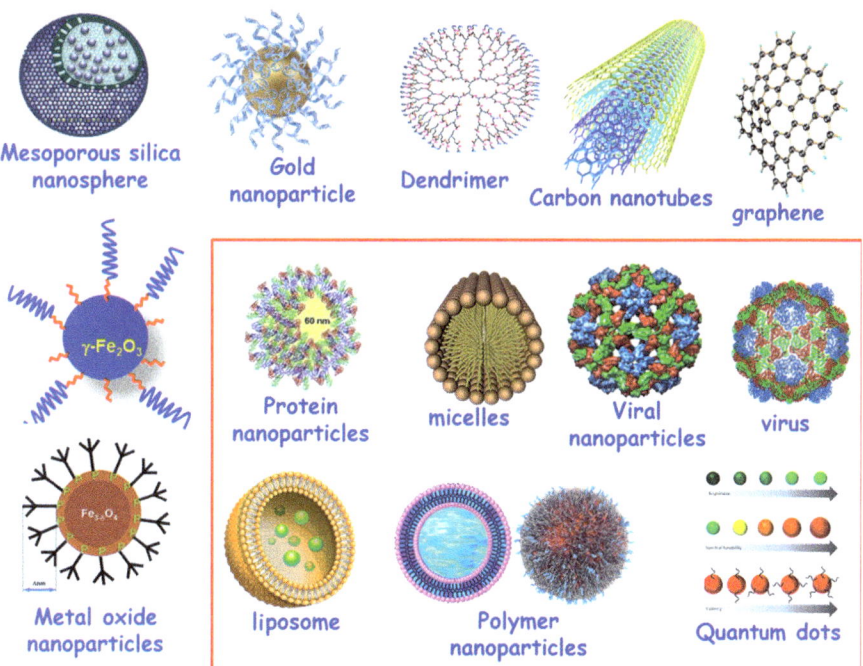

Figure 5.14 Different types of nanocarriers. All those depicted outside the box are inorganic.

nanoparticles are gaining increasing significance since they exhibit higher mechanical strength, chemical stability, biocompatibility and resistance to microbial attack than their organic analogs.[92,93] Moreover, the inorganic matrix efficiently preserves the chemical nature of entrapped molecules by protecting them from degradation or denaturation provoked by enzymatic attack or pH or temperature variations in the local environment.

In the framework of inorganic nanomaterials, mesoporous silica nanoparticles (MSNPs) are perhaps the most promising multifunctional platforms for nanomedicine. Since porosity and surface play a fundamental role in some biomedical applications,[94] MSNPs are receiving increasing research attention for their forthcoming applications in biotechnology and nanomedicine.[11,55,59,72–74,76–78,95–105]

MSNPs suitable for nanomedicine applications require synthesis methods that produce uniform nanoparticle sizes in the 30–300 nm range and minimize the self-aggregation of nanoparticles. Accordingly, the two main strategies used to synthesize MSNPs are the so-called *modified Stöber method* and *aerosol-assisted synthesis*. The former consists of silica condensation under a basic medium using cationic surfactants as template agents. Aggregation of nanoparticles is reduced by the alkaline and highly diluted conditions, which favor negatively charged and more fully-condensed surfaces.[106] Aerosol-assisted synthesis produces MSNPs by using not only cationic but also anionic and even non-ionic surfactants under basic or acid conditions.[107–111] Regardless of the synthesis route taken, the last stage consists of elimination of the surfactant, which usually leads to materials with cylindrical mesopores organised in a 2D hexagonal way, typical of MCM-41-type materials.[2,112] MSNPs present singular and valuable textural and structural characteristics, such as high surface area ($\sim$1000 m^2 g^{-1}) and pore volume ($\sim$1 cm^3 g^{-1}), stable mesostructure, narrow pore size distributions that can be tuned in the 2–10 nm range, two functional surfaces (outer particle and inner pore surfaces) and tunable particle size. When aiming at targeted intracellular delivery MSNP size must be within the 50–300 nm range. The synthesis of MSNPs with sizes <50 nm is not easily achievable due to the inherent mesoporosity of nanoparticles, whereas nanoparticles >300 nm cannot easily cross physical membranes in the body. The particle size has been proved to be one of the key factors that dictate the final fate of nanoparticles in living systems.[113–115] For this reason, narrow particle size distributions are desired and therefore, wet-chemical synthesis routes are normally preferred to other physical methods such as spray-drying, which normally lead to somewhat broad particle size distributions.

The high density of silanol groups on the surface, which can be functionalized with a wide array of entities, is undoubtedly one of the great milestones of MSNP research.[116–118] When the aim is to fulfill nanomedicine applications, the surface functional groups of MSNPs can play diverse roles, such as tailoring the surface charge of nanoparticles, allowing the grafting of functional molecules inside and outside the pores, or facilitating the capping of the nanopore openings to prevent premature release of entrapped drugs.

Functionalization of MSNPs is mainly achieved using three distinctive methods: co-condensation, post-synthesis and surfactant displacement. The co-condensation method involves the synthesis and functionalization in a unique step, *i.e.* the organosilanes are added during the synthesis stage together with the silica source.[107,109,118-123] The surfactant removal is carried out by ion exchange using either an ethanol solution of ammonium nitrate[124] or hydrochloric acid.[125,126] This straightforward synthesis method produces MSNPs displaying uniform distribution of functional groups. Nonetheless, depending on the solvent used, the surfactant extraction is sometimes incomplete, which can be a serious issue for any further application of the resulting nanoparticles in medicine. The post-synthesis or grafting method consists of grafting the functional groups after the surfactant removal, either by extraction or calcination, has been completed.[127] This method allows the allocation of diverse functional groups in MSNPs and also to chemically attach delicate organic functions prone to hydrolysis and elimination reactions.[128] One of the major issues of the post-synthesis method is the possible blocking of the mesopore openings by the functional groups, which would lead unavoidably to heterogeneous functionalization of the mesoporous matrix.[129] Finally, the surfactant displacement method consists of direct surface silylation with simultaneous surfactant removal using acidic alcohol as a solvent.[130,131] This method produces a homogeneous monolayer coating, where the amount of functional groups on the mesoporous silica surface can be precisely tuned.

It should be highlighted that the exclusive topology of MSNPs allows three well-defined domains to be distinguished that can be independently functionalized: the silica framework, the mesopores and the outermost surface of nanoparticles. Due to the remarkable features of MSNPs they are excellent nanoplatforms to integrate multiple functionalities for therapy and diagnosis of different pathologies, as represented in Figure 5.15.[11,59,77,78,105,132-137]

Therefore, different molecules can be incorporated into the silica matrix acting as contrast agents for optical imaging and magnetic resonance imaging (MRI). It is also feasible to entrap magnetic nanoparticles within the silica framework, which would play a dual role: contrast agents for MRI and thermoseeds for the treatment of tumors by hyperthermia. In addition, mesoporous channels can accommodate diverse organic molecules, such as drugs, proteins, nucleic acids for gene therapy or photosensitizers for photodynamic therapy. These molecules can be simply adsorbed or covalently grafted to the inner part of the mesopore walls. A ground-breaking approach consists of providing MSNPs of stimuli-responsive controlled delivery capability. To tackle this goal, molecular nanogates can be covalently anchored to the pore outlets, obstructing the mesopore entrances and therefore avoiding the premature release of entrapped cargo. A given stimulus would increase the aperture of the nanogates and trigger the release of molecules into the target cell or tissue. To optimize the performance of MSNPs for diverse nanomedicine applications, the outer surface of nanoparticles can be decorated with targeting agents for specific cellular uptake, functional groups

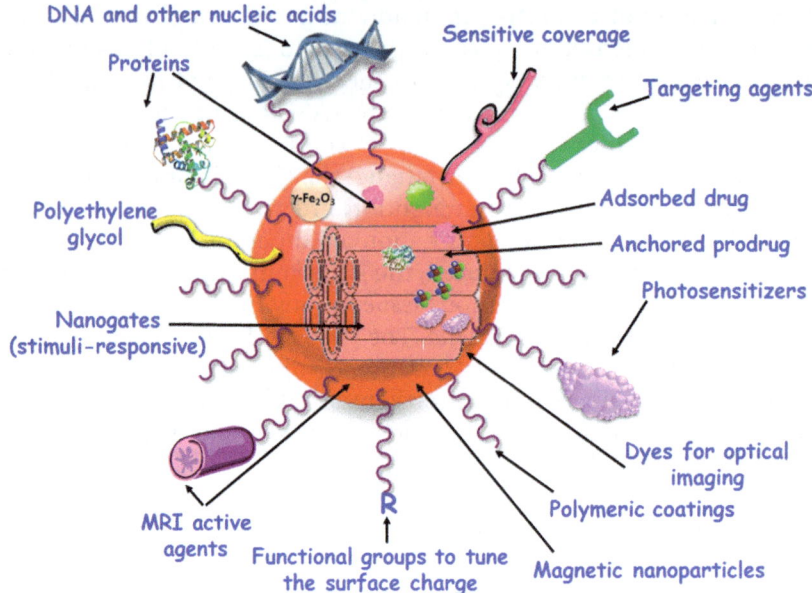

DNA and other nucleic acids

Sensitive coverage

Proteins

Targeting agents

Polyethylene glycol

Adsorbed drug

Anchored prodrug

Photosensitizers

Nanogates (stimuli-responsive)

Dyes for optical imaging

MRI active agents

Polymeric coatings

R

Functional groups to tune the surface charge

Magnetic nanoparticles

Figure 5.15   Multiple platforms can be designed from a silica mesoporous nanoparticle.

to modulate surface charge, polymeric coatings such as polyethyleneglycol to confer "stealth" properties, or stimuli-responsive polymers, MRI contrast agents, photosensitizers, nucleic acids, proteins, *etc.* The high versatility of MSNPs allows the design of multifunctional nanoplatforms for "theranostic", *i.e.* therapeutic and diagnostic nanomedicine.

5.7   From Bulk to Nano: Potential Use in Nanomedicine

As previously discussed in Section 5.6, all the analysis and descriptions for bulk mesoporous materials are also applicable to mesoporous nanoparticles. The only requirement is to find synthesis pathways that can preserve the inner ordered mesoporosity while producing the required morphology and size for the scaffold.

This step expands even more the field of applications and, in this particular case, moves from the clinical field to the particular realm of nanomedicine. Stimuli-responsive systems may be the solution for plenty of issues in the field of controlled drug delivery, ensuring that the release is only verified upon arrival at the final destination and, once there, release can be efficiently controlled. This is a very attractive feature in oncology: the use of cytotoxic species and the efforts to reduce their dosage and constrain their administration to the tumor area could be met using nanocarriers functionalized with selective chemical species akin to the tumor mass, preserving the confinement of the cytotoxic load until arrival at the tumor. Any potential generation of toxicity must be carefully studied beforehand.

Cytotoxicity generally increases as a function of the uptake grade and this grade is strongly size-dependent. Thus, particles with size >1 μm are slowly internalized by human cells and therefore present low toxicity. Particles around 50–100 nm are internalized in short periods of time (30 min) and can produce significant toxicity when high concentrations of particles are used.[138] The silanol groups present on the surface of mesoporous silica nanoparticles can interact with the lipids of the cellular membranes and some proteins altering their properties. Lin and co-workers studied in detail the interaction of MSNPs with the membraned of human red blood cells showing that MCM-41-type particles barely alter the membrane morphology while SBA-15 induces strong membrane deformations.[139] The degradability of mesoporous silica nanoparticles has been studied using SBF. The process takes place in three steps: fast bulk silica degradation (hour scale), followed by the deposition of magnesium/calcium silicate on the particle surface and finally, slow dissolution of the particles (day scale).[140] Tamanoi and co-workers studied the maximum safe dose for drug delivery applications using murine models, showing that the animals can tolerate without significant alterations doses up to 50 mg kg^{-1}.[141]

Calcium phosphate nanoparticles are usually very biocompatible due to their natural presence in the body, both dissolved and in solid state. In the form of carbonated hydroxyapatite they forms the structure of bones, teeth and tendons. The biological forms of these materials are usually nanocrystals that are precipitated under physiological conditions.[142] They dissolve into their ionic components Ca^{2+} and PO_4^{3-} which are naturally present in the body in concentrations ranging from 1 to 5 mM. One possible harmful effect is that their degradation increases the intracellular Ca^{2+}, concentration which could initiate protein aggregation.

References

1. T. Yanagisawa, T. Shimizu, K. Kuroda and C. Kato, *Bull. Chem. Soc. Jpn.*, 1990, **63**(4), 988.
2. C. T. Kresge, M. E. Leonowicz, W. J. Roth, J. C. Vartuli and J. S. Beck, *Nature*, 1992, **359**(6397), 710.
3. J. S. Beck, J. C. Vartuli, W. J. Roth, M. E. Leonowicz, C. T. Kresge, K. D. Schmitt, C. T. W. Chu, D. H. Olson, E. W. Sheppard, S. B. McCullen, J. B. Higgins and J. L. Schlenker, *J. Am. Chem. Soc.*, 1992, **114**(27), 10834.
4. S. Inagaki, S. Guan, T. Ohsuna and O. Terasaki, *Nature*, 2002, **416**(6878), 304.
5. Y. Sakamoto, M. Kaneda, O. Terasaki, D. Y. Zhao, J. M. Kim, G. Stucky, H. J. Shim and R. Ryoo, *Nature*, 2000, **408**(6811), 449.
6. S. Che, Z. Liu, T. Ohsuna, K. Sakamoto, O. Terasaki and T. Tatsumi, *Nature*, 2004, **429**(6989), 281.

7. S. Guan, S. Inagaki, T. Ohsuna and O. Terasaki, *J. Am. Chem. Soc.*, 2000, **122**(23), 5660.
8. H. X. Deng, S. Grunder, K. E. Cordova, C. Valente, H. Furukawa, M. Hmadeh, F. Gandara, A. C. Whalley, Z. Liu, S. Asahina, H. Kazumori, M. O'Keeffe, O. Terasaki, J. F. Stoddart and O. M. Yaghi, *Science*, 2012, **336**(6084), 1018.
9. M. E. Davis, *Nature*, 2002, **417**(6891), 813.
10. A. Corma, *Chem. Rev.*, 1997, **97**(6), 2373.
11. M. Vallet-Regi, F. Balas and D. Arcos, *Angew. Chem., Int. Ed.*, 2007, **46**(40), 7548.
12. A. Stein, *Adv. Mater.*, 2003, **15**(10), 763.
13. M. Vallet-Regí, I. Izquierdo-Barba and M. Colilla, *Philos. Trans. R. Soc., A*, 2012, **370**, 1400.
14. M. Vallet-Regí, Ceramics For Medical Applications, *J. Chem. Soc., Dalton Trans.*, 2001, 97–108, (portada de la revista).
15. M. Colilla, B. González and M. Vallet-Regí, *Biomater. Sci.*, 2013, **1**, 114.
16. J. S. Beck, J. C. Vartuli, W. J. Roth, *et al.*, A new family of mesoporous molecular sieves prepared with liquid crystal templates, *J. Am. Chem. Soc.*, 1992, **114**(27), 10834–10843.
17. Y. Sakamoto, M. Kaneda, O. Terasaki, *et al.*, Direct imaging of the pores and cages of three-dimensional mesoporous materials, *Nature*, 2000, **408**(6811), 449–453.
18. M. Manzano and M. Vallet-Regi, *Prog. Solid State Chem.*, 2012, **40**, 17.
19. M. Vallet Regí, M. Manzano and M. Colilla, *Biomedical Applications of Mesoporous Ceramics*, CRC Press, Taylor & Francis, Boca Raton, 2013.
20. M. Vallet Regí, *Bio-Ceramics with Clinical Applications*, John Wiley and Sons, India, 2014.
21. M. Vallet-Regí, J. C. Doadrio, A. L. Doadrio, I. Izquierdo-Barba and J. Pérez-Pariente, *Solid State Ionics*, 2004, **172**, 435.
22. D. Y. Zhao, J. L. Feng, Q. S. Huo, N. Melosh, G. H. Fredrickson, B. F. Chmelka and G. D. Stucky, *Science*, 1998, **279**(5350), 548.
23. B. González, M. Colilla, C. López de Laorden and M. Vallet-Regí, *J. Mater. Chem.*, 2009, **19**, 9012.
24. B. González, E. Ruiz, M. J. Feito, C. López, D. Arcos, C. Ramírez, C. Matesanz, M. T. Portolés and M. Vallet-Regí, *J. Mater. Chem.*, 2011, **21**(12), 4598.
25. S. Sánchez-Salcedo, M. Colilla, I. Izquierdo and M. Vallet-Regí, *J. Mater. Chem. B*, 2013, **1**, 1595.
26. X. X. Yan, C. Z. Yu, X. F. Zhou, J. W. Tang and D. Y. Zhao, *Angew. Chem., Int. Ed.*, 2004, **43**, 5980.
27. A. López-Noriega, D. Arcos, I. Izquierdo-Barba, Y. Sakamoto, O. Terasaki and M. Vallet-Regí, *Chem. Mater.*, 2006, **18**, 3137.
28. I. Izquierdo, A. J. Salinas and M. Vallet-Regí, *Int. J. Appl. Glass Sci.*, 2013, **4**, 149.
29. C. J. Brinker, Y. F. Lu, A. Sellinger and H. Y. Fan, *Adv. Mater.*, 1999, **11**, 579.

30. J. Zhong and D. C. Greenspan, *J. Biomed. Mater. Res.*, 2000, **53**, 694.
31. X. X. Yan, H. X. Deng, X. H. Huang, G. Q. Lu, S. Z. Qiao, D. Y. Zhao and C. Z. Yu, *J. Non-Cryst. Solids*, 2005, **351**, 3209.
32. X. X. Yan, X. H. Huang, C. Z. Yu, H. X. Deng, Y. Wang, A. D. Zhang, S. Z. Qiao, G. Q. Lu and D. Y. Zhao, *Biomaterials*, 2006, **27**, 3396.
33. H. S. Yun, S. E. Kim and Y. T. Hyeon, *Mater. Lett.*, 2007, **61**, 4569.
34. H. S. Yun, S. E. Kim and Y. T. Hyeon, *Solid State Sci.*, 2008, **10**, 1083.
35. M. Kaneda, T. Tsubakiyama, A. Carlsson, Y. Sakamoto, T. Oshuna, O. Terasaki, H. Joo and R. Ryoo, *J. Phys. Chem. B*, 2002, **106**, 125.
36. J. C. Doadrio, E. M. B. Sousa, I. Izquierdo-Barba, A. L. Doadrio, J. Pérez-Pariente and M. Vallet-Regí, *J. Mater. Chem.*, 2006, **16**, 462.
37. A. García, M. Cicuéndez, I. Izquierdo-Barba, D. Arcos and M. Vallet-Regí, *Chem. Mater.*, 2009, **21**, 5474.
38. I. Izquierdo-Barba, D. Arcos, Y. Sakamoto, O. Terasaki, A. López-Noriega and M. Vallet-Regí, *Chem. Mater.*, 2008, **20**, 3191.
39. D. Arcos, I. Izquierdo-Barba and M. Vallet-Regí, *J. Mater Sci. Mater. Med.*, 2009, **20**, 447.
40. M. Vallet-Regí and F. Balas, *Open Biomed. Eng. J.*, 2008, **2**, 1.
41. Y. Sakamoto, M. Kaneda, O. Terasaki, D. Y. Zhao, J. M. Kim, G. Stucky, H. J. Shim and R. Ryoo, *Nature*, 2000, **408**(6811), 449.
42. S. Che, Z. Liu, T. Ohsuna, K. Sakamoto, O. Terasaki and T. Tatsumi, *Nature*, 2004, **429**(6989), 281.
43. S. Guan, S. Inagaki, T. Ohsuna and O. Terasaki, *J. Am. Chem. Soc.*, 2000, **122**(23), 5660.
44. H. X. Deng, S. Grunder, K. E. Cordova, C. Valente, H. Furukawa, M. Hmadeh, F. Gandara, A. C. Whalley, Z. Liu, S. Asahina, H. Kazumori, M. O'Keeffe, O. Terasaki, J. F. Stoddart and O. M. Yaghi, *Science*, 2012, **336**(6084), 1018.
45. D. J. Derosier and A. Klug, *Nature*, 1968, **217**(5124), 130.
46. R. A. Crowther, D. J. Derosier and A. Klug, *Proc. R. Soc. London, Ser. A*, 1970, **317**(1530), 319.
47. M. Vallet-Regí, *Chem.–Eur. J.*, 2006, **12**, 5934.
48. M. Vallet-Regi, A. Ramila, R. P. del Real and J. Perez-Pariente, *Chem. Mater.*, 2001, **13**(2), 308.
49. M. Manzano, M. Colilla and M. Vallet-Regi, *Expert Opin. Drug Delivery*, 2009, **6**(12), 1383.
50. M. Manzano and M. Vallet-Regi, *J. Mater. Chem.*, 2010, **20**(27), 5593.
51. F. Balas, M. Manzano, P. Horcajada and M. Vallet-Regi, *J. Am. Chem. Soc.*, 2006, **128**(25), 8116.
52. M. Vallet-Regi, *Chem.–Eur. J.*, 2006, **12**(23), 5934.
53. M. Vallet-Regi, M. Manzano, J. M. Gonzalez-Calbet and E. Okunishi, *Chem. Commun.*, 2010, **46**(17), 2956.
54. F. Hoffmann, M. Cornelius, J. Morell and M. Froba, *Angew. Chem., Int. Ed.*, 2006, **45**(20), 3216.
55. M. Vallet-Regi, M. Colilla and B. Gonzalez, *Chem. Soc. Rev.*, 2011, **40**(2), 596.

56. J. C. Doadrio, E. M. B. Sousa, I. Izquierdo-Barba, A. L. Doadrio, J. Perez-Pariente and M. Vallet-Regi, *J. Mater. Chem.*, 2006, **16**(5), 462.

57. A. Nieto, F. Balas, M. Colilla, M. Manzano and M. Vallet-Regi, *Microporous Mesoporous Mater.*, 2008, **116**(1–3), 4.

58. A. Baeza, E. Guisasola, E. Ruiz-Hernández and M. Vallet-Regí, *Chem. Mater.*, 2012, **24**(3), 517.

59. M. Vallet-Regi and E. Ruiz-Hernandez, *Adv. Mater.*, 2011, **23**(44), 5177.

60. M. Vallet-Regi, *J. Intern. Med.*, 2010, **267**(1), 22.

61. B. González, M. Colilla and M. Vallet-Regí, *Chem.–Eur. J.*, 2013, **19**, 4883.

62. M. Cicuéndez, M. Malmsten, J. C. Doadrio, Mª T. Portolés, I. Izquierdo-Barba and M. Vallet-Regí, *J. Mater. Chem. B.*, 2014, **2**, 49.

63. M. Vallet-Regí, *J. Intern. Med.*, 2010, **267**(1), 22.

64. D. Arcos, I. Izquierdo-Barba and M. Vallet-Regi, *J. Mater. Sci.: Mater. Med.*, 2009, **20**(2), 447.

65. D. Arcos and M. Vallet-Regi, *Acta Biomater.*, 2010, **6**(8), 2874.

66. M. Vallet-Regí, I. Izquierdo-Barba and M. Colilla, *Philos. Trans. R. Soc., A*, 2012, **370**(1963), 1400.

67. C. G. Trejo, D. Lozano, M. Manzano, J. C. Doadrio, A. J. Salinas, S. Dapia, E. Gomez-Barrena, M. Vallet-Regi, N. Garcia-Honduvilla, J. Bujan and P. Esbrit, *Biomaterials*, 2010, **31**, 8564.

68. F.-M. Chen, Y. An, R. Zhang and M. Zhang, *J. Controlled Release*, 2011, **149**(2), 92.

69. A. Baeza, M. Colilla and M. Vallet-Regí, *Expert Opin. Drug Delivery*, 2015, **12**, 319.

70. A. Baeza, E. Guisasola, A. Torres-Pardo, J. M. González-Calbet, G. J. Melen, M. Ramirez and M. Vallet-Regí, *Adv. Funct. Mater.*, 2014, **24**, 4625.

71. J. Simmchen, A. Baeza, D. Ruiz and M. Vallet-Regí, *Nanoscale*, 2014, **6**, 8907.

72. J. L. Vivero-Escoto, I. I. Slowing, B. G. Trewyn and V. S. Y. Lin, *Small*, 2010, **6**, 1952.

73. Y. Zhao, J. L. Vivero-Escoto, I. I. Slowing, B. C. Trewyn and V. S. Y. Lin, *Expert Opin. Drug Delivery*, 2010, **7**, 1013.

74. J. M. Rosenholm, C. Sahlgren and M. Linden, *Nanoscale*, 2010, **2**, 1870.

75. M. Vallet Regí, E. Ruiz-Hernandez, B. Gonzalez and A. Baeza, *J. Biomater. Tissue Eng.*, 2011, **1**, 6.

76. J. Liu, X. Jiang, C. Ashley and C. J. Brinker, *J. Am. Chem. Soc.*, 2009, **131**, 7567.

77. C. E. Ashley, E. C. Carnes, G. K. Phillips, D. Padilla, P. N. Durfee, P. A. Brown, T. N. Hanna, J. W. Liu, B. Phillips, M. B. Carter, N. J. Carroll, X. M. Jiang, D. R. Dunphy, C. L. Willman, D. N. Petsev, D. G. Evans, A. N. Parikh, B. Chackerian, W. Wharton, D. S. Peabody and C. J. Brinker, *Nat. Mater.*, 2011, **10**, 389.

78. M. W. Ambrogio, C. R. Thomas, Y.-L. Zhao, J. I. Zink and J. F. Stoddartt, *Acc. Chem. Res.*, 2011, **44**, 903.

79. Z. X. Li, J. C. Barnes, A. Bosoy, J. F. Stoddart and J. I. Zink, *Chem. Soc. Rev.*, 2012, **41**, 2590.

80. J. Khandare, M. Calderon, N. M. Dagia and R. Haag, *Chem. Soc. Rev.*, 2012, **41**, 2824.
81. V. P. Torchilin, *Nat. Rev. Drug Discovery*, 2005, **4**, 145.
82. W. J. M. Mulder, G. J. Strijkers, G. A. F. Van Tilborg, D. P. Cormode, Z. A. Fayad and K. Nicolay, *Acc. Chem. Res.*, 2009, **42**(7), 904.
83. R. Haag and F. Kratz, *Angew. Chem., Int. Ed.*, 2006, **45**(8), 1198.
84. K. T. Oh, H. Q. Yin, E. S. Lee and Y. H. Bae, *J. Mater. Chem.*, 2007, **17**, 3987.
85. Y. Ma, R. J. M. Nolte and J. J. L. M. Cornelissen, *Adv. Drug Delivery Rev.*, 2012, **64**, 811.
86. P. Ghosh, G. Han, M. De, C. K. Kim and V. M. Rotello, *Adv. Drug Delivery Rev.*, 2008, **60**, 1307.
87. A. M. Smith, H. W. Duan, A. M. Mohs and S. M. Nie, *Adv. Drug Delivery Rev.*, 2008, **60**, 1226.
88. C. Sun, J. S. H. Lee and M. Q. Zhang, *Adv. Drug Delivery Rev.*, 2008, **60**, 1252.
89. E. J. Anglin, L. Y. Cheng, W. R. Freeman and M. J. Sailor, *Adv. Drug Delivery Rev.*, 2008, **60**, 1266.
90. C. Barbe, J. Bartlett, L. G. Kong, K. Finnie, H. Q. Lin, M. Larkin, S. Calleja, A. Bush and G. Calleja, *Adv. Mater.*, 2004, **16**, 1959.
91. Y. Piao, A. Burns, J. Kim, U. Wiesner and T. Hyeon, *Adv. Funct. Mater.*, 2008, **18**, 3745.
92. D. Avnir, T. Coradin, O. Lev and J. Livage, *J. Mater. Chem.*, 2006, **16**, 1013.
93. W. H. Tan, K. M. Wang, X. X. He, X. J. Zhao, T. Drake, L. Wang and R. P. Bagwe, *Med. Res. Rev.*, 2004, **24**, 621.
94. M. Vallet-Regi and A. Ramila, *Chem. Mater.*, 2000, **12**, 961.
95. N. Mas, D. Arcos, E. Aznar, S. Sánchez, F. Sancenón, A. García, M. D. Marcos, A. Baeza, M. Vallet-Regí and R. Martínez, *Small*, 2014, **10**(23), 4859.
96. M. Colilla, M. Martínez, S. Sánchez, M. L. Ruiz, J. M. González-Calbet and M. Vallet-Regí, *J. Mater. Chem. B*, 2014, **2**(34), 5639.
97. N. Knezevic, E. Ruiz-Hernández, W. Hennink and M. Vallet-Regi, *RSC Adv.*, 2013, **3**, 9584.
98. A. L. Doadrio, J. Sánchez, J. C. Doadrio, A. Salinas and M. Vallet-Regi, *Micropor. Mesopor. Mat.*, 2014, **195**, 43.
99. O. Prymak, S. Ristig, W. Meyer-Zaika, A. Rostek, L. Ruiz, J. M. Gonzalez-Calbet, M. Vallet-Regi and M. Epple, *Russ. Phys. J.*, 2013, **10**, 5.
100. A. Baeza, D. Arcos and M. Vallet-Regí, *J. Phys. Condens. Matter*, 2013, **25**, 484003.
101. M. Cicuendez, M. Portoles, I. Izquierdo-Barba and M. Vallet-Regí, *Chem. Mater.*, 2012, **24**, 1100.
102. D. Molina-Manso, M. Manzano, J. C. Doadrio, G. Del Prado, A. Ortiz-Pérez, M. Vallet-Regí, E. Gómez-Barrena and J. Esteban, *Int. J. Antimicrob. Agents*, 2012, **40**, 252.
103. Z. X. Li, J. C. Barnes, A. Bosoy, J. F. Stoddart and J. I. Zink, *Chem. Soc. Rev.*, 2012, **41**, 2590.

104. D. Lozano, C. G. Trejo, E. Gómez-Barrena, M. Manzano, J. C. Doadrio, A. J. Salinas, M. Vallet-Regí, N. García-Honduvilla, P. Esbrit and J. Buján, *Acta Biomater.*, 2012, **8**, 2317.

105. S. H. Wu, Y. Hung and C. Y. Mou, *Chem. Commun.*, 2011, **47**, 9972.

106. M. Grun, I. Lauer and K. K. Unger, *Adv. Mater.*, 1997, **9**, 254.

107. D. Arcos, A. Lopez-Noriega, E. Ruiz-Hernandez, O. Terasaki and M. Vallet-Regi, *Chem. Mater.*, 2009, **21**, 1000.

108. C. Boissiere, D. Grosso, A. Chaumonnot, L. Nicole and C. Sanchez, *Adv. Mater.*, 2011, **23**, 599.

109. C. J. Brinker, Y. F. Lu, A. Sellinger and H. Y. Fan, *Adv. Mater.*, 1999, **11**, 579.

110. M. Colilla, M. Manzano, I. Izquierdo-Barba, M. Vallet-Regi, C. Boissiere and C. Sanchez, *Chem. Mater.*, 2010, **22**, 1821.

111. Y. F. Lu, H. Y. Fan, A. Stump, T. L. Ward, T. Rieker and C. J. Brinker, *Nature*, 1999, **398**, 223.

112. I. Izquierdo, S. Sánchez, M. Colilla, M. J. Feito, C. Ramírez, M. T. Portolés and M. Vallet-Regi, *Acta Biomater.*, 2011, **7**, 2977.

113. O. C. Farokhzad and R. Langer, *ACS Nano*, 2009, **3**, 16.

114. Y. S. Lin and C. L. Haynes, *J. Am. Chem. Soc.*, 2010, **132**, 4834.

115. W. H. Suh, Y. H. Suh and G. D. Stucky, *Nano Today*, 2009, **4**, 27.

116. F. Hoffmann, M. Cornelius, J. Morell and M. Froba, *Angew. Chem., Int. Ed.*, 2006, **45**, 3216.

117. D. Bruhwiler, *Nanoscale*, 2010, **2**, 887.

118. F. Hoffmann and M. Froba, *Chem. Soc. Rev.*, 2011, **40**, 608.

119. F. Hoffmann, M. Cornelius, J. Morell and M. Froba, *Angew. Chem., Int. Ed.*, 2006, **45**(20), 3216.

120. D. Arcos, V. Fal-Miyar, E. Ruiz-Hernández, M. García-Hernández, M. L. Ruiz-González, J. G. Calbet and M. Vallet-Regí, *J. Mater. Chem.*, 2012, **24**, 64.

121. M. Colilla, I. Izquierdo-Barba and M. Vallet-Regí, *Micropor. Mesopor. Mater.*, 2010, **135**, 51.

122. D. Bruhwiler, *Nanoscale*, 2010, **2**, 887.

123. M. Colilla, I. Izquierdo-Barba, S. Sánchez-Salcedo, J. L. G. Fierro, J. L. Hueso and M. Vallet-Regí, *Chem. Mater.*, 2010, **22**, 6459.

124. N. Lang and A. Tuel, *Chem. Mater.*, 2004, **16**(10), 1961.

125. M. C. Burleigh, M. A. Markowitz, M. S. Spector and B. P. Gaber, *Chem. Mater.*, 2001, **13**(12), 4760.

126. S. Huh, J. W. Wiench, J. C. Yoo, M. Pruski and V. S. Y. Lin, *Chem. Mater.*, 2003, **15**(22), 4247.

127. M. H. Lim and A. Stein, *Chem. Mater.*, 1999, **11**(11), 3285.

128. C. Zapilko, M. Widenmeyer, I. Nagl, F. Estler, R. Anwander, G. Raudaschl-Sieber, O. Groeger and G. Engelhardt, *J. Am. Chem. Soc.*, 2006, **128**(50), 16266.

129. H. Ritter and D. Bruhwiler, *J. Phys. Chem. C*, 2009, **113**(24), 10667.

130. H.-P. Lin, L.-Y. Yang, C.-Y. Mou, S.-B. Liu and H.-K. Lee, *New J. Chem.*, 2000, **24**(5), 253.

131. Y. H. Liu, H. P. Lin and C. Y. Mou, *Langmuir*, 2004, **20**(8), 3231.

132. D. Arcos and M. Vallet-Regí, *Acta Mater.*, 2013, **61**, 890.

133. M. Colilla, B. González and M. Vallet-Regí, *Biomater. Sci.*, 2013, **1**, 114.

134. J. M. Rosenholm, C. Sahlgren and M. Linden, *Nanoscale*, 2010, 2(10), 1870.

135. A. Baeza and M. Vallet-Regí, Ceramic Smart Drug Delivery Nanomaterials, in *Bio- and Bioinspired Nanomaterials*, ed. D. Ruiz-Molina, F. Novio and C. Roscini, Wiley-VCH Verlag GmbH & Co. KGaA, Weinheim, Germany, 2015.

136. M. Colilla, A. Baeza and M. Vallet-Regí, Mesoporous silica nanoparticles for drug delivery and controlled release applications, in *The Sol-Gel Handbook (3 volumes)*, ed. D. Levy and M. Zayat, Wiley-VCH Verlag GmbH & Co. KGaA, Weinheim, 2015.

137. M. Colilla and M. Vallet-Regí, Smart Drug Delivery from Silica Nanoparticles, in *Smart Materials for Drug Delivery (Volume 2)*, ed. C. Alvarez-Lorenzo and A. Concheiro, Royal Society of Chemistry, UK, 2013.

138. I. I. Slowing, B. G. Trewyn and V. S.-Y. Lin, *J. Am. Chem.*, 2006, **128**, 14792.

139. Y. Zhao, X. Sun, G. Zhang, B. G. Trewyn, I. I. Slowing and V. S.-Y. Lin, *ACS Nano*, 2011, **5**, 1366.

140. Q. He, J. Shi, M. Zhu, Y. Chen and F. Chen, *Microporous Mesoporous Mater.*, 2010, **131**, 314.

141. J. Lu, M. Liong, J. I. Zink and F. Tamanoi, *Small*, 2010, **6**, 1794.

142. S. V. Dorozhkin and M. Epple, *Angew. Chem., Int. Ed.*, 2002, **41**, 3130.

CHAPTER 6

Mesoporous Nanoceramics for Drug Delivery

6.1   What is Nanomedicine?

In recent years we have witnessed the appearance of nanoscience and nano-technology as emerging fields of outstanding interest in research.

The term nanoscience refers to all research focused on understanding the structure and properties of materials and devices within the nanoscale size range. Nanotechnology uses this knowledge to produce materials, structures and devices with new properties emanating from this minute nanometer size.

The nano- prefix is also used in previously existing fields of research, where the aim is to highlight their current drift towards the study of nanoscale phenomena. Hence, we now have nanomedicine, nanoelectronics, nanochemistry, nanophotonics, nanobiotechnology, *etc.*

The application of nanotechnology to the treatment, diagnosis, monitoring and control of biological systems is known as nanomedicine. Currently, the field of nanomedicine encompasses four areas:

- drug release with focalized therapy;
- diagnostics;
- theranostics, which is the combination of therapy and diagnostics; and
- nanodevices.

Drug release with focalized therapy attempts to improve the pharmacological profile of a drug, to selectively release the drug on the target

RSC Nanoscience & Nanotechnology No. 39
Nanoceramics in Clinical Use: From Materials to Applications, 2nd Edition
By María Vallet-Regí and Daniel Arcos Navarrete
© M. Vallet-Regi and D. Arcos Navarrete 2016
Published by the Royal Society of Chemistry, www.rsc.org

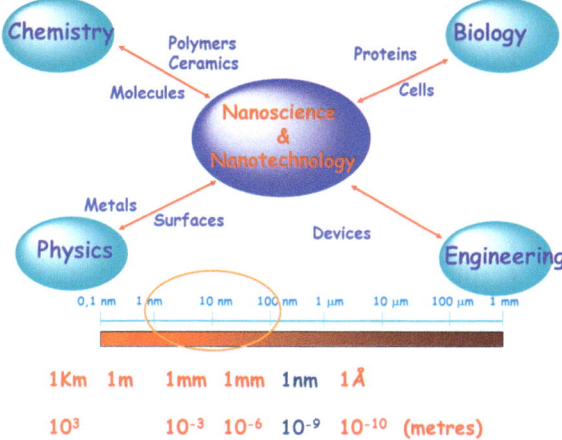

Figure 6.1 The interdisciplinary field of nanoscience.

tissue, to overcome biological barriers allowing the insertion of drugs and/ or genes in sick cells and to diminish the secondary effects of traditional treatments.

In diagnostics, nanomedicine attempts to detect sick tissues with higher sensitivity, faster and more precisely, to diagnose in the early stages of a disease and to specifically detect pathological markers.

In theranostics, the aim of nanomedicine is to detect and simultaneously provide focalized therapy on a given sick tissue, as well as to visualize and assess the performance of a treatment.

And finally, the nanodevices of interest in nanomedicine are predominantly biosensors with higher precision and sensitivity than traditional sensors, and nanorobots able to detect and repair tissue at cellular level.

Nanotechnology is a cross-disciplinary meeting point for chemists, biologists, physicists, physicians and engineers. With such a diverse collection of knowledge and contributions, it is possible to tackle any problem from new and original perspectives and to obtain prototypes that bring new solutions to previously unanswered issues. This joint effort of various disciplines is the only way to fruitfully develop new focalized therapies, better diagnostics, the combination of therapy and diagnostics and the development of new nanodevices. Figure 6.1 shows the interaction of disciplines responsible for great innovations in nanoscience and nanotechnology.

Nano means small in Greek; the nano- prefix, as in nanometer, equals the billionth part of a meter. Nanoscale has been arbitrarily defined as the size and length scale ranging from 1 to 100 nm. In this nanoscale we find atoms, molecules, proteins, viruses, DNA chains, carbon nanotubes, graphene, nanoparticles, *etc.* All these nano- objects are of special interest in nanoscience and nanotechnology. Figure 6.2 depicts the relative positions of these items on the nanoscale.

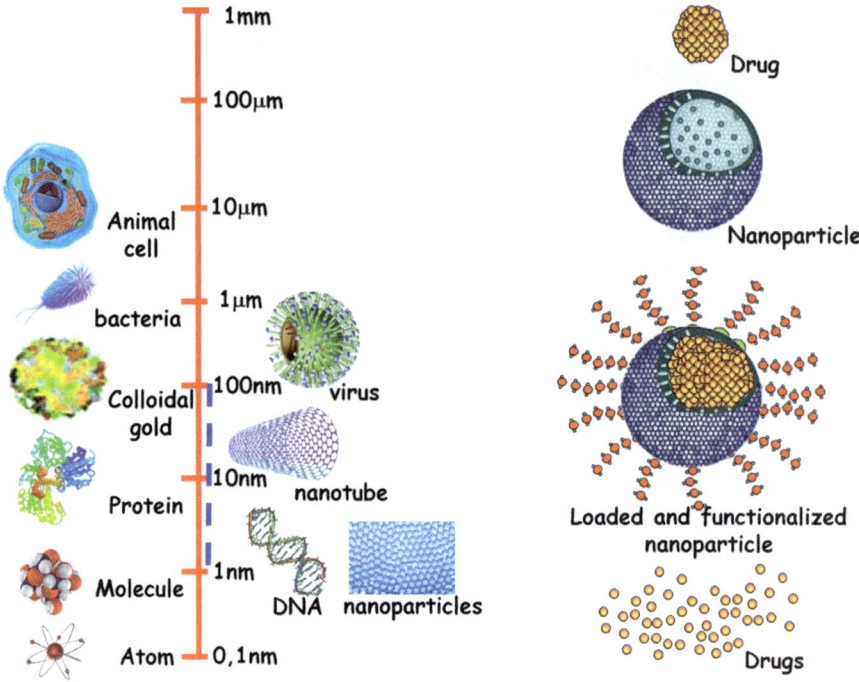

Figure 6.2   The size scale of various species. Silica mesoporous nanoparticles are within the nanometer range, as well as many loaded drugs and functionalizing chemical species.

Ten hydrogen atoms in line measure 1 nm. A DNA molecule is ~2.5 nm wide, while the thickness of a human hair is nearly 80 000 nm. Only atomic-resolution microscopes can deal with the nanoscale.

Both nanoscience and nanotechnology developed in a very quiet fashion until the 1990s, 30 years after Feynman in his visionary talk made scientists dream about the possibility to understand and manipulate matter at the nanometer scale; this is why he can be regarded as the father of nanoscience. At the annual conference of the American Physical Society in 1959, he was the first to mention the limits of miniaturization and to foresee the ability to manipulate and control matter at the nanoscale. However, the first scanning tunneling microscope was not developed until 1981, allowing experimental work in this area. It was discovered by Heinrich Rohrer and Gerd Binnig, awarded the Nobel Prize in Physics in 1986 for this technological break-through which boosted the nano world. These microscopy techniques are the "eyes" and "hands" used to study and manipulate nanostructures.

Scanning tunneling microscopy provides images at a very small scale, down to individual atom resolution. Besides, this technology also characterizes magnetic domains at the atomic level, and allows the manipulation of atoms as single isolated elements on a surface, hence designing new structures. This discovery triggered the design of several additional techniques, such as the atomic force microscope. This tool, based on the interaction between the probe and the surface through van der Waals forces, enables the detection of interactions in the field of picoNewton (10^{-12} N). Such sensitivity allows the performance of studies on binding forces, molecular interactions such as antigen–antibody, DNA hybridization processes or protein folding, among others.

Simultaneously with the development of characterization techniques, new and different nanostructures arose with promising technological applications. Among the allotropic forms of carbon, nanotubes for instance exhibit remarkable mechanical, electrical and thermal properties. These materials have been proposed in the biomedical field for several applications and have opened the door for another member of this family: the graphene, potentially useful in similar applications and with the benefit of all previous testing and experimentation with carbon nanotubes. Similarly, metal and metal oxide nanoparticles have been developed, exerting control on their shape, size, magnetic and electrical properties using different synthesis methods.

6.1.1   The Link Between Biotechnology and Medicine

The mutual relationship of these two disciplines is opening or about to open new possibilities. It is gaining momentum in very important aspects of our current society, such as:

- New analysis techniques for biological phenomena thanks to equipment such as atomic force microscopes.
- New diagnosis techniques based on the use of functionalized nanoparticles, that is, coated with a material capable of providing new properties such us detecting the focus of a disease. Certain nanoparticles, used as optical or magnetic markers in early disease detection, can also act as therapeutic elements destroying sick tissue upon adhesion.
- Disease diagnosis with the development of new nanoelectronic sensors to carry out a precise and exact monitoring of our health.
- Development of nanoelectronic devices to restore visual and audible capabilities in a patient.
- Development of drug dosage and release systems, able to work with precision, at the required moment and at the predetermined location.
- Within the field of biomaterials, biocompatible nanomaterials are already satisfactorily in use, helping to obtain more resilient prostheses.
- Development of nanoparticles which can be autonomously used as biomarkers in high-sensitivity image techniques for biomedicine, *e.g.* particles for the early detection of carcinogenic tumors.

Several physical systems are already in use in bio or nanomedicine, based on metal semiconducting nanoparticles, for early detection techniques. Wavelengths of these nanoparticles, in order to be used as biomarkers, must be within the 600–1000 nm range, where human tissue is most transparent.

The use of heavy metals for these applications should not be hazardous in principle, since most of the nanoparticles involved in the assay can be eliminated. Those nanoparticles not attached to infected cells must be released through the excretion system, while those successfully attached will be eliminated during the therapy phase, together with the cells of interest.

All these strategies are difficult to achieve by following a conventional molecular architecture, hence the increase in use of nanoparticles in medicine.

6.2   What is a nanocarrier?

A nanocarrier is basically a nanoparticle which can be loaded with drugs, contrasting agents or other species, in order to be subsequently transported to a specific tissue.

In the previous chapter, several types of nanoparticles which could be used as nanocarriers are reviewed. There are plenty of different types that can load and unload drugs. Some of them are soft in texture, such as liposomes or polymers, and there may exist the risk of premature release of the enclosed drug. In this sense, the use of inorganic nanoparticles could be the solution, since generally speaking, they are harder and environment-resistant systems; the risk of unwanted release could be significantly decreased. This is the main reason behind the current use of inorganic type nanoparticles, with sizes ranging from 10 to 200 nm, as vectors for the insertion of cytotoxics inside tumor cells.

Moreover, if conveniently functionalized, the specificity of the particle can be achieved or boosted. In this chapter, the focus will be on ceramic nanoparticles with these features, highlighting the possibilities of silica mesoporous nanoparticles, described in Chapter 5.

Obviously, a nanocarrier has to accept the payload that must be transported, and its surface must also accept decoration with different species such as dispersing agents – to avoid nanoparticle buildup; polymers that render it undetectable by macrophages – to increase its travel time in the bloodstream; and targeting ligands, which guide them to the tissues where the load has to be released. All these issues are reviewed in this chapter. Figure 6.3 depicts two nanocarriers; the first one is a silica mesoporous nanoparticle and the second is a carbon nanotube, both are loaded with drugs, contrast agents and magnetic material, and with surface treatment to favor their binding with a dispersing agent (polyethylene glycol) and different targeting agents.

Nanoparticles used as drug carriers are submicrometer-sized systems which should have a size <100 nm in at least one orthogonal direction. In recent years, these nanocarriers have offered a promising alternative to conventional therapy in oncology and have been prepared using a wide range of materials such as polymers, lipids, viruses and ceramics.[1] Conventional antitumoral drugs usually present small sizes and they are removed from systemic circulation very fast, so higher dosages are needed. This increases the risk of severe side effects caused by the high toxicity of these antitumoral agents towards healthy tissues. The encapsulation of cytotoxic drugs in nanometric carriers improves their pharmacokinetic profile, protecting them against enzymatic degradation

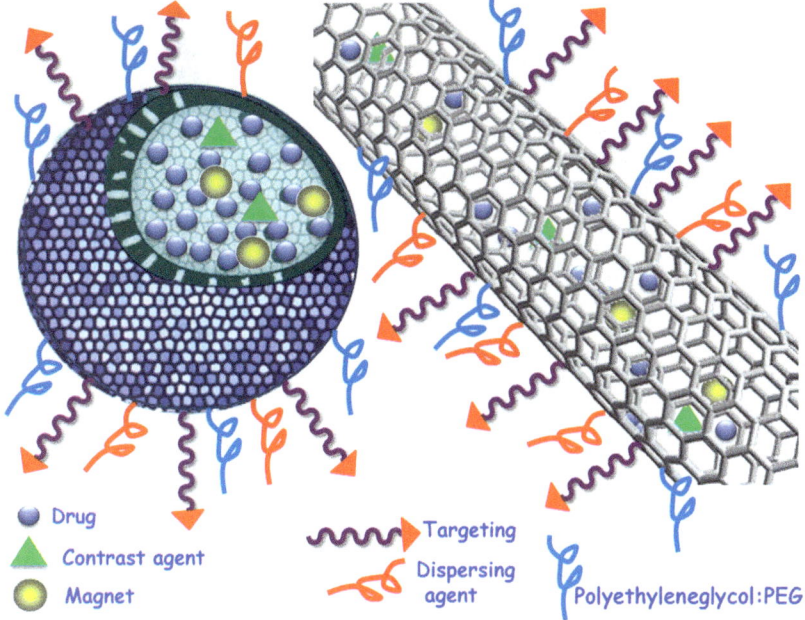

● Drug		〜〜▶	Targeting
▲ Contrast agent		∿∿	Dispersing agent
● Magnet		ⵜ	Polyethyleneglycol:PEG

Figure 6.3 Layout of a silica mesoporous nanoparticle and a carbon nanotube. Both can be loaded with drugs, contrast agents or tiny magnets. They can be functionalized with dispersing agents, targeting and other species.

and allows their selective accumulation into the diseased tissue owing to the *enhanced permeation and retention effect* (EPR), as mentioned below. Moreover, using nanoparticles it may be possible to obtain other important advantages such as: improving the delivery of hydrophobic drugs or macromolecular-type therapeutic agents (proteins, enzymes or oligonucleotides), to transport and deliver multiple therapeutic agents at the same time, which is enormously important for combined therapy, to cross tight epithelial or endothelial barriers and even to visualize in real time the drug delivery process using imaging agents attached to or encapsulated into the nanocarrier. The nanocarrier selectivity can be improved attaching targeting agents on their surface able to be specifically recognized by cancer cells or by the tumor microenvironment. Additionally, there is clear commercial interest in the re-formulation of drugs that have not passed clinical trials due to poor solubility or extensive side effects.

The maximum size of these nanocarriers is limited by the primary immunogenic system. Macrophages and other specialized cells are designed for the elimination of foreign entities and operate in all tissues. The clearance mechanisms orchestrated by these cells are generally not effective for objects with sizes <100–200 nm, although other factors such as particle shape or surface charge must be taken into account. Meanwhile, particles or macromolecules with sizes <5–6 nm are rapidly excreted by kidney. Therefore, the optimal size of nanocarriers for biomedical purposes is 10–200 nm.

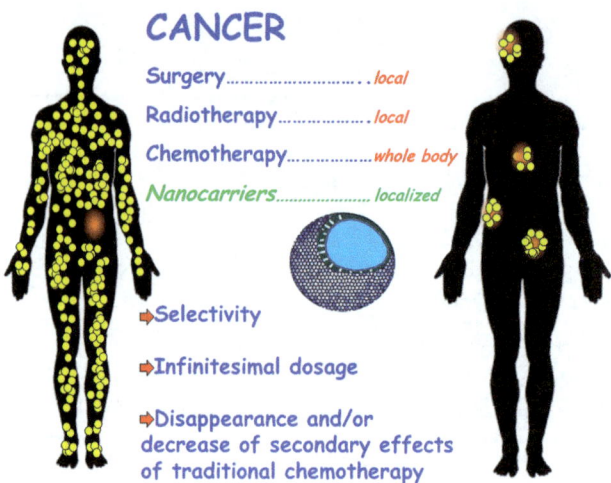

Figure 6.4 List of cancer-fighting strategies and areas of improvement on current therapies.

The specific application of these nanocarriers in cancer treatment could be extremely beneficial compared with current treatments such as surgery, radiotherapy and chemotherapy. Surgery and radiotherapy are locally applied when the tumor can be operated upon and no metastasis is present. Therefore it is not always possible, either because there is no easy access to the tumor or because there is metastasis. The solution applied in these cases is chemotherapy, which is non-selective and reaches all tissues in the human body.

Advances in the field of nanosystems with applications in medicine have now produced new smart materials which could solve new clinical requirements.

A pivotal concern in medicine is the administration of therapeutic agents to a patient through the most physiologically acceptable route. In too many cases, drug doses are deliberately excessive, to ensure that the minimum adequate dose reaches the critical region. But most of the administered dose – actually almost the full dose – is acting throughout the body, affecting undesired areas and organs. Hence large doses are needed because the drug is released everywhere, in a non-specific fashion, and in places where it is not needed. This is an acute problem in oncology, where the risk–benefit ratio associated with chemotherapy complicates the adoption of the right decision, given the high toxicity of the drugs involved. It is generally accepted that drug adsorption by the body is favored by smaller drug sizes and by the type of coating or encapsulating material.

The purpose of a nanocarrier is to selectively reach the damaged tissues only, and to release there the transported cytotoxin, hence dosage would be considerably lower than in traditional chemotherapy, decreasing or even suppressing side effects. This scenario is currently in the early research and development stages. We describe several specific examples throughout the chapter. Figure 6.4 expresses these ideas.

6.3   Ceramic Nanoparticles in Medicine

Among ceramic nanoparticles having applications in medicine, the main types are mesoporous silica, gold, various metal oxides, graphene and carbon nanotubes. Figure 6.5 exhibits these nanoparticles.

The extremely small size of these structures implies a remarkable ability to penetrate biological tissues. Nanoparticle behavior depends on their size, shape and reactivity with tissue surfaces. Given the large specific surface of these nanosystems, the contact with fluids and tissues may trigger the adsorption of different macromolecules.

The use of nanoparticles in the medical field goes hand in hand with the development of molecular labels capable of sick cell recognition, as well as other methods of anchoring to the detected sick cell.

When any nanoparticle is designed as a drug release system, several factors have to be taken into account.

It must be ensured that the nanoparticles are non-toxic, that the drug will remain stable and protected within the nanostructure, that it will be physically, chemically and biologically stable, and also that it shows good compatibility with the receptor.

Besides these factors, and considering that they must be applied in clinical practice, it is important to ensure that the product can be easily and safely sterilized, stored and administered.

These nanoparticles will be in contact with blood fluids; hence they must exhibit hemocompatibility, a long circulation time in the bloodstream, adequately controlled drug release, and they must not form clusters, they cannot

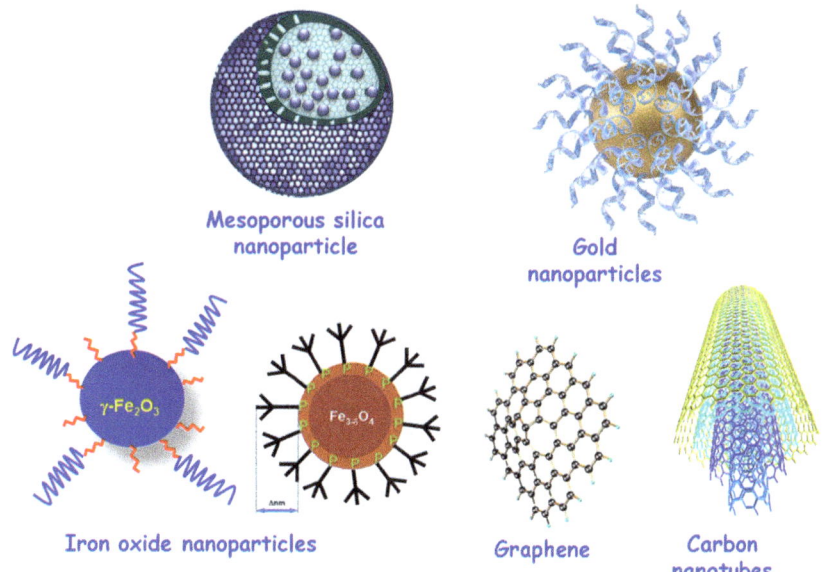

Mesoporous silica nanoparticle

Gold nanoparticles

Iron oxide nanoparticles

Graphene

Carbon nanotubes

Figure 6.5   Inorganic nanoparticles with applications in cancer therapy.

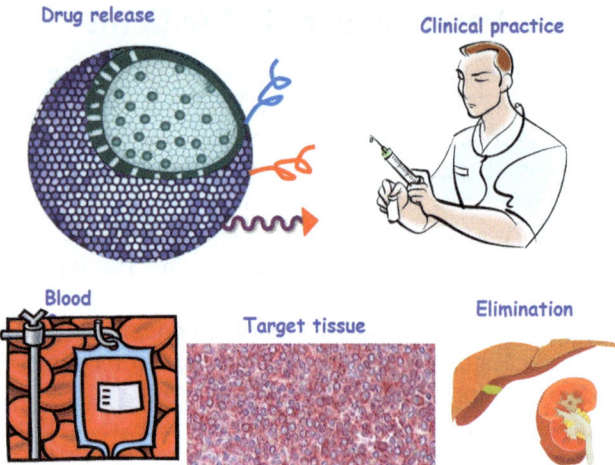

Figure 6.6   Desired features of a nanocarrier to ensure its correct clinical application. The nanocarrier should be easy to functionalize with specific targeting agents, injectable, stable in blood, able to reach the target tissue
and easy to remove *via* liver metabolism or urinary system.

activate phagocytes and it is critical that they are not eliminated *via* the
pulmonary route and do not cause embolism.

Regarding vectorization to the target tissue, the nanocarrier must be selective, capable of crossing biological barriers, directed to the internment in
cells or organs where it is due while ensuring its biodegradability, biocompatibility and non-cytotoxicity.

Finally, it is necessary also to foresee their elimination, bearing in mind
that they must have low molecular weight, that their decomposition products must be non-toxic, and also that there is no accumulation process in any
organ; they must be safely excreted by liver and kidneys. Figure 6.6 illustrates
these ideas.

6.3.1   Mesoporous Silica Nanoparticles

Mesoporous silica nanoparticles (MSNs) comprise one of the most promising
materials for antitumoral purposes among the inorganic nanocarriers, and their
application has reached *in vivo* trials using mainly murine models.[2] In 2011, the
first silica-based nanoparticle was approved by the US Food and Drug Administration for the first in-human clinical trial in oncologic diagnosis, which constitutes a great step forward in the clinical acceptance of this material.[3]

As it mentioned in Chapter 5, MSNs present very interesting properties for
biomedical uses, such as high specific surface area, robustness, easily tunable size, shape and pore diameter, among others. These properties provide
unique advantages to encapsulate different drugs, from small molecules to
therapeutic macromolecules.[4–6]

The production of mesoporous silica particles is simple, controllable, cost-effective and easily scalable. Moreover, this material presents a high proportion of silanol groups on its surface, which provide many options for the development of multifunctional materials through covalent grafting using the well-established silane chemistry. The higher research efforts in the field of antitumoral drug delivery applications of these materials have been performed using the MCM-41 and SBA-15 types, the first being the most studied because it material can be easily obtained in form of nanoparticles <200 nm, which is a perfect size for drug delivery purposes.[7] Mesoporous silica particles with more exotic morphologies, such as hollow or rattle-type spheres have also received great attention.[8]

The application of the MSNs as drug carriers in antitumoral therapy requires their effective suspension in biological solutions with high salt and protein contents. Bare mesoporous silica particles show a tendency to aggregate due to the formation of interparticle hydrogen bonds caused by the silanol groups. The particle aggregation is even more intense when they are exposed to biological fluids. In order to avoid this undesirable behavior, different strategies have been described that involve the attachment of hydrophilic moieties on the particle surface, such as phosphonates or polyethyleneglycol-derived phospholipids (PEG-lipid).[9]

This last method provides particles with significantly lower non-specific protein absorption due to the PEG coating. Small mesoporous nanoparticles (<50 nm) with their surface decorated with polyethyleneimine (PEI)-PEG copolymer have been evaluated *in vivo* using a murine xenograft model of human squamous carcinoma showing high passive accumulation (12%) within the tumoral mass and an improved therapeutic efficacy, higher than the free drug.[10]

PEGylated silica nanorattles loaded with docetaxel, which is a potent hydrophobic antitumoral drug, have demonstrated their ability to inhibit the tumoral growth 15% more than the clinical formulation of docetaxel, Taxotere, and also showing less systemic toxicity.[11]

MSNs can also be employed for the capture of important biomolecules within the tumoral cells, provoking inhibition of their cellular growth. Thus, Lin and co-workers have decorated the external surface of MSNs with phenanthridinium molecules, a group able to bind to cytoplasmic oligonucleotides, as messenger RNAs, interfering with the normal development of the cell.[12]

The shape of the particles is an important aspect in terms of cellular internalization and presents cell-type dependence. Lin and co-workers have reported that spherical MSNs exhibit a faster endocytic rate than rod-shaped particles when Chinese hamster ovary (CHO) cells were employed, whereas the internalization rate for both was similar using human fibroblast cells.[13]

Thus, the shape design of the carrier depends upon the type of target cell. Targeting molecules able to be selectively internalized by tumoral cells can be attached onto the mesoporous silica surface in order to improve the efficacy of the therapy. Thus, the surface of MSNs has been conjugated with different ligands such as sugars,[14] folic acid,[15,16] transferrin or cyclic RGD-peptides[17] as well as antibodies[18] and DNA aptamers.[19]

6.3.2   Calcium Phosphate Nanoparticles

Calcium phosphates are the most important inorganic components of hard tissues and present excellent biocompatibility. This material is dissolved in acidic environments (pH ≈ 4–5), which can be naturally present in lysosomes after internalization or in tumoral tissues. Therefore, cytotoxic drugs can be encapsulated during the formation of calcium phosphate nanoparticles and released once the nanocarrier is internalized by the target cell or when it reaches the tumoral environment. Hollow thin nanoparticles have demonstrated their potential application as stimuli-responsive carriers able to release their cargo in response to ultrasound stimulus.[20]

Cisplatin has been trapped in calcium phosphate nanoparticles and showed a sustained release of the cytotoxic agent and an improved effectivity against the resistant A2780cis human ovarian cancer cell line.[21]

Ceramide is a lipid-derived drug which induces apoptotic death of melanoma tumoral cells. Unfortunately, the potential application of this molecule is hampered by its low solubility in water. Recently, ceramide has been encapsulated into calcium phosphate nanoparticles of 20 nm diameter stabilized by a PEG coat.[22]

This material is easy to disperse in buffered solutions at physiological pH, maintaining stable in solution for long periods, which is an important issue for drug delivery purposes. The biological evaluation of this material was performed using different human tumoral cell lines, showing its efficacy in delivering hydrophobic drugs to these malignant cells. The same material has also demonstrated its ability to transport cytotoxic drugs and organic dyes in order to combine drug delivery and imaging within the same carrier.[23]

Photosensitizers can be also transported by calcium phosphate nanoparticles for photodynamic therapy. Thus, methylene blue and 5,10,15,20-tetrakis (3-hydroxyphenyl)porphyrin (*m*THPP) have been trapped in polymeric layers attached to the surface of calcium phosphate nanoparticles.[24]

These particles have demonstrated their capacity to destroy tumoral cells with high efficacy following light irradiation, showing very low toxicity under dark conditions.[25]

Another strategy to destroy tumors is the encapsulation of superparamagnetic iron oxide nanocrystals into calcium phosphate particles. Thus, magnetic hydroxyapatite nanoparticles have been synthesized by the coprecipitation process using different concentrations of Fe^{2+} present in the media.[26]

The as-formed particles present round shapes, 20–50 nm in size and having good magnetic properties ~3–20 emu g^{-1}. The ability to produce local hyperthermia of these particles under the exposition of alternative magnetic fields, with temperature increase from 38 to 40 °C in mice models was confirmed showing a significant tumor volume reduction. Magnetic nuclei have also been trapped within dicalcium phosphate particles showing excellent biocompatibility and promising capacity to destroy tumoral cells under magnetic stimulus without affecting healthy cells.[27]

6.3.3 Carbon Allotropes

The broad range of applications of carbon materials in medicine is the consequence of the high number of preparation methods. Moreover, the structural versatility of the different allotropes also facilitates functionalization and other surface modification techniques to tailor/engineer the already outstanding properties, such as mechanical properties, high electrical conductivity, high thermal conductivity, optical absorbance, *etc.* An element existing in more than one structural form with different molecular configurations is said to exhibit allotropy.[28,29] Twenty-five years ago, the world of carbon allotropes mostly focused on diamond- and graphite-based materials, fibers and compounds. But the field was significantly increased with the discovery of the C_{60} buckminsterfullerene, which became the principal turnover in our way of thinking about materials at the nanoscale when Kroto, Smalley and Curl won the Nobel Prize in Chemistry in 1996. Following them, the discoveries of carbon nanotubes by Iijima and the Nobel Prize in 2010 for the isolation of graphene by Geim and Novoselov made these the materials with the most potential in the new era of nanotechnology.

The development of novel graphene-based drug delivery nanocarriers has received great attention in the recent years, due to the ultra-high surface area available for drug loading.[30]

The application of these types of materials in drug delivery applications requires their surface functionalization with organic groups, polymers or biomolecules, which allow their suitable dispersion in aqueous solutions. Thus, PEG has been attached to the graphene surface in order to avoid agglomeration in serum media.[31]

The delocalized electron of the graphene sheets allows the retention of different cytotoxic drugs by π-stacking, such as doxorubicin. Surfactants such as Pluronic F127 have been employed for the fabrication of water-stable graphene carriers which present high loading capacity of doxorubicin (up to 280% w/w).[32] Moreover, this carrier is able to release its cargo in response to pH changes, achieving higher release in mildly acidic conditions (pH 5). Camptothecin is a quinolone alkaloid which shows a very potent cytotoxic activity against several human tumoral cell lines. However, due to its low solubility in water and high toxicity, the clinical application of this agent is compromised. Nanoparticles of graphene oxide (NGO) covered with chitosan have demonstrated their ability to retain this drug by intermolecular forces guided by π-stacking attractions, acting as drug-delivery carrier to different tumoral cell lines.[33]

One additional advantage of this type of material is its intrinsic photoluminescence in the visible and infrared regions, which can be used in order to localize the particles as they travel around the body.

As mentioned above, functional groups present in graphene oxide allow the attachment of targeting molecules able to guide the graphene-carriers to the tumoral cells. As in the previous materials, folic acid has been attached onto the NGO surface in order to provide selectivity.[34]

NGOs functionalized with folic acid on the surface can also deliver multiple drugs at the same time, such as doxorubicin and camptothecin, increasing the efficacy of the nanodevice by a synergistic effect.[35]

As in the case of mesoporous silica particles, a very interesting approach is the development of NGO devices able to release their cargo in response to external or internal stimuli. A redox-responsive NGO carrier has been synthesized attaching PEG onto the grapheme surface using disulfide cross-linkers.[36]

The presence of a PEG layer hampers the release of the loaded drugs and therefore the release rate is higher when the polymeric layers are detached from the surface, which happens if the nanoparticle reaches the intracellular space. Finally, pH has also been exploited in order to create NGO-responsive systems. Thus, NGOs combined with superparamagnetic iron oxide nanoparticles have been synthesized by a chemical precipitation method.[37]

Targeting molecules, in this case folic acid, were attached onto the iron oxide surface and doxorubicin was loaded onto the graphene, showing a very high loading capacity (as high as 0.387 mg mg^{-1}). This material exhibits strong pH dependence, presenting higher release in a mild acidic environment (pH 5).

Single-walled carbon nanotubes (SWNTs) have also been employed in drug delivery for cancer treatment using similar strategies to graphene oxide, due to their very similar chemical nature.[38]

As in the previous case, they require the external functionalization with hydrophilic moieties in order to increase their dispersability in water solutions. PEG-functionalized SWNT has demonstrated its ability to transport and release doxorubicin in a pH-dependent manner.[39]

Although SWNTs present the potential to encapsulate doxorubicin or other small drugs, recent studies indicate that the drugs are usually attached to the surface by π-stacking forces, because the average diameter of the SWNT increases after loading.[40]

The release rate of doxorubicin is related to the diameter and surface charge of the SWNT, being slow when the SWNTs present lower diameter and surface charges. Drug release is also accelerated with higher temperatures, due to the increase of the molecular motion and therefore it can be also employed as trigger stimulus. Other cytotoxic drugs have been loaded on the SWNTs and studied using *in vivo* models. Paclitaxel was chemically conjugated *via* a cleavable ester bond to branched PEG chains placed on the SWNT surface.[41]

This material has demonstrated high efficiency in suppressing tumor growth in a murine 4T1 breast cancer model.

Both SWNTs and NGOs can be employed in photothermal therapy against cancer because they are able to generate heat under light irradiation. The *in vivo* application of PEGylated NGO labeled with a fluorescent group was studied using different xenograft tumor mouse models.[42]

This material showed a high passive tumor accumulation and relatively low retention by the reticuloendothelial system. Under laser irradiation in the near infrared region, the particles were able to generate enough heat to destroy the tumoral cells, which indicates the suitability of these materials in this therapy. On other hand, SWNTs functionalized with PEGylated phospholipids present the capacity to destroy tumors under near-infrared irradiation at low doses up to 3.6 mg kg^{-1} without toxic side effects.[43]

6.3.4 Iron Oxide Nanoparticles

Iron oxide ceramic nanoparticles are also used in nanotechnology, but given their special relevance in stimulus-response systems, they are fully described in Chapter 7.

6.4 Administration

Nanoparticles can be introduced into the body by two routes, local or systemic, as pictured in Figure 6.7.

When using local administration, nanoparticle size can be in the 2–5 nm range. If the systemic route is used, nanoparticles must be 10–200 nm in size, although this is still under discussion. These ideas are discussed in the following sections.

The local route is applied when dealing with a very localized tumor, and the nanoparticles can be directly injected into the tumor mass.

The systemic route follows a longer path. Nanoparticles are injected into the bloodstream and circulate until they reach the tumor. This will only happen if the nanoparticles exhibit a clear selectivity towards these tissues. This procedure is at present performed *via* two different routes, *passive targeting* and *active targeting*. Both options are reviewed in Sections 6.6 and 6.7, respectively.

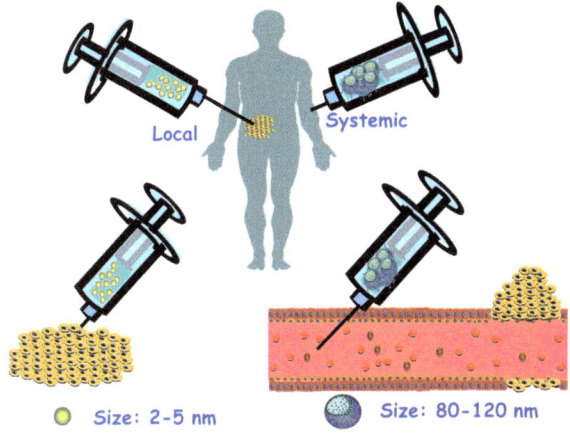

Figure 6.7   Routes of administration for a nanocarrier.

6.5   Design of Nanocarriers

This section deals with the design of nanocarriers for drug loading, transport and cell internalization.

In the design illustrated in Figure 6.8, the main players are the tumor cells and the nanoparticles loaded with cytotoxics. The cells have receptors on their surface and this has to be acknowledged when designing the nanoparticles in order to functionalize the particle surface with targeting agents, able to direct preferably to the receptors of the tumor cells.

If we attach to the nanoparticle surface molecules that can be selectively recognized by the tumoral cell, we can improve the internalization of the nanoparticles by the tumoral cells and therefore their therapeutic effect. The epithelial cell layer in the tumoral zone is not well distributed and there are fenestrations and pores with diameters up to few hundred micrometers. Thus, the nanoparticles will be able to extravasate from the blood vessels in this area. The nanoparticles, loaded with cytotoxic drugs, are injected into the bloodstream, but they are detected very quickly by macrophages. Therefore they must be rendered invisible to macrophages, which is why they must be functionalized with PEG. Now, the nanoparticles loaded with cytotoxic drugs and functionalized with targeting molecules are injected into the bloodstream. Once they have reached the tumoral mass, the targeting molecules of the nanoparticle allow their internalization (Figure 6.9). At this point, an external stimulus must be applied to produce the release of the cytotoxic drug. This cytotoxic agent induces apoptosis in the tumoral cells.

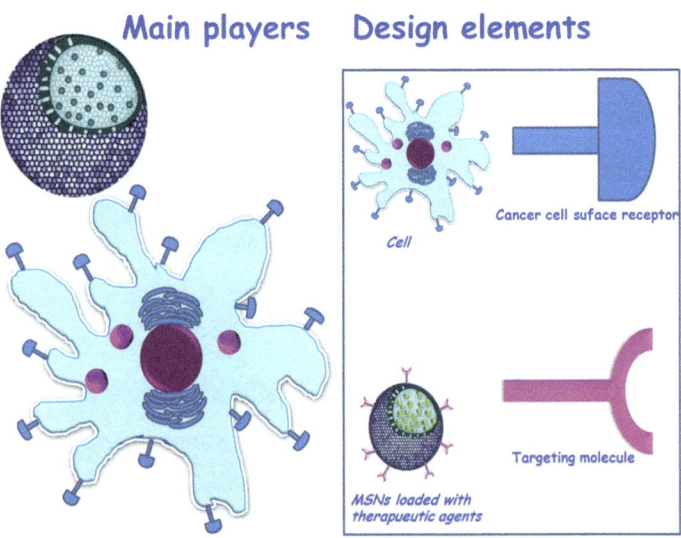

Figure 6.8   The cell dictates the design of an effective nanocarrier. MSN: mesoporous silica nanoparticles.

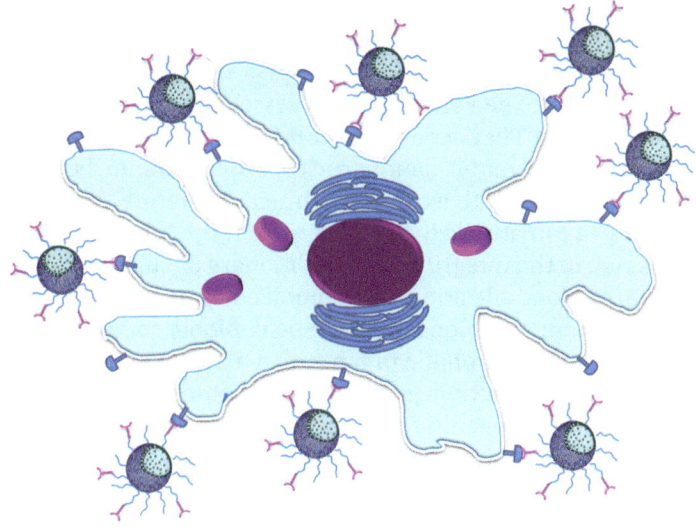

Figure 6.9 Layout of a cell receiving nanocarriers attracted by its receptors.

Therefore, this is the roadmap to follow:

- synthesis of nanoparticle;
- load it with cytotoxics;
- functionalize the nanoparticle with PEG, rendering it invisible to macrophages;
- functionalize with a locating agent to direct it to the tumor cell;
- upon arrival at the tumor cell, release the cytotoxin; and
- perform a biodistribution study to confirm the effectiveness of the treatment.

In order to discuss these topics, we present some specific examples.

6.5.1 Biodistribution and Excretion/Clearance Pathways

The effective and safe application of a nanomaterial for drug delivery requires a comprehensive view of the interactions that may take place with the biological system.[44]

Biocompatibility and biodistribution of a given material depends upon its particular features (composition, size, shape and surface charge), but also on other factors, such as route of administration (oral or intravenous), target body localization or internalization route into the cell. In this section, general concepts about the introduction of nanoparticles into the human body will be briefly presented considering the particularities of the ceramic particles. Nanoparticles can be introduced into the body by different routes; in antitumor therapy the most useful way is by peripheral intravenous injection, which is be the only way considered here.

Once the nanoparticles are injected in the bloodstream, their period of circulation through the body will involve several organs that must be carefully considered.[45]

Lung capillaries are the smallest blood vessel in the body, with diameters between 2 and 13 μm. They act as some kind of sieving constraint, especially in the case of rigid ceramic nanoparticles, which cannot be deformed in order to navigate through these tight channels. Therefore, ceramic particles with diameters >2 μm are likely to be trapped in the lungs, hence constituting a serious risk to the integrity of the pulmonary circulation. Furthermore, kidneys perform blood filtration. Macromolecules or particles <5–6 nm are easily removed from the bloodstream. Spent blood components are then stored in the spleen, a lymphatic organ which also takes care of lymphocyte maturation. Particle retention in this organ should always be avoided due to the potential immunogenic reactions involved. The interaction between nanoparticles and the spleen is responsible for the accelerated blood clearance effect.[46]

This effect can appear after repeated administration of nanoparticles, inducing a rapid elimination of these carriers. The accelerated blood clearance effect is divided into two phases: the first is the induction phase or sensitization, due to the interaction of the nanoparticles with B cells in the spleen. These cells recognize the particles as foreign entities and start to produce specific antibodies. Between 2 and 4 days after the first injection, the amount of specific antibodies against the nanoparticle increases. This is when the second phase, or effectuation phase, takes place. In this phase, the nanoparticles are rapidly covered by these antibodies and cleared by the liver. Thus, the blood circulation times of the nanocarriers are shortened with each injection.

Whether this accelerated process takes place or not, finally all blood components – and consequently the injected nanoparticles – reach the liver. This organ contains almost 90% of the total body population of macrophages. Once there, specific phagocytic cells known as Kupffer cells engulf the nanoparticles and eliminate them either by degradation or by retention in residual bodies within the cell. The interaction of smaller nanocarriers (<1 μm) is less favorable and these particles can escape the liver retention. However, other characteristics of the carriers such as shape or surface charge can strongly influence the nanoparticle capture by these phagocytic cells and it is necessary to study in detail each type of particle. Therefore, liver, spleen and lungs usually retain the major amount of injected nanoparticles. It is necessary to design these carriers to increase their chances of avoiding retention in these organs and also to avoid renal clearance.

Another issue that must be tackled in order to preview the outcome and distribution of nanomedicines is their interaction with blood components. When nanoparticles are exposed to this biological fluid, they are in contact with more than 3700 different proteins and many other complex biomolecules, which can bind competitively with the nanocarrier surface. After a given time since the injection, the particle surface will be coated by several

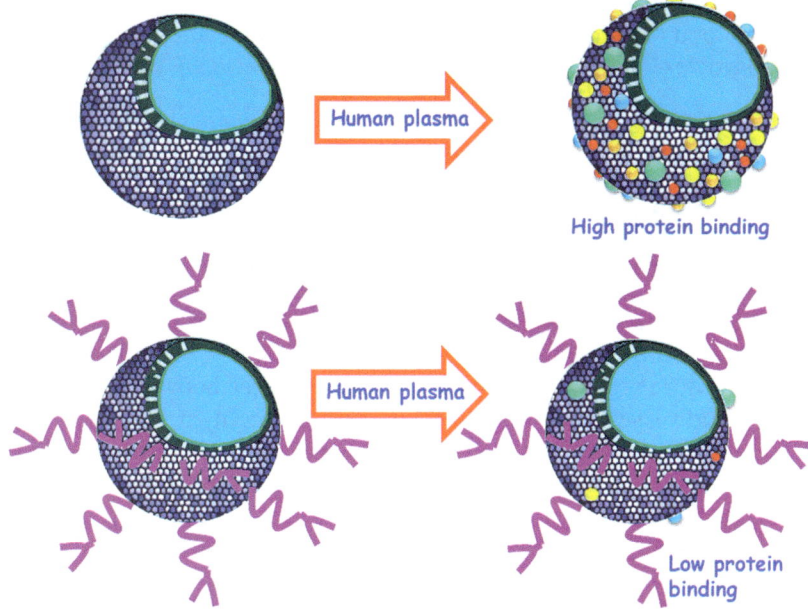

Figure 6.10 Potentially high protein binding onto the nanocarrier surface in biological fluids. Surface protection avoids this phenomenon.

proteins forming the *corona* of the nanoparticle, which constitutes a nanoparticle–biomolecule interface "readable" by the cells.[47] See Figure 6.10.

Among blood proteins, a very important type is the opsonin family, which are proteins designed to mark certain antigens or foreign bodies and induce the internalization and degradation of these entities by the mononuclear phagocytic system. This process is called opsonization. The macrophages of the mononuclear phagocytic system have the capacity to eliminate the nanocarriers from the bloodstream within seconds of their intravenous administration. However, these phagocytic cells cannot directly identify the nanoparticles by themselves, but rather recognize the opsonins bound to the particle surface. Thus, the injected nanoparticles that are covered by opsonins become visible for the macrophages and are rapidly removed from the bloodstream. In order to avoid this detection, several methods have been reported. These methods are usually based on the decoration of the external surface of the nanoparticle with molecules that interfere with the binding of opsonins. One of the most employed strategies is the grafting of shielding groups which block or interfere with the electrostatic or hydrophobic interactions that allow the protein's absorption onto the nanoparticle surface. These groups are usually long hydrophilic polymeric chains, such as PEG chains of different lengths.[48]

Anchoring PEG chains to the surface of nanoparticles in order to increase their half-life in the blood circulation is a well-established procedure known as PEGylation. Thus, PEGylated nanoparticles are able to avoid macrophage detection due to the lack of opsonin adhesion, preventing their clearance from the blood stream for longer periods; for this reason they are known as stealth particles.

6.6 Passive Targeting

6.6.1 Enhanced Permeation and Retention Effect

The EPR effect was first reported by the Japanese researchers Matsumura and Maeda in 1986.[49] They discovered that macromolecules >40 kDa, which is the threshold of renal clearance, selectively leak out from tumor vessels and accumulate within tumor tissues.[50]

This behavior is caused by the dramatic differences between tumoral and healthy blood vessels. Tumoral blood vessels are highly irregular, heterogeneous and tortuous, leaving unperfused spaces within the tumor. Moreover, their vessel wall structure shows specific characteristics such as wide inter-endothelial junctions, pericyte deficiencies, aberrant basement membranes, large numbers of fenestrations and transendothelial pores with diameters as large as several hundred nanometers. Consequently, vascular permeability is significantly higher in tumors than in healthy tissues. This enhanced permeability ensures sufficient supply of nutrients and oxygen to tumor tissues to enable their fast growth rate, but it could also be their weak spot. Drug-loaded nanoparticles with sizes up to a few hundred nanometers would be able to extravasate selectively through the tumoral blood vessel fenestrations, releasing their payloads within solid tumors without affecting other tissues. In addition, macromolecules, waste products or excess fluid in the interstitial space of healthy tissues are effectively recovered by the lymphatic system. In contrast, the rapidly growing tumoral cells compress the lymphatic vessels, especially at the center of the solid tumors, provoking their collapse. Thus, clearance *via* the lymphatic drainage is greatly compromised in neoplastic tissues, causing an additional retention of colloidal nanomedicines. The combination of high permeability and enhanced retention is the basics of the EPR effect and the reason why this phenomenon has become the "gold standard" in nanoparticle-based antitumoral therapy (Figure 6.11). It should be highlighted that this particular vascular architecture has been observed even in tumor nodules <0.2 mm.

As been previously mentioned, nanoparticles exploit the EPR effect in order to be selectively accumulated into the tumoral tissue. However, it is necessary to take into account that this effect is a highly heterogeneous phenomenon which shows significant variations from one tumor to another, between different patients, and even within the same tumor. Thus, in order to improve the therapeutic effectiveness of nanomedicines several strategies focused on the augmentation of EPR effect can be employed. The administration of angiotensin II before the nanoparticle treatment produces systemic hypertension which leads to a higher nanoparticle retention within the tumoral mass. The reason is that tumor blood vessels show very little response to this

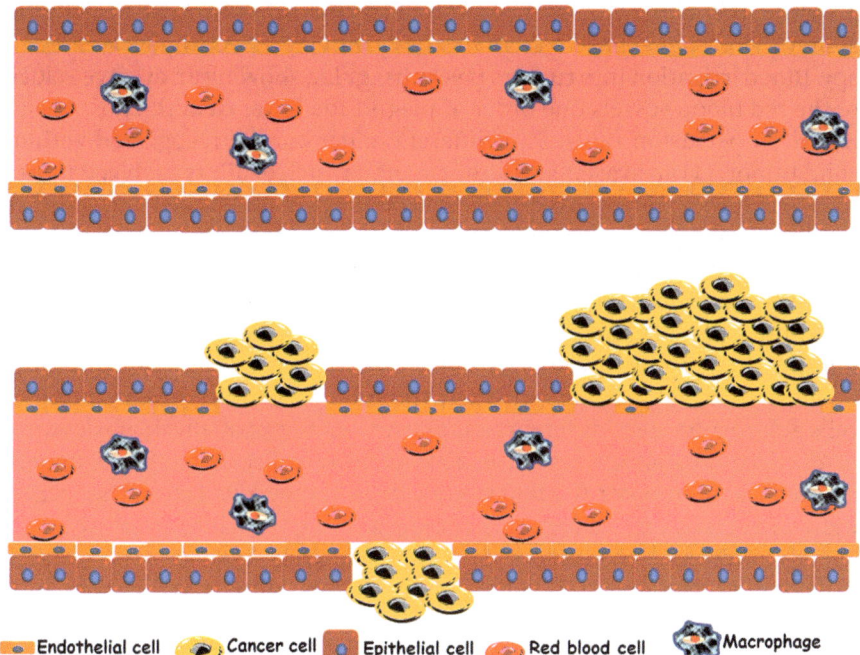

Endothelial cell | Cancer cell | Epithelial cell | Red blood cell | Macrophage

Figure 6.11 Healthy blood vessel (top). Three tumors growing on a blood vessel (bottom).

hormone due to the lack of a smooth muscle layer or pericytes required for vasoconstriction, whereas healthy vessels show significant constriction.

Therefore, the administered nanoparticles are pushed out into the tumoral space due to the different pressures between healthy or tumoral blood vessels. Nitric oxide (NO) is one of the main factors that sustain the EPR effect and is produced by the tumoral cells in higher amounts than normal cells in order to increase the blood flux and maintain their high nutrient requirements. Recently, it has been verified that the administration of nitroglycerine and other NO-releasing agents, which can be converted to NO under hypoxic conditions, increase the efficacy of the nanomedicine therapy.[51]

6.6.2 Tumor Microenvironment

The lack of an effective lymphatic system, particularly at the core of the tumor, in association with fluid leakage from the tumoral blood vessels produces interstitial hypertension, which constitutes an important barrier for the transport of nanoparticles within the tumor.[52]

Due to the increased interstitial fluid pressure, the major mechanism of mass transport within a tumor is diffusion, a process highly dependent upon the molecular weight, and therefore very slow for nanoparticles. In some cases, interstitial fluid pressure inside tumors can exceed the vascular pressure, provoking the intravasation of the nanocarriers back to the blood system. This could cause systemic toxicity and low effectiveness of the therapy. Blood irrigation into tumors is very irregular, showing an average velocity of the red blood vessels one order of magnitude lower than the host vessels. Based on perfusion rates, four different zones can be recognized within a solid tumor: (1) an avascular necrotic region which exhibits high cell mortality due to low content of oxygen and nutrients; (2) a transition zone called the seminecrotic region; (3) a stabilized microcirculation region which exhibits normalized blood perfusion; and (4) an advancing front.[53]

The presence of these unperfused regions with low partial oxygen pressure and acidic pH stimulates the formation of multiresistant tumoral cells immune to radiotherapy and chemotherapeutic agents, due to evolutionary forces in this highly hostile microenvironment. Moreover, mutated tumoral cells able to migrate to different tissues usually show up in this hypoxic region.

An effective nanocarrier must be homogeneously distributed throughout the affected tissue in order to eliminate all the tumoral cells before they acquire these drug-resistant and metastatic abilities. Thus, the nanoparticle should be able to diffuse into the tumoral extracelular matrix. This matrix is composed of a complex mixture of different proteins, mainly collagen and glycoproteins, which form a dense network that hampers the diffusion of the nanocarriers. Particles with negative charges on their surface are strongly retained by the collagen fibers due to the positive charge of collagen at physiological pH. Tumors that exhibit higher amounts of negatively charged proteins such as sulfated glycosaminoglycans can obstruct the penetration of positive carriers. One obvious rule is that the smaller the nanocarrier, the higher the diffusion through the extracellular matrix. However, if the particle is too small, the discrimination between healthy and tumoral tissues by the EPR effect is diminished. It is also compulsory to consider other aspects, such as the luminal surface of the blood vessels, which presents a negative charge and therefore the positively charged particles can be non-specifically attached to their surface and suffer rapid elimination from the bloodstream. For this reason, the nanocarrier should be carefully engineered with specific size, shape and surface charge according to each type of tumor. Figure 6.12 depicts a blood vessel invaded by three tumor masses. On the lower left side, one of the tumor masses shows the presence of nanocarriers loaded with cytotoxics being released, starting the cell death process; the lower right side displays the final situation, where the whole tumor necroses.

Nanoparticles administrated intravenously tend to accumulate into tumoral lesions due to the EPR effect. This effect constitutes a passive targeting effect of nanomedicines towards tumoral tissues and is one of the main reasons for their use in antitumoral therapy.

6.7 Active Targeting

Solid tumors are composed of a heterogeneous mixture of tumoral cells and healthy cells and therefore it is very important to discriminate between them. The external surface of the nanoparticles can be decorated

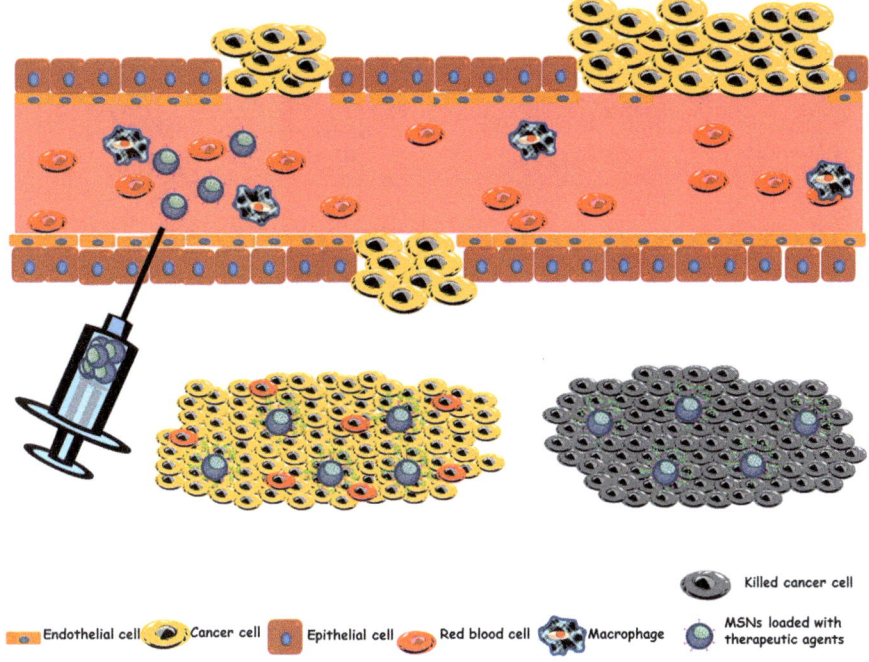

Figure 6.12 Layout of a blood vessel with three tumors; nanoparticles loaded with cytotoxic agents have been inserted. The arrival and load release of nanocarriers to the tumors are depicted on the bottom left side. The bottom right side illustrates the final scenario with necrosed tumors. MSN: mesoporous silica nanoparticles.

with molecules able to be specifically recognized by tumoral cells, thus providing the capacity to selectively kill tumoral cells in the presence of healthy ones. Therefore, these functionalized nanoparticles could attack specifically the tumoral cells present in the target tissue without affecting the neighboring healthy ones, producing a strong decrease of the side effects usually associated with antitumoral chemotherapy. This strategy is the basis of active targeting, and it also explains why nanoparticles are so promising in cancer treatment. Engineered nanoparticles can satisfy the requirements of the "magic bullet" principle postulated by Ehrlich in 1906.[54]

This concept is based on three principles: to find a proper target for a particular disease, to find the drug that effectively treats the disease, and finally to find how to carry the drug to the desired place.

The basis of active targeting consists of the use of targeting agents, specifically selected for each pathology, placed on the periphery of the nanocarriers. In the case of cancer, the targeting ligand is usually chosen to bind to a receptor overexpressed by tumor cells or tumor vasculature and not expressed by healthy cells or vessels. This receptor should be homogeneously expressed by all diseased cells (Figure 6.13).

Active targeting

Tumoral cell

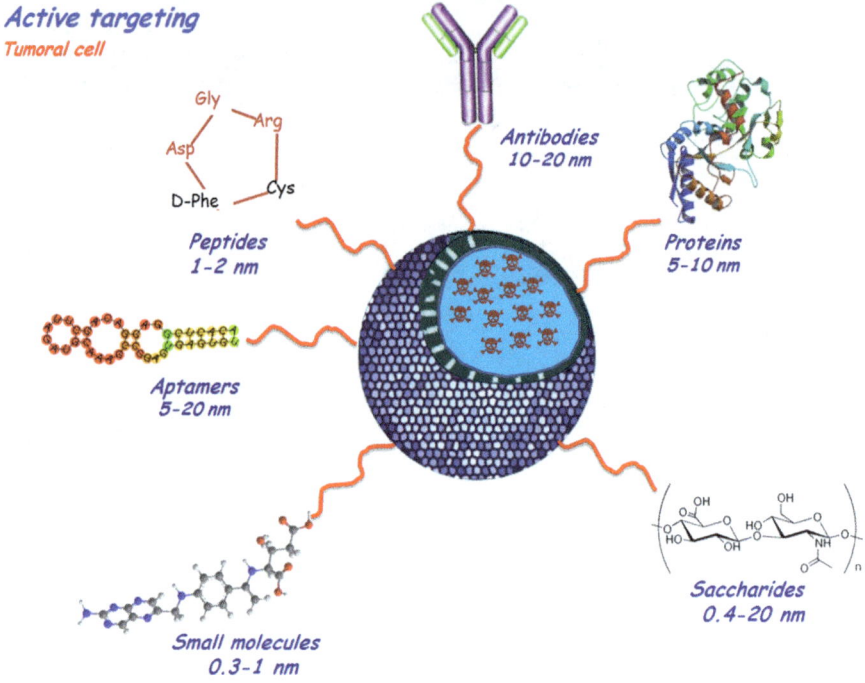

Figure 6.13   Active targeting.

One of the most useful families of targeting molecules is the antibodies. An antibody is a protein-based macromolecule fabricated by living organisms that is able to bind selectively to certain regions of foreign entities or pathogens. Once labeled by the antibody, the macrophages and other cells of the immune system are able to detect the foreign body and destroy it. Conjugation of antibodies on the surface of a nanoparticle could be performed randomly using chemical reagents such as carbodiimides, which create robust amide bonds between carboxylic acid groups placed on the nanoparticle surface and the free amino groups of the antibody, provided by lysine or the N-terminal group. Using this strategy, the activity of the attached antibodies is usually lower because the antibody is bonded to the surface in multiple ways, some of which can block the recognition region of the macromolecule. Alternatively, antibodies can be conjugated using maleimide-type cross-linkers using thiol groups present in known regions of the antibody (provided by cysteine) or using engineered antibodies with thiol groups placed away from the recognition site. Other strategies for the grafting of antibodies onto inorganic nanoparticles are the use of biomolecules as cross-linkers, such as biotin–streptavidin bridges[55] or protein A.[56]

The first strategy involves the biotinylation of the antibody and the nanoparticle followed by the coupling to each other using streptavidin, acting as a bridge between them. Streptavidin is a 60 kDa protein isolated from the bacteria *Streptomyces avidinii*, which presents a strong affinity for the biotin moieties, producing one of the strongest non-covalent intermolecular interactions in nature. This protein has four pockets able to bind a biotin molecule and act as a bridge between the antibody and the nanoparticle. The second strategy involves the use of protein A as linker. Protein A is a cell-wall-associated protein produced as a defensive mechanism by the bacteria *Staphylococcus aureus*. It binds to antibodies secreted by the host *via* their Fc region, avoiding the recognition of the pathogen. Thus, the use of this molecule for antibody grafting presents a very important advantage because the recognition zone of the antibody is not involved in the bond.

Some antibodies conjugated with nanoparticles are currently used in clinical applications, such as trastuzumab (anti-human epidermal growth factor receptor-2) against breast cancer or panitumumab (anti-epidermal growth factor receptor) against colorectal cancer, among others. It is also possible to employ small antibody fragments that contain only the recognition site.[57] The antibody fragment retains the capacity to recognize the antigen while lacking the constant Fc effector region which is responsible for binding to immune cells and can induce complement activation, resulting in the premature clearance of the conjugated nanocarrier.

Tumoral cells present a fast growth which that requires larger amounts of nutrients than normal cells, and for this reason they usually overexpress receptors to capture some essential molecules. Thus, the attachment of these molecules to the surface of nanocarriers could improve the grade of internalization and therefore the effectiveness of the therapy. For instance, transferrin receptors are overexpressed 100-fold more than the average expression in healthy cells due to the higher requirements for iron of the tumoral cells. Folic acid is an essential vitamin required for the biosynthesis of DNA, among other important roles. Thus, the folate receptor is usually upregulated in many tumoral cells (up to 40% of human cancers), in some cases more than two orders of magnitude.[58]

Due to the essential role of this molecule, the cell internalizes the folic acid into a vesicle and releases it into the cytosol. Folic conjugation is widely used in nanomedicine because this molecule is inexpensive, non-toxic, stable, non-immunogenic and easy to conjugate with different substrates while retaining its binding capacity to the desired surface, depending on the designed application.[59]

The cellular membrane is usually decorated with different glycoproteins, which are a distinctive label of each cell type. Tumoral cells present their own glycoproteins on their surfaces and this can be exploited in order to guide specifically the nanocarriers. Lectins are non-immunogenic proteins able to recognize the glycoproteins or glucides present in these cells and therefore they can be used as targeting agents. It is also possible to employ the opposite strategy, which consists of the attachment of certain glucides to the nanoparticle surface, which can be recognized by the glycoproteins present in the tumoral cells.[60]

This active targeting is limited by the receptor capacity. The number of receptors located on the cell surface limits the number of targeting molecules that can be bonded to the tumoral cell. Under ideal conditions we could assume an infinite binding affinity, the number of ligands that can be bound by the tumors equaling the number of available receptors (considering a $1:1$ binding ratio). However, the real situation is that only a fraction of the ligands are able to bind the receptors. The rest of the targeting conjugates suffer the same fate as non-targeted substrates.[61]

Using targeting ligands that show a high affinity for the receptors we can overcome this limitation, but a new problem appears in this case. The high affinity of the ligand to the receptor causes the conjugated nanocarrier to be strongly retained by the first cellular line after extravasation, leading to a poor penetration capacity and therefore producing only local effects. This effect is also known as *binding-site barrier*. Nanoparticles can be decorated with multiple copies of one single targeting agent, enabling the use of ligands with a low affinity constant with the receptor. This property is called "multivalency" and has been applied in particular with peptides that show moderate affinity for the cellular receptors.[62] The multivalency effect generally shows enhancements of $10-10^4$ orders of magnitude in the binding capacity of the nanocarriers. Unfortunately, the presence of a large number of receptors can also increase recognition by macrophages of the reticuloendothelial system, leading to faster clearance.

6.7.1 Angiogenesis-Associated Active Targeting

The tumoral mass induces the formation of new blood vessels (angiogenesis), secreting different growth factors, such as vascular endothelium growth factor (VEGF) or platelet-derived growth factor in order to support their fast growth rates. Attacking the tumoral vasculature can destroy the tumoral mass due to the lack of an efficient blood supply providing nutrients and oxygen to the tumor. This strategy presents some interesting advantages, as follows.

(1) The extravasation of the nanocarriers from the blood vessels is not required.
(2) Tumoral blood vessels usually overexpress certain receptors, which are easily accessible by the functionalized nanoparticles.
(3) Endothelial cells that compose the tumoral vessels are less susceptible to mutation due to their more stable environment, compared with the tumoral cells housed inside the solid tumoral mass which are exposed to harder conditions (lower pH values, low O_2 pressure, *etc.*). This reduces the risk of multidrug resistance under prolonged treatments.
(4) The endothelial cell markers are common in different tumors.

The main angiogenic markers that have been widely explored in nanomedicines are:

(1) *VEGF*: when the tumoral cells are exposed to hypoxic conditions they increase the production of VEGF. This results in an upregulation of the VEGF receptors (VEGFR-1 and VEGFR-2) in the endothelial cells. Thus, it is possible to exploit the presence of these receptors using VEGF as a

targeting ligand, inducing the endocytotic pathway. Also, the opposite strategy can be employed using anti-VEGF attached to the nanocarrier as targeting ligands, in order to inhibit ligand binding to VEGFR-2.

(2) $\alpha_v\beta_3$ *integrin*: an endothelial cell receptor for extracellular matrix proteins, which is highly overexpressed in neovascular endothelial tumoral cells, but is scarcely present in healthy cells. Oligopeptides harboring the RGD sequence (Arg-Gly-Asp) bind selectively to this receptor. Ruoslahti and co-workers discovered that the use of a cyclic peptide with a hidden RGD motif (CRGDK/RGPD/EC) attached to the nanoparticle surface allows not only the targeting of tumoral blood vessels, but also improves the penetration of the nanoparticle within the tumoral mass.[63] This peptide sequence binds to the $\alpha_v\beta_3$-integrins of the tumor endothelium and then, after a proteolytic cleavage, the RGD motif is exposed and interacts with neuropilin-1 receptors, which promote the internalization of the nanoparticle.

(3) *Vascular cell adhesion molecule-1*: an immunoglobulin transmembrane protein that is expressed only on the surface of tumoral blood vessels and during inflammation. This protein is responsible for inducing cell-to-cell adhesion, an important requisite of the angiogenesis process. The attachment of antibodies specifically designed to bind to this molecule could improve the antitumoral activity of the nanocarrier, theoretically without side effects.

(4) *Matrix metalloproteinases*: proteins that degrade the extracellular matrix, mainly by hydrolyzing the collagen network. They play a determinant role in angiogenesis and metastasis. Some antibodies capable of binding selectively to these proteins have been conjugated to different nanocarriers, especially antibodies that recognize the membrane type-1 metallo-proteinase, which is present on the endothelial tumoral cells of a large number of malignancies.

Nanocarrier design for applications in active targeting must follow the steps illustrated in Figure 6.14, starting with loading the cytotoxin, followed by surface functionalization.

6.8 Preparation of Nanocarriers

6.8.1 Synthesis

There are plenty of chemical methods to synthesize nanoparticles. We describe some of the most important methods.

6.8.1.1 Modified Stöber Method

The most common method preparing MSNs is the so-called *modified Stöber method*.[64] The Stöber method consists of the precipitation of silica nanoparticles from ethanol solution under basic pH. Starting from soluble species

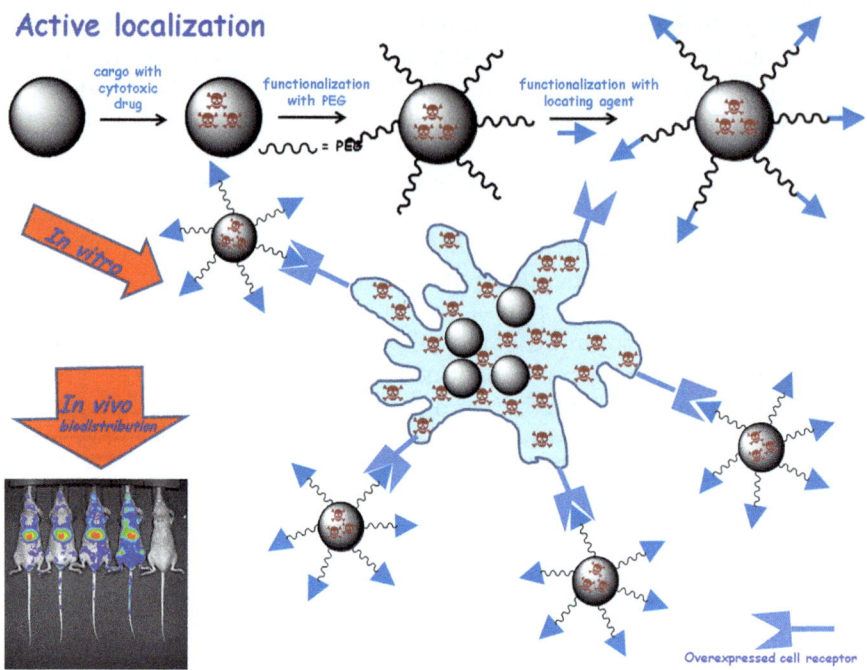

Figure 6.14   Steps in active targeting.

of silica precursors, generally alkoxisilanes such as tetraethyl orthosilicate (TEOS), the fast hydrolysis and condensation under alkaline conditions leads to the formation of negatively charged surface nanoparticles in the range of a few nanometers. The modified Stöber method involves incorporating a structure-directing agent (SDA), especially of cationic nature like cetyl trimethylammonium bromide (CTAB), so that preformed micelles are present during nanoparticle precipitation. When the synthesis conditions such as temperature, stirring, *etc.* are well established, the precipitated nanoparticles maintain the mesostructure imposed by the SDA. Finally, by removing the surfactant by calcination or extraction techniques, ordered MSNs with a narrow size distribution are obtained.[65]

Figure 6.15 shows a scheme of the synthesis using the modified Stöber method.

MSNs with additional functionalities can be prepared *via* the modified Stöber method. For instance, the preparation of magnetic MSNs has been one of the most pursued goals in this field.[66] Generally, magnetic MSNs have been prepared by adding colloidal iron oxide nanoparticles (commonly of Fe_3O_4 or $\gamma\text{-}Fe_2O_3$) to the reaction media.[67] However, the presence of ferrofluids leads to phase separation and hinders the ordering of the mesophase. For this reason, strategies involving the functionalization of the magnetic nanoparticle or aqueous/oil phase exchange have been developed to prepare these composites as non-aggregated nanoparticles with narrow size

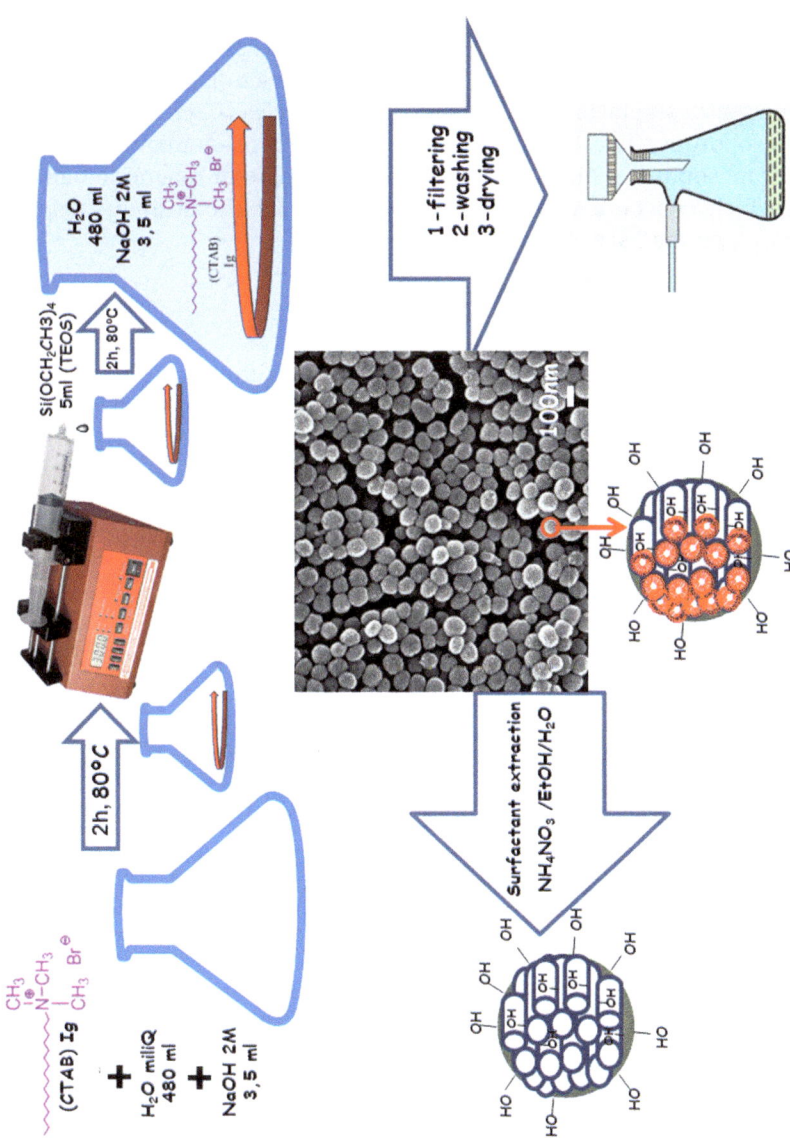

Figure 6.15 Synthesis of nanoparticles by the modified Stöber method.

distribution.[68] In any case, we will begin by synthesizing the nanoparticle, load it, and prepare its surface. In addition, as previously discussed, magnetic nanoparticles and contrast agents can be placed in the inner region, if the application so requires, not only drug molecules. It is important to note that the chosen method must provide homogeneous nanoparticles, both in terms of size and shape, with inner pores to load the drug, and with a high number of OH groups to effect different functionalizations.

In Figure 6.16 we can see another method with similar results, although here the synthesis starts with magnetic nanoparticles which are then coated with mesoporous silica. In this system, besides the already-mentioned advantages of mesoporous silica, there is the added value of magnetic nanoparticles, which can be a very important element for the stimulus-response actions, as we shall see later.

6.8.1.2 Aerosol-Assisted Methods

Mesoporous microspheres can be synthesized through aerosol-assisted routes. This strategy is based on preparing micro- or nanoparticles by forming an aerosol from a solution containing the material precursors.[69,70] Each droplet of the aerosol behaves as a micro- or nanoreactor, in such a way that the final particle size is strongly dependent on the size of the aerosol droplets. MSNs can be prepared by incorporating an amphiphilic copolymer as a

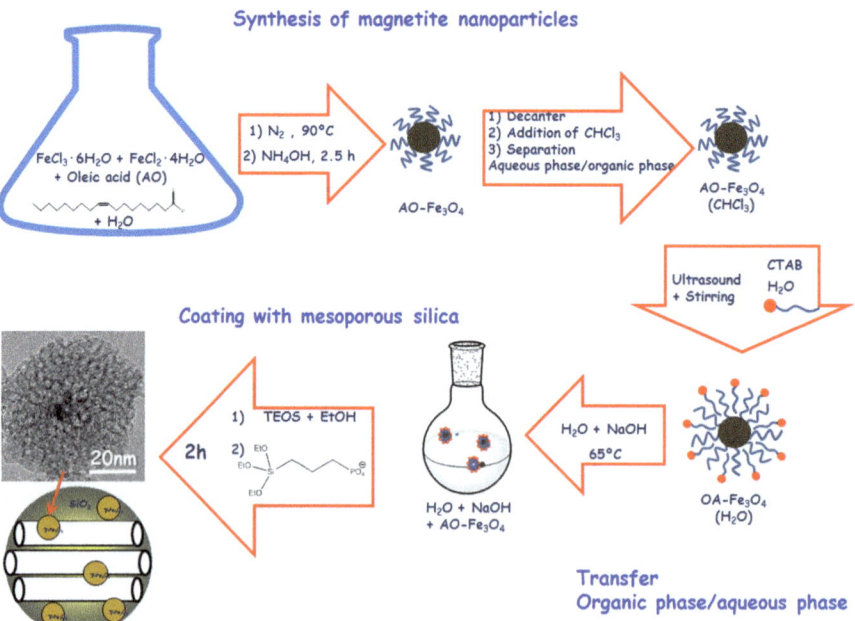

Figure 6.16 Synthesis of mesoporous silica nanoparticles with magnetic nanoparticles inside.

structure-directing agent, together with silica hydrolyzed species, thus leading to the formation of a mesostructure through a self-assembly mechanism. After calcination or ethanol/acid extraction, the surfactant is removed, leading to highly ordered mesoporous nanoparticles with high surface area and pore volume. The aerosol-assisted method leads to spherical shapes with controlled sizes in the micro- and nanoscale, instead of irregular and difficult to control bulky grains, as those obtained by standard hydrothermal methods.

The aerosol can be generated by several techniques, such as Venturi-effect atomizers,[71] radiofrequencies,[72] *etc.* Moreover, the aerosol droplets can be treated in different ways to obtain the final products (laser and thermal pyrolysis, vapor deposition over substrates, *etc.*). Figure 6.17 represents a device where the aerosol is generated by ultra-high-frequency spraying of the solution.[73] When the piezoelectric transducer is excited near its own resonance frequency (~850 kHz for the transducer used), a geyser is formed at the surface of the liquid. This geyser produces ultrafine droplets, which form an aerosol. N_2 or Ar gases are generally used as carrier to convey the aerosol to the pyrolysis zone, which consists of a long tubular furnace. The residence time of the particles in the high-temperature zone is controlled by the gas flow. Finally the dried particles are collected outside the furnace with an electrostatic filter, which is a thin tungsten wire suspended in the center of a tubular stainless steel collection plate.

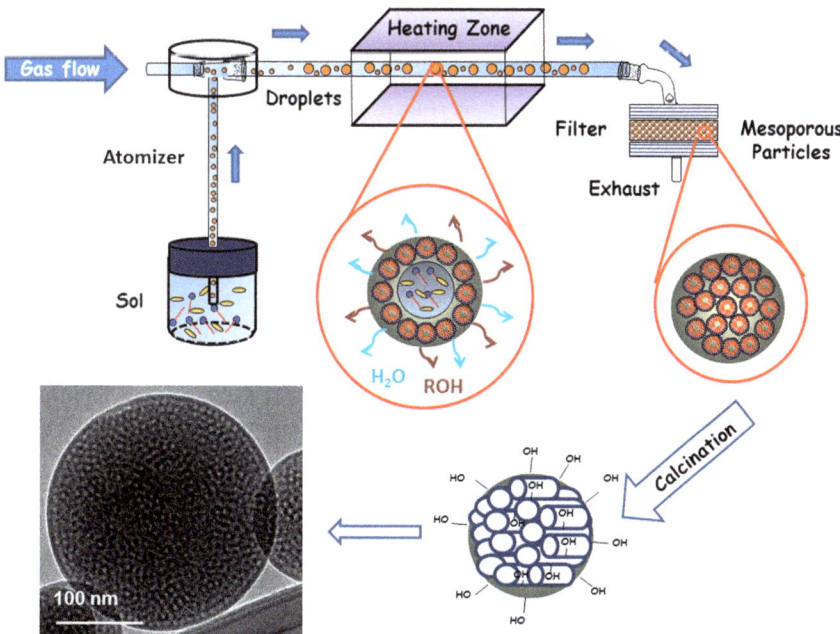

Figure 6.17 Pyrosol method.

Magnetic mesoporous nanoparticles can be also prepared by this method.[74] In fact, by adding certain amounts of colloidal iron oxide nanoparticles (generally magnetite or maghemite) as ferrofluid, these nanoparticles are easily incorporated to the aerosol droplets. Certainly, high amounts of ferrofluid within the precursor solution hinder the preparation of an ordered mesoporous structure. However, significant amounts of magnetite nanoparticles have been incorporated into mesoporous particles, providing good magnetic properties while keeping the ordered mesoporous structures.[75]

6.8.2   Loading

Once prepared, the mesoporous nanoparticles must be loaded with the adequate drug for a given therapy. The incorporation of the payload is achieved using impregnation techniques. In this sense, drug solutions at the highest possible concentrations are prepared and the mesoporous particles are soaked in them. Loading parameters such as time and temperature are strongly dependent on the solvent. The hydrophobic/hydrophobic character of the drug must be taken into account. This is because the drug will attach to the mesoporous matrix or will remain in the solution, depending on the affinity of these molecules with the solvent of the chemistry surface of the material.[76,77]

These particles loaded with drugs and magnetic nanoparticles inside, can be used in hyperthermia treatments, controlling time and temperature, in order to eliminate malignant cells; or treated as a drug delivery system, with the adequate cytotoxic products to eliminate cancerous tissue. However, they are not smart systems, because they start to release the product as soon as they are in contact with the medium, which is far from what was expected at the design stage. Figure 6.18 shows how the drugs are loaded in the nanoparticle pores but, as soon as they are in contact with the medium, they rapidly release >50% of the load, with a subsequent sustained release for the remaining load. The purpose in these systems is to achieve zero release before reaching the target, so these do not comply with this requirement.

This acute initial release is a very serious issue, particularly in systems treating carcinogenic tumors and hence loaded with cytotoxins. Therefore it is compulsory to find a solution against premature release. Several options are available, and they are described throughout this chapter; but perhaps the most evident method, which can be taken as starting point, is the design of stimulus-response systems.

6.8.3   Stimulus-Response Systems

Stimulus-response systems, also known as smart systems, exhibit the ability to transport efficiently a load along a zero-release trajectory unless they receive a predesigned stimulus that triggers the release. The stimulus might be internal or external, and it is chosen depending on the requirements of

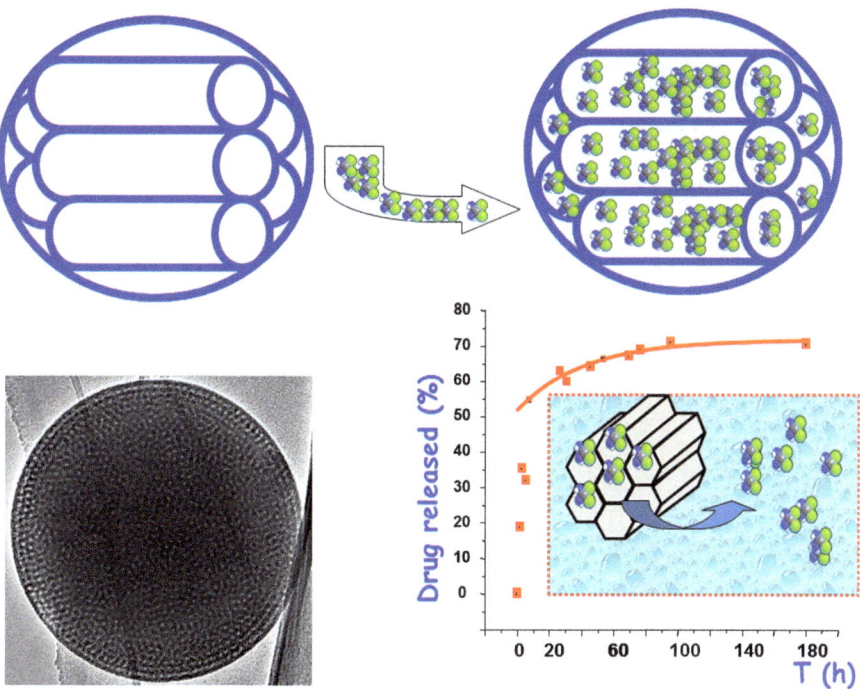

Figure 6.18 Schematics of drug loading of a porous nanoparticle. The transmission electron microscope image (bottom left) shows the pores. Drug release graphics (bottom right) reveal that a massive release – more than 50% – is verified in the first hours, before reaching a sustained regime.

the particular system. The *external stimuli*, remotely applied by the clinician, or *internal stimuli*, usually defined by the treated pathology, is the key to start releasing the load. Both will trigger the release of the trapped drugs, achieving a better control of the administered dose. Smart nanoparticles that employ internal stimuli present the advantage of not requiring external apparatus to trigger the release. However, the control of the administrated dosage is lower than in the case of devices that employ an external stimulus. In any case, each type of system presents pros and cons which should be evaluated taking into account their potential clinical application.

Such systems are successfully in use at present in different fields, but this chapter focuses on drug-release applications, with emphasis on cytotoxic drugs, due to the serious problem created in the body if any load is released before reaching the tumor mass.

Silica mesoporous nanoparticles can be used for such a system if, after loading the cytotoxic drug, their pores are closed by molecular gates which cannot open until the corresponding command is received. Hence, controlled release of the encapsulated species could be achieved.

The design of stimulus-response systems could solve the problem of drug release before the nanocarrier reaches its destination. What is required is

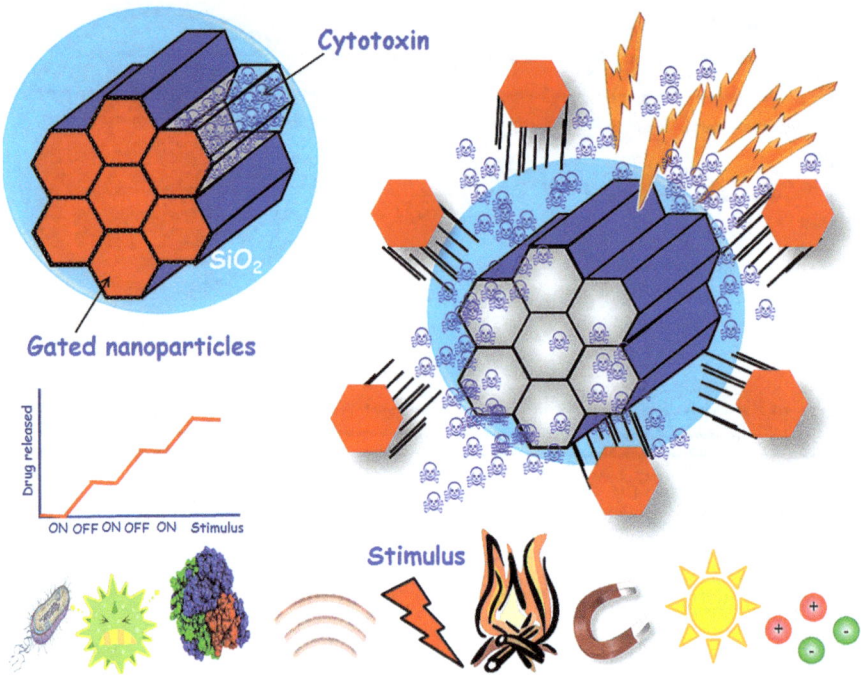

Figure 6.19   A loaded nanoparticle; the load is kept inside thanks to molecular gates. Release is only achieved when an external stimulus is applied (right). The graphic (left) shows the characteristic behavior of stimuli-responsive systems. Different stimuli used are also shown (bottom).

for the nanoparticles to be loaded and for them to retain cytotoxin inside them, until arrival at the target. In that moment, thanks to an external stimulus, the release can occur. So we need to design smart systems having zero release on the way, and then starting to release the drug under an external stimulus once the targeted tissue has been reached. A first solution to this problem would be to design stimulus-response systems. We can choose among many different stimuli. There are several examples of them all in the literature. Figure 6.19 shows the purpose of a stimulus-response system, and the different stimuli already tested in different studies, which are reviewed later on.

The important issue here is to ensure that the particle reaches its target, and that the chosen stimulus acts efficiently at the desired moment. Nanoparticles can be loaded with a drug; after loading, we could fit locks preventing the drug release and, with a stimulus, send a command to open those locks and start the release (see Figure 6.20).

Another possibility is to coat the loaded nanoparticle with a cleavable shell that preserves the drug until it reaches its destination.

And a third possibility would be to transport pro-drugs which, in the case of unwanted premature release, do not pose any risk to the body. When the

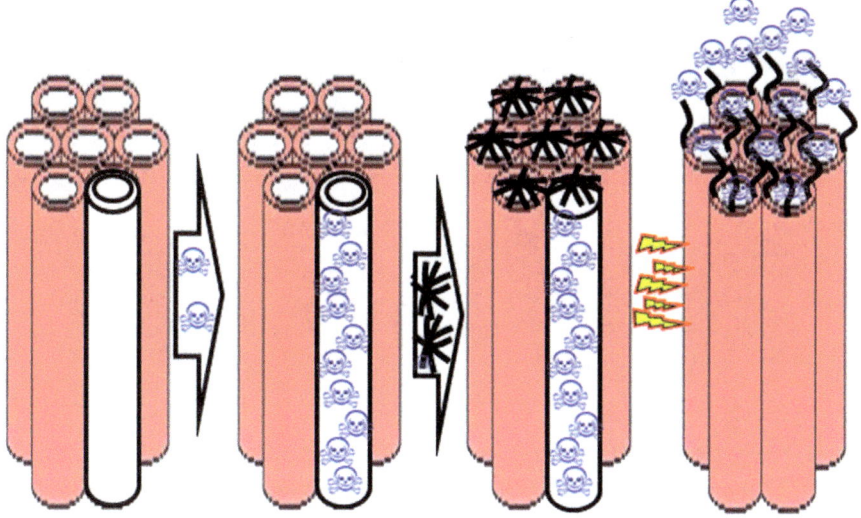

Figure 6.20 Steps to follow in a stimuli-responsive system. From left: micro-organisms, enzymes/biomolecules, ultrasound, electric field, heat, magnetic field, light and pH.

target is reached, a certain mechanism would be activated (for instance, a particular enzyme activity) that transforms the pro-drugs into the active cytotoxic drug.

These three possibilities are reviewed below; none of them is totally effective, but their outcome is infinitely better than the systems discussed in Section 6.8.2.

Among these three options, pro-drug systems are initially advantageous, because any small premature release would not affect other tissues, since the load is non-toxic prior to its activation.

Figure 6.21 depicts these three solutions to premature drug release.

We present a proof-of-concept study of these systems, to elucidate if these three options improve cytotoxin retention inside the nanocarriers.

The idea is to obtain a smart drug release system using nanocaps, magnetic nanogates in this particular case. With this purpose, we performed the synthesis of mesoporous nanoparticles with single DNA strands attached to their surface and then we loaded the nanoparticles with the cytotoxic drug.[78] Next, we synthesized the magnet-sensitive caps; in this case we used superparamagnetic iron oxide nanoparticles. Their size should be equivalent to the pore size in order to achieve a better capping effect. We attached complementary DNA strands to those employed in mesoporous silica particles on the surface of the iron oxide particles.

When in contact, both complementary strands are hybridized and the pores are closed.

Under an external stimulus, a magnetic field in this case, the maghemite particles acting as gates loosen by DNA dehybridization, due to the heat

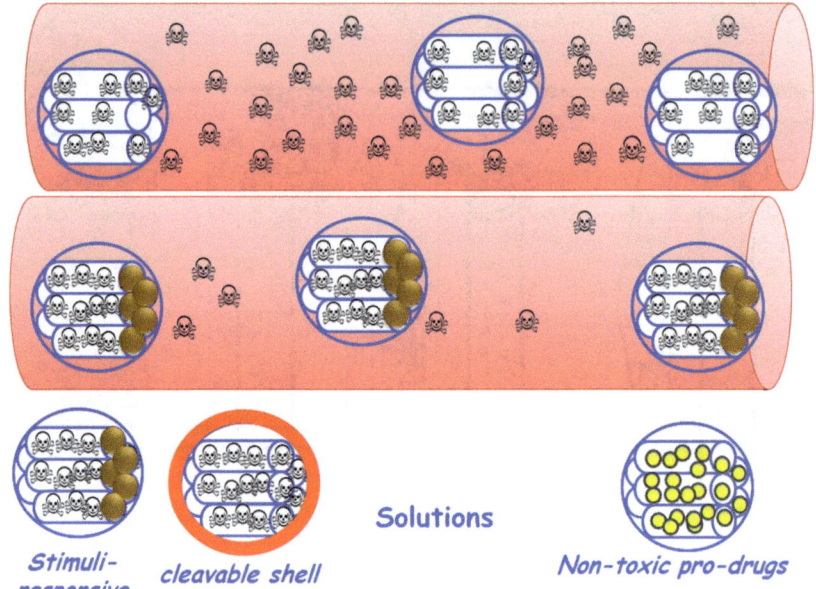

Figure 6.21   Loaded nanocarrier (top): when in contact with the bloodstream, the load starts to exit. Loaded nanocarrier with closed pores (center): it may release part of the load, but in a much smaller amount. Three potential solutions to unwanted load release in open nanocarriers (bottom): stimulus-responsive, cleavable shell and non-toxic pro-drugs.

produced, and the drug loaded inside is gradually released. Figure 6.22 schemes this process.

This is a reversible system, because the gates open under the applied field, and if the field is turned off the gates close again. It is in fact a stimulus-response system with on–off behavior. The proof of concept is passed. Now, what is needed is an adequate functionalization of these nanoparticles, so that they are preferentially directed towards the tumor cells. In this first assay, drug retention before reaching the target was improved.

The second solution can be to coat the nanocarrier with a layer (*e.g.* polymer). Following the previous example and adding an additional step, we use a thermosensitive polymer shell able to respond to temperature changes. The aim is to design a system that can release two different drugs with two different release rates; it could deliver different drugs to specific cancer cell lines, and different macromolecules with the aim to improve its therapeutic response. When we apply the magnetic field, the iron oxide particles cause an increase in the temperature of the surrounding tissue, and then, the polymer collapses and the macromolecules and drugs trapped inside diffuse out following different kinetics: the macromolecule exhibits a fast

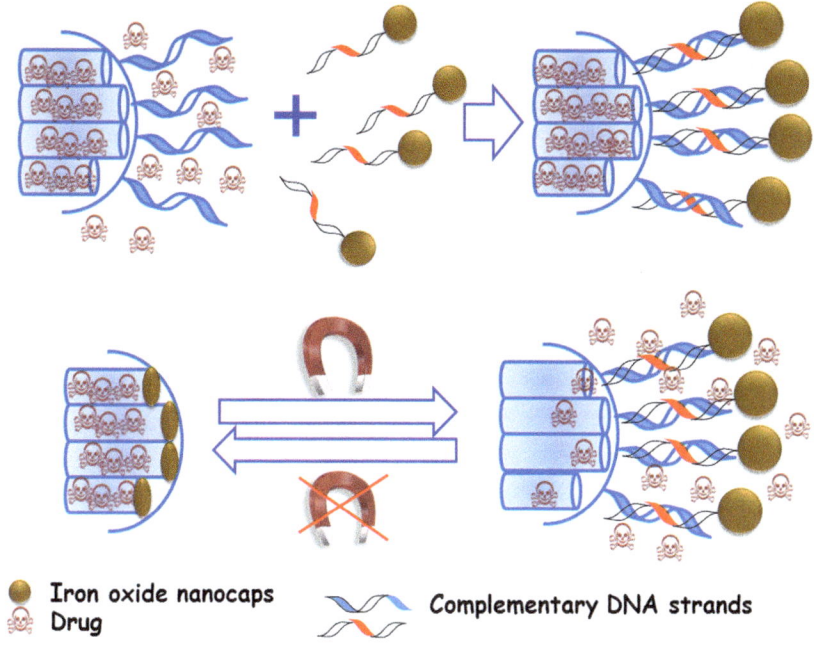

Figure 6.22 Schematics of a stimulus-responsive system.

release while the more toxic drug diffuses slowly.[79] Figure 6.23 schemes this idea.

The most important factor is that we can control on demand the substrate release by controlling the external magnetic field.

As already mentioned, the third solution relies on the use of pro-drugs. Using non-toxic pro-drugs the system would work in a similar way, but in this case, since pro-drugs are non-toxic, their release away from the target would not be a problem. Therefore, nanoparticles loaded with pro-drugs would travel in the bloodstream and reach the tumoral tissue. Once at the destination, the stimulus is applied to release them. Following their release and when in contact with enzymes, these pro-drugs are activated, producing the cytotoxin that will attack the tumor cells. In this design, we aim to transport nanoparticles loaded with pro-drugs which have been previously bound to encapsulated enzymes.

The idea is that, as soon as they reach the target, the stimulus to release pro-drug is activated, the enzymes will be immediately contacted, so that the pro-drugs are activated and the cytotoxin is generated. In this work indol-3-acetic acid is used as the cytotoxin and the enzyme horseradish peroxidase is protected with a polymeric capsule.

In addition, the inner surface of the particle must be functionalized to maximize loading capacity. Enzymes oxidize quickly in presence of proteases, temperature or oxidation, producing hydrogen peroxide. If this happens, the enzyme will be useless for our purpose. Therefore, the enzyme must be protected with a polymer capsule which avoids oxidation. In this way, we can use the enzymes for our purpose.

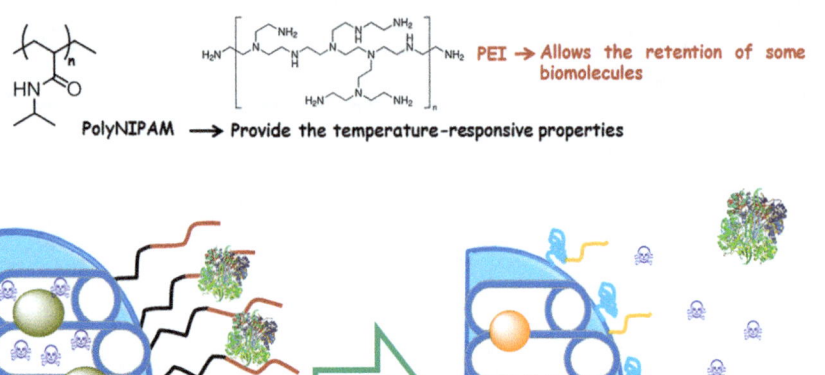

PolyNIPAM ⟶ Provide the temperature-responsive properties

Figure 6.23 Operation mode of a nanocarrier with both nanogates and a cleavable shell.

Here we can see a possible chemical process to encapsulate these enzymes. Therefore, the first step is to synthesize an amino-functionalized mesoporous silica carrier which will then be loaded with the pro-drug. The next step is to bind the pro-drug-loaded nanoparticle to the encapsulated enzyme. And finally, we have the full system that must travel to the tumoral mass.

Figure 6.24 shows nanoparticles used to load the pro-drug, the nanocapsules, where the enzyme is confined, and the complete system, nanoparticle bound to nanocapsules with confined enzymes. It is important to note that enzyme functionality is preserved, even after being encapsulated by the polymer; cytotoxicity studies have been performed with human neuroblastoma cell lines.

Studies have been undertaken *in vitro* to verify cell viability evaluation.

Having described these three methods to retain the load until the destination is reached, we discuss different options to design a stimulus-response system with silica mesoporous nanoparticles acting as carriers, making use of different stimuli.

6.8.4   Different Kinds of Stimuli

In this section, different stimuli such as magnetic field, redox, light, enzymes and small molecules are described. Figure 6.25 shows the different stimuli that can be applied over the mesoporous silicon nanoparticles. Figure 6.26 depicts three of these stimulus-response systems.

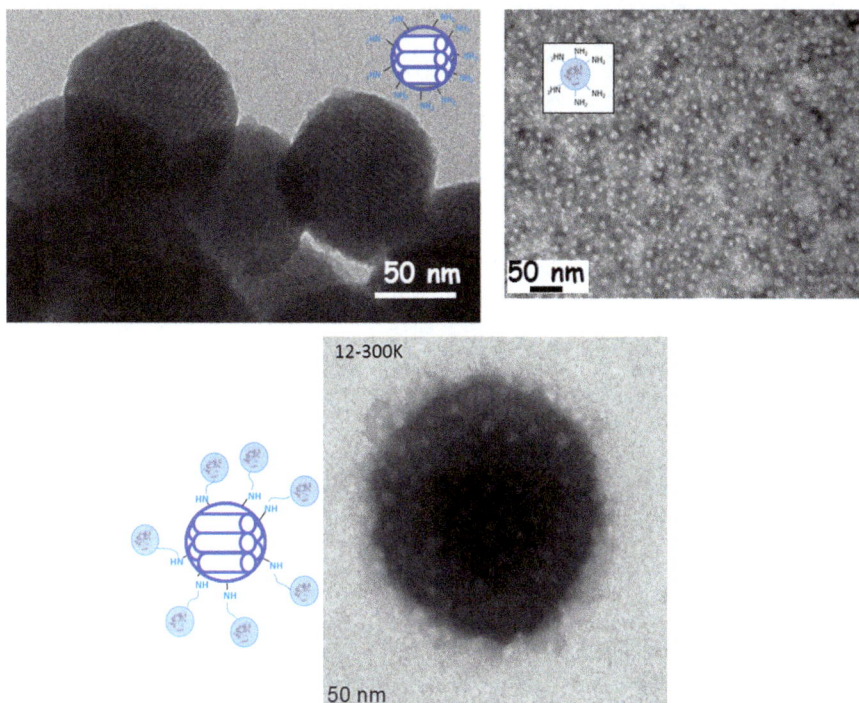

Figure 6.24 Transmission electron microscope images of mesoporous silica nanoparticles with attached nanocapsules containing enzymes.

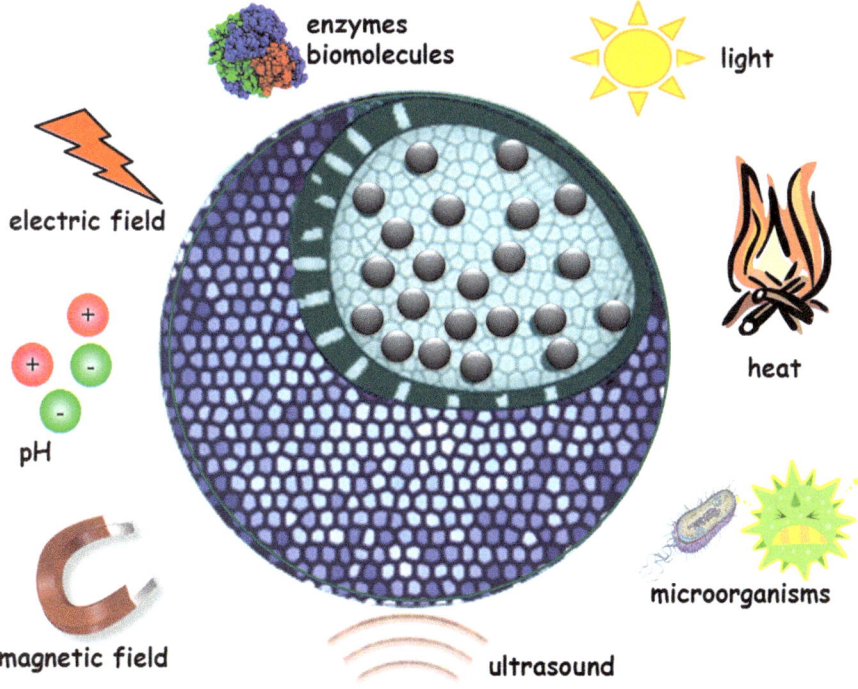

Figure 6.25 Mesoporous silica nanoparticle and different stimuli eligible to open its nanogates.

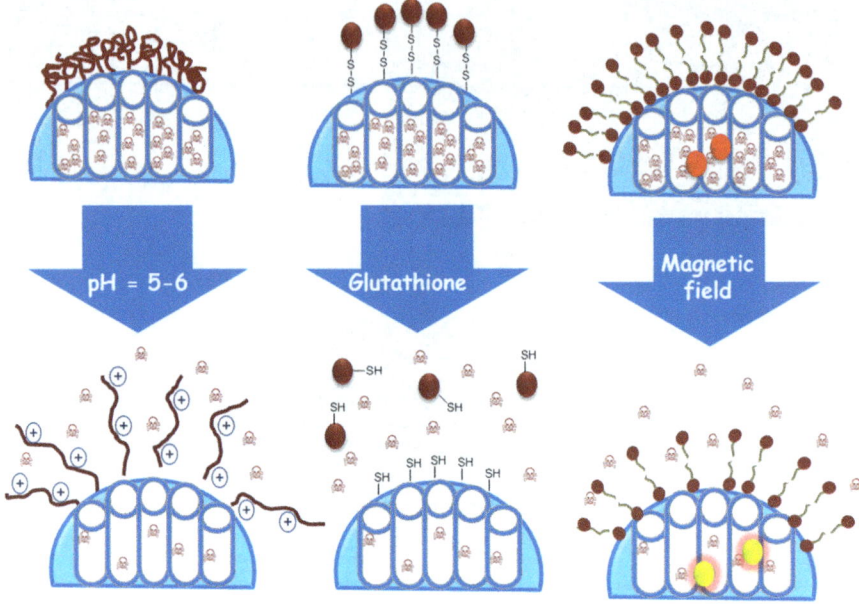

Figure 6.26   Depiction of three different stimulus-response systems: pH, glutathione and magnetic field.

6.8.4.1   pH

Some tissues of the body show alterations in their pH level when they are in a pathological state. Thus, tumors and inflamed tissues usually present lower pH values (up to pH 5.5–6) than blood or healthy tissues (at pH ≈ 7).[80]

   This pH gradient can be employed as trigger event in MSNs functionalized with pH-sensitive gates attached onto the pores. Moreover, the pH value of the endosomes or lysosomes becomes even lower and therefore the internalized nanocarrier functionalized with these pH gates can release the cargo once there, avoiding the premature release of the drug outside the cell. Different nanocaps, such as β-cyclodextrins[81] or gold nanoparticles[82] can be attached onto the pore outlets through acid-cleavable linkers. Another strategy is based on coating the mesoporous surface with a polymeric shell which undergoes a physicochemical transformation in acid media.[83]

   Inorganic pH-sensitive coatings such as calcium phosphates can be also applied to the MSN surface.[84] This shell is dissolved into non-toxic ions under acidic media (pH 5), allowing drug release.

6.8.4.2   Temperature

The artificial production of hyperthermia or hypothermia by physical means is a well-established procedure in clinical practice. Also, some tissues exhibit higher temperature values when affected by different diseases. One of the most common approaches to exploit this in order to produce stimulus-responsive carriers, is the attachment of thermosensitive polymers, generally poly-N-isopropylacrylamide (PNIPAM) and its derivatives onto the external surface of the MSNs.[85] This polymer is in the hydrated form below the lower critical solution temperature of 32 °C, which prevents the release of the drugs trapped inside, whereas it suffers a collapse if the temperature exceeds this value, producing the pore opening.

6.8.4.3 *Magnetic Field*

MSNs do not have magnetic properties and therefore it is necessary to attach superparamagnetic moieties into their structure in order to make the material sensitive to magnetic fields. When exposed to alternative magnetic fields, the presence of these superparamagnetic nanocrystals provoke a temperature increase of the media due to the rapid rotation of the magnetic nuclei (Brownian fluctuations) and the fluctuation of the magnetic moment[86] (Néel fluctuations).

The production of magnetic MSNs can be developed by trapping iron oxide nanocrystals into the silica matrix using aerosol-assisted[87] or modified Stöber methods.[88]

These materials, under the exposition of alternative magnetic fields are able to generate enough heat to lead to the destruction of tumoral cells (magnetic hyperthermia).[89]

The external surface of these magnetic particles can be decorated with thermosensitive gates which open or close the pores in response to the temperature changes. A very interesting thermosensitive molecule is the polymer of DNA. A single strand of DNA is able to bind selectively with its complementary strand through hydrogen bonds between the nitrogenated bases; adenine (A) is complementary with thymine (T) and guanidine (G) is complementary with cytosine (C). The two strands will be separated if the temperature reaches a certain value, which depends on the G/C content and the length of the strand, in a process called dehybridization.[90] The surface of magnetic MSNs have been decorated with a specific 15-base single-pair oligonucleotide sequence, selected to display a melting temperature of 47 °C. After loading the particles with a drug model, the pores of these particles were capped with iron oxide nanocrystals (size ~5 nm) previously functionalized with complementary DNA strands. Under alternative magnetic fields, the presence of the iron oxide particles produced a temperature increase, reaching the melting temperature of the DNA strands, with consequent drug release. Thus, the system showed zero-release behavior at room temperature, but it released the drug when the temperature reached the melting temperature.

Thermosensitive polymers have been also attached onto the magnetic MSN surface in order to control release. A thermoresponsive copolymer of poly(ethyleneimine)-b-poly(N-isopropylacrylamide) (PEI/NIPAM) has been used as a temperature-sensitive coating of mesoporous silica nanoparticles which carried superparamagnetic iron oxide nanocrystals into their structure.[79]

The reason to use this polymer coat is because it can combine two interesting properties: the temperature-responsive gatekeeper behavior of the poly-NIPAM and the protein or DNA retention capacity of the PEI. This material has proven its ability to transport and release two different molecules (small molecules retained within the porous channels and macromolecules trapped within the polymeric shell) in response to the application of an external magnetic field.

6.8.4.4   Redox

There is a remarkable difference in the redox potential between the mildly oxidizing extracellular space and the reducing intracellular space, which can be exploited in order to produce redox-responsive MSNs. The concentration of glutathione is 1000 times higher in the cytosol than outside the cell. Thus, different moieties have been attached on the pore outlets through a redox-cleavable bond (disulphide), such as inorganic nanocaps[91,92] biomolecules such as collagen[93] and also polymers which contain these breakable bonds.[94]

However, when the carriers are taken up by the cells they are confined in endosome vesicles, where the presence of reductive species is lower than necessary, or even where mild oxidative agents are present.[95] Therefore, designing a system able to accelerate the endosomal escape, for instance disrupting the endosome membrane by irradiation of a photosensitizer attached to the carrier,[96] is of paramount importance in this type of nanodevice.[97]

6.8.4.5   Light

A pioneering group in the use of light as a trigger stimulus in drug release with MSNs is Fujiwara and co-workers. They demonstrated that functionalization with coumarin on the pore outlets can trigger the release of the drugs trapped in the mesoporous structure by light irradiation.[98]

Additionally, this system was improved by functionalization of the internal pore structure with azobenzene molecules. Azobenzene act as an impeller because it suffers a continuous rotation–inversion movement under ultraviolet/visible irradiation, which accelerates the release rate.[99] Also based on the photo-isomerization of azobenzene, Zink and co-workers have reported a mesoporous silica nanocarrier able to release its cargo in response to UV irradiation.[100]

In this system, cyclodextrins (β-CD) are used as pore caps due to their binding affinity for the *trans*-azobenzene, which decorates the pore outlets. UV light irradiation at 351 nm causes azobenzene isomerization from the *trans* to the *cis* conformation and consequently the β-CD caps are removed from the surface with subsequent drug release. As in the previous cases, it is also possible to use nanocaps attached through sensitive linkers. Thus, gold nanoparticles can be anchored to the mesoporous surface using thioundecyl-tetraethyleneglycoesteronitrobenzylethyldimethylammonium bromide (TUNA), which is converted to the negative form under UV irradiation. This provokes repulsion of the gold nanocaps, causing the drug to be released.[101]

All these systems have been designed to respond in the presence of UV/visible light, which exhibits poor penetration depth in living tissues. The application of near infrared (NIR) light could improve the effectiveness of these devices, since the transmission of this radiation through blood and soft tissue is optimal thanks to its low energy absorption, and therefore allows deep penetration in body tissues. Recently, a mesoporous silica carrier has been designed to exploit this energy to trigger the release of an antitumoral drug.[102]

In this system, a gold nanorod is trapped within the silica particle in order to provide the NIR-responsive behavior, because these gold rods are able to capture the NIR radiation, transforming it into thermal energy. The pores are capped with a double DNA strand which suffers thermal dehybridization when the material is irradiated with NIR light. The DNA dehybridization produces the pore opening and drug release.

6.8.4.6 Enzymes

Some diseases are characterized by the overexpression of certain enzymes at levels higher than the normal values, and therefore this characteristic can be exploited to trigger drug release.[103]

Thus, as in the previous cases, different nanocaps can be covalently attached on the pore entrances using cleavable linkers sensitive to the presence of one particular enzyme, such as esterases[104] or amylases.[105] Dysregulation in the amount of proteases have been described in different pathologies and can also be exploited as an internal triggering stimulus. Avidin molecules were attached on MSN surfaces acting as caps using their well-known affinity for the biotin moieties, which were previously grafted onto the pore outlets.[106]

These caps were eliminated by the action of a specific protease (trypsin) allowing the release of the loaded drug. Singh *et al.*[107] covered the MSN surface with an enzyme-sensitive polymeric shell which prevents the premature drug release until certain proteases are present. In this case, the enzymes responsible for the polymer degradation are matrix metalloproteinases known to be highly overexpressed in tumoral environments.

6.8.4.7 Small Molecules

The presence of specific molecules in the target tissue can trigger release in MSNs provided with their corresponding sensitive gatekeepers. Mártinez-Mañez and co-workers[108] have described MSNs capped with an antibody that recognizes selectively the presence of a certain molecule. When this molecule is present in the media, the antibody is detached from the mesoporous surface, leading to drug release. Recently, MSNs have been capped with gold particles modified with aptamers able to bind to adenosine triphosphate (ATP) molecules.[109]

Aptamers are single-stranded oligonucleotide polymers able to recognize selectively certain molecules. They are more stable to degradation and easy to prepare than antibodies. Thus, when these nanocarriers are exposed to the presence of ATP, the gold caps are removed by displacement reaction.

6.9 Other Applications of Nanoparticles in Nanomedicine: Imaging and Theranostic Applications

As already mentioned, the external surface of nanoparticles can be decorated with different molecules or polymers in order to provide interesting properties such as targeting capacity, improved solubility in water solutions and stimulus-responsive properties. Additionally, fluorescent dyes, superparamagnetic crystals or radioactive compounds can be incorporated, making them multifunctional nanodevices which combine drug delivery capacity with imaging properties, receiving the name of theranostics (therapeutics and diagnostics devices).[110]

If both drugs and imaging agents are incorporated within the same carrier, it is possible to consider that they present the same biodistribution and targeted accumulation, and the efficacy of the therapy can be evaluated in real time.

Nowadays, nanoparticles can play an important role in some imaging techniques applied in oncology, such as optical imaging, magnetic resonance imaging (MRI), positron emission tomography (PET) and single photon emission computed tomography (SPECT), each of which has its own advantages and disadvantages (Figure 6.27). These techniques employ different agents (dyes, magnetic particles or radioactive compounds) which are preferentially accumulated into the target organ. One single nanoparticle can transport large amounts of these agents, obtaining enhanced signals. Also, the targeting capacity of the nanoparticles usually leads to better selectivity towards the target tissue. Moreover, different imaging agents can be incorporated on the same nanocarrier in order to overcome the limitations of each technique, obtaining an improved diagnosis agent. For instance, it can be possible to combine PET with fluorescence or PET with MRI in the same particle. These multimodal imaging agents can provide more reliable and accurate detection of disease sites. Finally, if the nanocarrier is also loaded with therapeutic compounds, it would be possible to treat the disease and collect information on the pathology at the same time. In this section, a brief summary of the recent advances on the use of ceramic nanoparticles for theranostic applications is described.

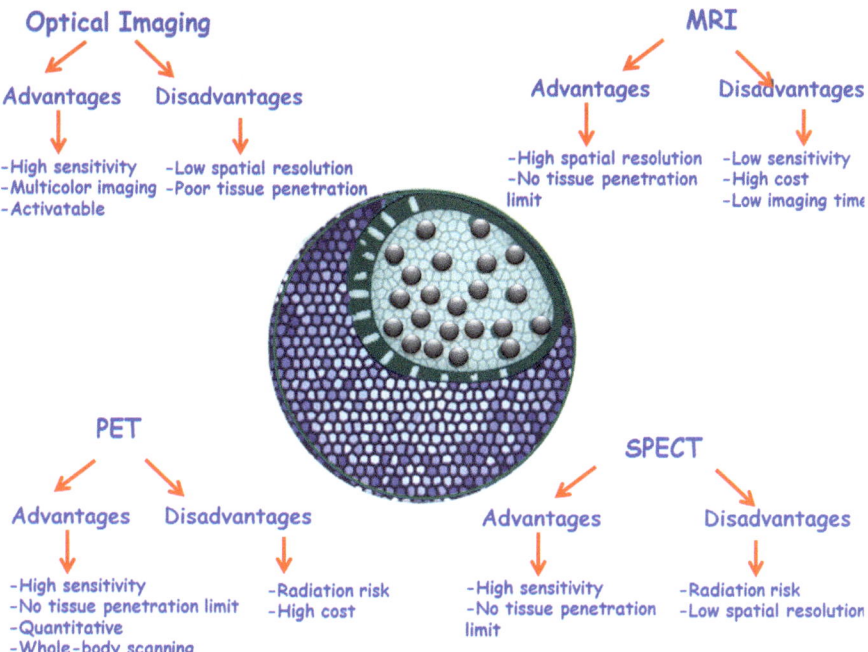

Optical Imaging

Advantages — Disadvantages

-High sensitivity
-Multicolor imaging
-Activatable

-Low spatial resolution
-Poor tissue penetration

MRI

Advantages — Disadvantages

-High spatial resolution
-No tissue penetration limit

-Low sensitivity
-High cost
-Low imaging time

PET

Advantages — Disadvantages

-High sensitivity
-No tissue penetration limit
-Quantitative
-Whole-body scanning

-Radiation risk
-High cost

SPECT

Advantages — Disadvantages

-High sensitivity
-No tissue penetration limit

-Radiation risk
-Low spatial resolution

Figure 6.27 Imaging fields of interest for the use of mesoporous silica nanoparticles. MRI: magnetic resonance imaging; PET: positron emission tomography; SPECT: single-photon emission computed tomography.

6.9.1 Mesoporous Silica Nanoparticles

Different dyes can be encapsulated into silica particles. Thus, fluorescein has been covalently attached to the surface of mesoporous silica particles previously functionalized with PEI.[111]

The presence of the fluorescein allowed tracking of the carrier destination by fluorescence microscopy. The external surface of this carrier was also functionalized with folic acid and the resulting system could be selectively recognized by human tumoral cells (HeLa). Fluorescein-labelled hollow MSNs have been employed as carriers of doxorubicin and photodynamic therapy agents in order to improve the therapeutic response by synergic effect and to trace the particle during the process.[112]

The use of conventional fluorophores, such as fluorescein or rhodamine in fluorescence microscopy is limited due to the attenuation of photon propagation and poor signal-to-noise ratio caused by living tissue autofluorescence. In order to overcome these limitations, fluorophores excited with NIR radiation (usually 750–900 nm) can be employed, since this radiation presents higher penetration in living tissues.[113]

An additional advantage is that MSNs protect the dyes from degradation. Lee *et al.*[114] have described a MSN-based theranostic device composed of NIR dyes (ATTO 647N) and photosensitizers (*meso*-tetratolylporphyrin-Pd) attached within the silica matrix, and cyclic RGD moieties anchored to PEG chains placed on the external surface of the particles. This device is able to selectively destroy human tumoral cells (breast and glioblastoma) under irradiation at 552 nm. Recently, He *et al.*[115] developed a novel mesoporous silica material based on a post-calcination and heating process, which presents luminescent properties without the addition of fluorophores.

MRI is a very useful non-invasive diagnostic technique which provides high resolution images using highly penetrating magnetic fields. Gadolinium is one of the best-known agents for this imaging tool, but usually presents low selectivity and unsatisfactory image contrast enhancement, due to its small size. This molecule can be incorporated to MSNs using chelates covalently anchored to their external surface. This improves the sensitivity due to the enhanced relaxation as a result of reduced tumbling rates and large payloads of active magnetic nuclei.[116]

One of the main limitations of Gd-conjugated MSNs is their long-term tissue accumulation which can produce excessive toxicity. In order to reduce the toxicity, the Gd-chelate can be attached using a redox-cleavable linker which allows the rapid excretion of the metal after imaging.[117] Iron oxide nanocrystals can be encapsulated within fluorescein-labelled MSNs obtaining multifunctional nanodevices which combine imaging by MRI and fluorescence microscopy, with the capacity to transport and release therapeutic compounds housed within the pores.[118] Gold nanorods have also been trapped in MSNs, taking advantage of the fact that this metal combines surface plasmon resonance property, which allows its use as imaging probe, with the ability to produce heat under laser irradiation.[119]

6.9.2   Carbon Allotropes

Carbon nanotubes show interesting properties for imaging applications such as high absorption in the NIR region, strong Raman shift or photoacoustic properties.[120] Different radioisotopes such as ^{64}Cu, ^{125}I and ^{66}Ga can be attached to the carbon nanotube surface in order to obtain *in vivo* information about their distribution by PET.[121–123]

The *ex vivo* organ distribution of the particles can be evaluated by Raman spectroscopy thanks to the intrinsic Raman shift of the carbon nanotube. As in the previous material, fluorescent labels can be attached to its surface in order to trace the material by fluorescence imaging.[124] Graphene can produce fluorescence under an excitation of 400 nm. Thus, the cell internalization of drug-loaded graphene sheets can be easily monitored by fluorescence microscopy.[125,126]

However, the visible fluorescence of graphene competes with the intrinsic auto-fluorescence of living tissue, which makes its application in animal models difficult. As an alternative, graphene can be decorated with NIR fluorophores as Cy7 in order to avoid this fluorescent quenching. Also, gold nanoparticles with NIR photoluminescence[126] and quantum dots with strong fluorescence[127] have been attached to the graphene surface in order to improve the detection of these systems. Finally, superparamagnetic graphene oxide has been produced by growing iron oxide nanoparticles on the graphene oxide surface in order to employ this system as a MRI contrast agent.[128]

6.9.3 Iron Oxide Nanoparticles

Iron oxide nanoparticles (IONPs) provide a large T_2 relaxation effect and are widely used as contrast agents in MRI.[129] Hyeon and co-workers have developed a simple methodology to synthesize water-dispersible iron oxide particles for dual imaging (MRI and optical imaging) using a PEG-derivatized phosphine oxide (PO-PEG) moiety with fluorescein attached to the end.[130] The surface of these particles can also be decorated with NIR dyes such as Cy.5.5 in order to obtain *in vivo* information combining two imaging techniques.[131]

The combination of IONPs with PET or SPECT probes is currently receiving special attention. The most common approach to create these multifunctional devices consists of the conjugation of radionucleotides such as ^{64}Cu, ^{111}In and [124]I with different chelates (DOTA or DTPA) anchored on the iron oxide surface. Thus, ^{64}Cu–DOTA complexes conjugated to IONPs with RGD-functionalized IONPs allow non-invasive imaging of tumor targeting using PET and MRI.[132] Trimodal imaging probes have been synthesized by the attachment of PET probes to dual MRI/optical probes.[133]

The attachment of therapeutic compounds to the surface of IONPs by physical adsorption or covalent grafting enables the combination of imaging with drug delivery. Thus, methotrexate, an antitumoral drug, can be anchored to the iron oxide surface through an amide bond. Once the particle is internalized by the cancer cells, the drug is released due to the presence of intracellular proteases.[134]

Other antitumoral drugs such as doxorubicin[135] or glucocorticoid prednisolone[136] have been loaded onto the surface of polymer-coated-IONPs achieving excellent antitumoral activities. Proteins, *e.g.* human serum albumin, have been absorbed into the drug-loaded IONPs in order to improve biocompatibility while providing tumor target capacity.[137]

The cytotoxic drugs can be attached through sensitive bonds that undergo hydrolysis or other chemical transformations in the presence of certain stimulus. Thus, doxorubicin has been grafted to PEI-coated-IONP *via* a pH-sensitive hydrazone bond.[138]

In mildly acidic conditions, which are usually present in the endosomal or lysosomal intracellular compartments, this hydrazone bond undergoes hydrolysis, producing the release of the drug. These types of devices combine the capacity to release drugs in a controlled manner with their imaging properties.

6.10 Some Thoughts on Toxicity

According to the European Society for Biomaterials, the term biocompatibility can be defined as "the ability of a material to perform with an appropriate host response in a specific application". Regardless of the different nature of each type of inorganic nanoparticle (silica, calcium phosphates, carbon, *etc.*), nanoparticles can cause several adverse effects due to their small size in the range of the cellular organelles and compartments. Their characteristic high surface area over volume ratio greatly increases their interaction with the biological entities. One of the most common effects caused by the exposition to nanoparticles is the generation of reactive oxygen species (ROS), which can originate adverse side effects. In general, cells can tolerate small and transient increases of ROS species. However, under high or prolonged nanoparticle exposition, cells can suffer membrane damage or DNA alterations (genotoxicity). The ROS species can be generated as a consequence of exposure to the acidic endosomal environment, interaction of nanoparticle with the mitochondria and also through the activation of certain intracellular signaling pathways. The kinetics of the ROS formation depends on the total surface area and the coating stability and therefore, should be studied for each type of nanoparticle. In some cases, the surface functionalization plays an important role, as in the case of iron oxide nanoparticles coated by a dextran shell. These nanoparticles diminish the intracellular ROS levels inducing cellular proliferation instead of adverse effects.[139] Also, cellular internalization of rigid nanoparticles can provoke disruption of the cytoskeleton network, which can originate secondary side effects.

Mesoporous silica nanoparticles (SBA-15 and MCM-41) exhibit low toxicity at low concentrations. Cellular uptake is strongly size-dependent, the particles of 50 nm being the most capable of being internalized by human cells.[140] The capture of the mesoporous particles by macrophages is favored by the size increase, while smaller particles exhibit longer circulation times. Cytotoxicity is related to the uptake grade, thus, micrometric particles with sizes ~1 μm show lower toxicity than nanoparticles with sizes up to 200 nm. A general principle is that cationic nanoparticles induce more immunogenic response and cytotoxicity than their neutral or anionic counterparts. The silanol groups of the naked mesoporous silica nanoparticles can interact with the lipids of the cellular membranes and some proteins, altering their properties. Without any coating, these particles present negative surface charge at physiological pH. Due to this fact, the particles are rapidly covered by opsonins, captured by the macrophages and accumulated in liver and spleen shortly after entering the bloodstream. As mentioned above, the use of different coatings (PEG or other polymers) can increase the circulation time and reduce the toxicity of this material. The degradability of these particles has been studied using simulated body fluid. The process occurs in three steps: fast bulk silica degradation (hour scale), followed by the deposition of magnesium/calcium silicate on the particle surface, and finally, slow dissolution of the particles (day scale).

Regarding *in vivo* toxicity, Kohane and co-workers have demonstrated that these materials exhibit high toxicity when larger doses are employed (up to

1.2 g kg^{-1} in a single injection), whereas no toxicity was detected when the dose was below 40 mg kg^{-1}.[141]

The research group of Tamanoi has established 50 mg kg^{-1} as the safe dose for drug delivery purposes.[142] The intravenous administration of carbon nanotubes and fullerenes can activate the immune system, resulting in inflammation and granuloma formation.[143]

Meanwhile, the biocompatibility and cytotoxicity of another carbon allotrope, graphene oxide, has been recently studied in detail showing that this material can be safely employed in the treatment of ocular tumors.[144]

Iron oxide nanoparticles (mainly magnetite Fe_2O_3 and its oxidized and more stable form maghemite Fe_3O_4) are widely used as contrast agents in MRI and also in magnetic hyperthermia for cancer therapy. Once internalized into the cells, these particles are degraded into iron ions by several hydrolyzing enzymes.[145]

As in the other cases, the biodistribution of these particles depends on their charge, size, shape, external coatings and route of administration, usually being captured by the macrophages of the reticuloendothelial system if the nanoparticles are administered without any coating. Iron oxide nanoparticles are able to produce oxidative stress within the body by the Fenton reaction,[146] which is one of the main sources of ROS in living systems.

Moreover, the presence of iron oxide nanoparticles can produce a homeostasis imbalance within the body which can affect different organs. It is known that the accumulation of redox-active metals in the brain induces multiple degenerative disorders, such as Parkinson's and Alzheimer's diseases. This is particularly important because magnetic nanoparticles are able to penetrate into the body through the skin and can also cross the blood–brain barrier after exposure by inhalation.[147] It is important to remark that in all the cases the toxicity of these materials is a dose-dependent phenomenon and the biocompatibility should be determined with each case and treatment.

It has been verified in animal models that nanoparticles aimed towards a given organ can also be found in different organs.[148] Figure 6.28 depicts a nanocarrier which, as a consequence of its contact with biological fluids, carries certain proteins adhered onto its surface; these can then be internalized in cells, perhaps not as selectively as the materials scientist would have desired when designing the system.

Therefore, this is a controversial issue which requires a considerable research effort to ensure that these new nanocarriers are exclusively aimed at the targeted organ.

Another consideration worth remarking about toxicity is that the point of view of a material scientist is different from that of a toxicologist when dealing with this subject. This should not be the case in terms of biology. In this sense, the materials scientist is much more tolerant regarding toxicity

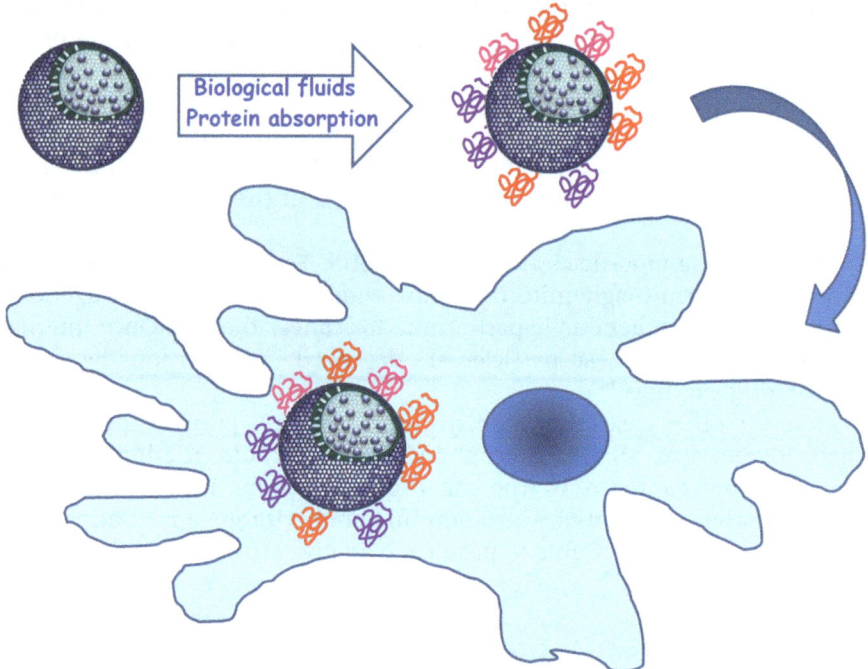

Figure 6.28   Cells and nanocarrier internalization.

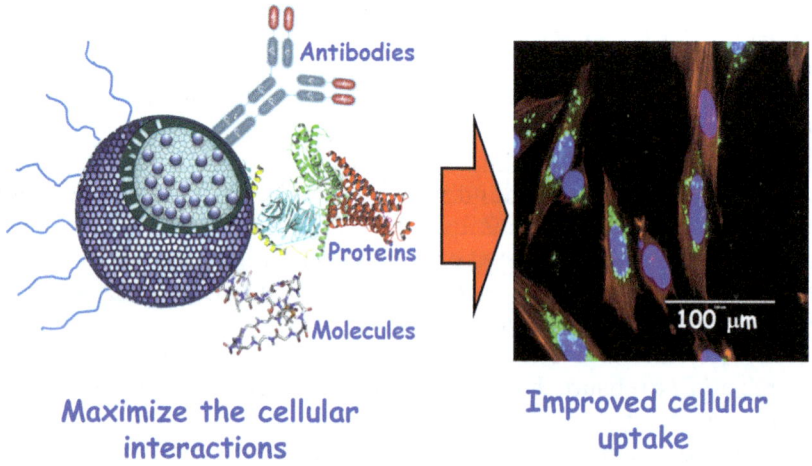

Figure 6.29   Nanoparticles internalized in cells.

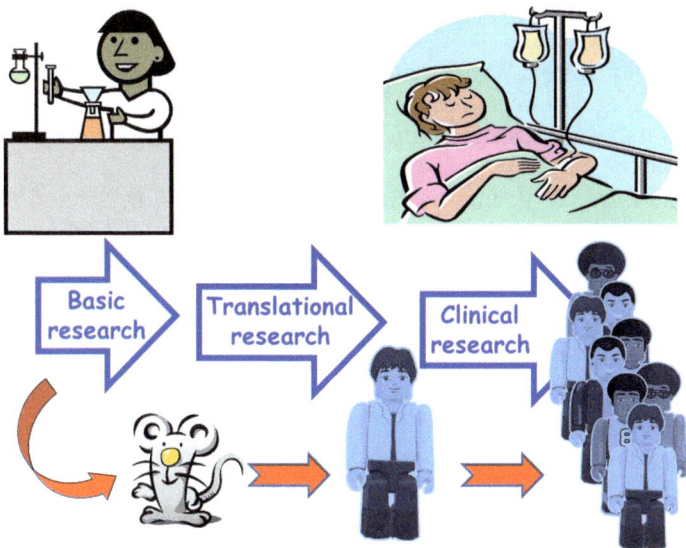

Figure 6.30 The route from basic research to the patient's bedside.

because the main focus of his/her work is to ensure that the damaged organ becomes a preferential target for the nanocarrier. The toxicologist, however, enjoys a wider and more comprehensive view of the potential effects of these nanocarriers on different organs and on the patient as a whole. Needless to say, materials scientists must "run the extra mile" to decrease as much as possible concentrations administered through nanocarriers, ensuring that they fulfill their mission while having the minimum possible effect on the rest of the body.

This implies that materials and biology scientists must cooperate in a joint effort to maximize cell interactions, improving nanocarrier selectivity and internalization in cells. It is critical to ensure an optimum reciprocal communication between both fields of research in order to reach valid solutions and the best and safest cytotoxic administration possible. Figure 6.29 tries to express this idea.

The basic research needs a so-called translational approach to reach the clinical field. In this sense, optimum cooperation between researchers and clinicals is the critical factor for success.

This is another important side to the research currently undertaken in the field of biomaterials, and more precisely, in nanocarriers. *In vitro* assays are not enough; it is essential to carry out animal testing, subsequently transferred to human tests, which require the close cooperation of researcher and clinician. The former needs to be fully familiarized with clinical needs in order to solve a particular healthcare issue, and the latter has to understand precisely the potential solutions put forward by the researcher. This whole process must be strictly controlled by all applicable health regulations (Figure 6.30).

References

1. K. Riehemann, S. W. Schneider, T. A. Luger, B. Godin, M. Ferrari and H. Fuchs, *Angew. Chem., Int. Ed.*, 2009, **48**, 872.
2. V. Mamaeva, C. Sahlgren and M. Lindén, *Adv. Drug Delivery Rev.*, 2013, **65**, 689.
3. M. Benezra, O. Penate-Medina, P. B. Zanzonico, D. Schaer, H. Ow, A. Burns, E. DeStanchina, V. Longo, E. Herz, S. Iyer, J. Wolchok, S. M. Larson, U. Wiesner and M. S. Bradbury, *J. Clin. Invest.*, 2011, **121**, 2768.
4. M. Vallet-Regi, A. Ramila, R. P. del Real and J. Perez-Pariente, *Chem. Mater.*, 2001, **13**(2), 308.
5. M. Vallet-Regí, *Chem.–Eur. J.*, 2006, **12**, 5934.
6. Z. Li, J. C. Barnes, A. Bosoy, J. F. Stoddart and J. I. Zink, *Chem. Soc. Rev.*, 2012, **41**, 2590.
7. B. G. Trewyn, S. Giri, I. I. Slowing and V. S.-Y. Lin, *Chem. Commun.*, 2007, 3236.
8. F. Tang, L. Li and D. Chen, *Adv. Mater.*, 2012, **24**, 1504.
9. M. Liong, J. Lu, M. Kovochich, T. Xia, S. G. Ruehm, A. E. Nel, F. Tamanoi and J. I. Zink, *ACS Nano*, 2008, **2**, 889.
10. H. Meng, M. Xue, T. Xia, Z. Ji, D. Y. Tarn, J. I. Zink and A. Nel, *ACS Nano*, 2011, **5**, 4131.
11. L. Li, F. Tang, H. Liu, T. Liu, N. Hao, D. Chen, X. Teng and J. He, *ACS Nano*, 2010, **4**, 6874.
12. J. L. Vivero-Escoto, I. I. Slowing and V. S.-Y. Lin, *Biomaterials*, 2010, **31**, 1325.
13. B. G. Trewyn, J. A. Nieweg, Y. Zhao and V. S.-Y. Lin, *Chem. Eng. J.*, 2008, **137**, 23.
14. D. Brevet, M. Gary-Bobo, L. Raehm, S. Richeter, O. Hocine, K. Amro, B. Loock, P. Couleaud, C. Frochot, A. Morère, P. Maillard, M. Garcia and J.-O. Durand, *Chem. Commun.*, 2009, 1475.
15. J. M. Rosenholm, A. Meinander, E. Peuhu, R. Niemi, J. E. Eriksson, C. Sahlgren and M. Lindén, *ACS Nano*, 2009, **3**, 197.
16. J. Lu, Z. Li, J. I. Zink and F. Tamanoi, *Nanomedicine*, 2012, **8**, 212.
17. D. P. Ferris, J. Lu, C. Gothard, R. Yanes, C. R. Thomas, J. C. Olsen, J. F. Stoddart, F. Tamanoi and J. I. Zink, *Small*, 2011, **7**, 1816.
18. C. P. Tsai, C. Y. Chen, Y. Hung, F. H. Chang and C. Y. Mou, *J. Mater. Chem.*, 2009, **19**, 5737.
19. X. He, Y. Zhao, D. He, K. Wang, F. Xu and J. Tang, *Langmuir*, 2012, **28**, 12909.
20. Y. Cai, H. Pan, X. Xu, Q. Hu, L. Li and R. Tang, *Chem. Mater.*, 2007, **19**, 3081.
21. X. Cheng and L. Kuhn, *Int. J. Nanomed.*, 2007, **2**, 667.
22. M. Kester, Y. Heakal, T. Fox, A. Sharma, G. P. Robertson, T. T. Morgan, E. I. Altinoglu, A. Tabakovic, M. R. Parette, S. M. Rouse, V. Ruiz-Velasco and J. H. Adair, *Nano Lett.*, 2008, **8**, 4116.
23. T. T. Morgan, H. S. Muddana, E. I. Altinoglu, S. M. Rouse, A. Tabakovic, T. Tabouillot, T. J. Russin, S. S. Shanmugavelandy, P. J. Butler, P. C. Eklund, J. K. Yun, M. Kester and J. H. Adair, *Nano Lett.*, 2008, **8**, 4108.

24. J. Schwiertz, A. Wiehe, S. Gräfe, B. Gitter and M. Epple, *Biomaterials*, 2009, **30**, 3324.

25. J. Klesing, A. Wiehe, B. Gitter, S. Gräfe and M. Epple, *J. Mater. Sci.: Mater. Med.*, 2010, **21**, 887.

26. C.-H. Hou, S. M. Hou, Y. S. Hsueh, J. Lin, H. C. Wu and F. H. Lin, *Biomaterials*, 2009, **30**, 3956.

27. C. H. Hou, C.-W. Chen, S.-M. Hou, Y.-T. Li and F.-H. Lin, *Biomaterials*, 2009, **30**, 4700.

28. R. B. Heimann, S. E. Evsyukov and Y. Koga, *Carbon*, 1994, **35**, 289.

29. E. Yasuda, M. Inagaki, K. Kaneko, M. Endo, A. Oya and Y. Tanabe, *Carbon alloys. New Concepts to develop carbon science and technology*, Elsevier, Amsterdam, 2003.

30. K. Yang, L. Feng, X. Shi and Z. Liu, *Chem. Soc. Rev.*, 2013, **42**, 530.

31. X. Sun, Z. Liu, K. Welsher, J. T. Robinson, A. Goodwin, S. Zaric and H. Dai, *Nano Res.*, 2008, **1**, 203.

32. H. Hu, J. Yu, Y. Li, J. Zhao and H. Dong, *J. Biomed. Mater. Res., Part A*, 2012, **100A**, 141.

33. H. Bao, Y. Pan, Y. Ping, N. G. Sahoo, T. Wu, L. Li, J. Li and L. H. Gan, *Small*, 2011, **7**, 1569.

34. Y. Yang, Y. M. Zhang, Y. Chen, D. Zhao, J. T. Chen and Y. Liu, *Chem.–Eur. J.*, 2012, **18**, 4208.

35. L. Zhang, J. Xia, Q. Zhao, L. Liu and Z. Zhang, *Small*, 2010, **6**, 537.

36. H. Wen, C. Dong, H. Dong, A. Shen, W. Xia, X. Cai, Y. Song, X. Li, Y. Li and D. Shi, *Small*, 2012, **8**, 760.

37. X. Yang, Y. Wang, X. Huang, Y. Ma, Y. Huang, R. Yang, H. Duan and Y. Chen, *J. Mater. Chem.*, 2011, **21**, 3448.

38. L. Meng, X. Zhang, Q. Lu, Z. Fei and P. J. Dyson, *Biomaterials*, 2012, **33**, 1689.

39. Z. Liu, X. Sun, N. Nakayama-Ratchford and H. Dai, *ACS Nano*, 2007, **1**, 50.

40. X. K. Zhang, L. J. Meng, Q. H. Lu, Z. F. Fei and P. J. Dyson, *Biomaterials*, 2009, **30**, 6041.

41. Z. Liu, K. Chen, C. Davis, S. Sherlock, Q. Cao, X. Chen and H. Dai, *Cancer Res.*, 2008, **68**, 6652.

42. K. Yang, S. Zhang, G. Zhang, X. Sun, S. T. Lee and Z. Liu, *Nano Lett.*, 2010, **10**, 3318.

43. J. Y. Robinson, K. Welsher, S. M. Tabakman, S. P. Sherlock, H. Wang, R. Luong and H. Dai, *Nano Res.*, 2010, **3**, 779.

44. S. Naahidi, M. Jafari, F. Edalat, K. Raymond, A. Khademhosseini and P. Chen, *J. Controlled Release*, 2013, **16**, 182.

45. N. Bertrand and J. C. Leroux, *J. Controlled Release*, 2012, **161**, 152.

46. T. Ishihara, M. Takeda, H. Sakamoto, A. Kimoto, C. Kobayashi, N. Takasaki, K. Yuki, K. I. Tanaka, M. Takenaka, R. Igarashi, T. Maeda, N. Yamakawa, Y. Okamoto, M. Otsuka, T. Ishida, H. Kiwada, Y. Mizushima and T. Mizushima, *Pharm. Res.*, 2009, **26**, 2270.

47. M. Lundqvist, J. Stigler, T. Cedervall, T. Berggard, M. B. Flanagan, I. Lynch, G. Elia and K. Dawson, *ACS Nano*, 2011, **5**, 7503.

48. A. S. Karakoti, S. Das, S. Thevuthasan and S. Seal, *Angew. Chem., Int. Ed.*, 2011, **50**, 1980.

49. Y. Matsumura and H. Maeda, *Cancer Res.*, 1986, **46**, 6387.

50. H. Maeda, H. Nakamura and J. Fang, *Adv. Drug Delivery Rev.*, 2013, **65**, 71.

51. J. Fang, H. Nakamura and H. Maeda, *Adv. Drug Delivery Rev.*, 2011, **63**, 136.

52. R. J. Jain and T. Stylianopoulos, *Nat. Rev. Clin. Oncol.*, 2010, **7**, 653.

53. R. K. Jain, *Adv. Drug Delivery Rev.*, 2001, **46**, 149.

54. P. Ehrlich, *The collected papers of Paul Ehrlich*, Pergamon, London, 1960, p. 3.

55. R. Narain, M. Gonzales, A. S. Hoffman, P. S. Stayton and K. M. Krishnan, *Langmuir*, 2007, **23**, 6299.

56. S. Mazzucchelli, M. Colombo, C. De Palma, A. Salvadè, P. Verderio, M. D. Coghi, E. Clementi, P. Tortora, F. Corsi and D. Prosperi, *ACS Nano*, 2010, **4**, 5693.

57. K. L. Vigor, P. G. Kyrtatos, S. Minogue, K. T. Al-Jamal, H. Kogelberg, B. Tolner, K. Kostarelos, R. H. Begent, Q. A. Pankhurst, M. F. Lythgoe and K. A. Chester, *Biomaterials*, 2010, **31**, 1307.

58. F. Danhier, O. Feron and V. Préat, *J. Controlled Release*, 2010, **148**, 135.

59. P. S. Low and A. C. Antony, *Adv. Drug Delivery Rev.*, 2004, **56**, 1055.

60. T. Minko, *Adv. Drug Delivery Rev.*, 2004, **56**, 491.

61. E. Ruoslahti, S. N. Bathia and M. J. Sailor, *J. Cell Biol.*, 2010, **188**, 759.

62. S. W. Reulen, P. Y. Dankers, P. H. Bomans, E. W. Meijer and M. Merkx, *J. Am. Chem. Soc.*, 2009, **131**, 7304.

63. K. N. Sugahara, T. Teesalu, P. P. Karmali, V. R. Kotamraju, L. Agemy, O. M. Girard, D. Hanahan, R. F. Mattrey and E. Ruoslahti, *Cancer Cell*, 2009, **16**, 510.

64. M. Grün, I. Lauer and K. K. Unger, *Adv. Mater.*, 1997, **9**, 254.

65. N. Gómez-Cerezo, I. Izquierdo-Barba, D. Arcos and M. Vallet-Regí, *J. Mater. Chem. B*, 2015, **3**, 3810.

66. M. Vallet-Regí, M. Colilla and B. González, *Chem. Soc. Rev.*, 2011, **40**, 596.

67. D. Arcos, V. Fal-Miyar, E. Ruiz-Hernández, M. Garcia-Hernández, M. L. Ruiz-González, J. González-Calbet and M. Vallet-Regí, *J. Mater. Chem.*, 2012, **22**, 64.

68. Y.-S. Lin and C. L. Haynes, *Chem. Mater.*, 2009, **21**, 3979.

69. C. Boissiere, D. Grosso, A. Chaumonnot, L. Nicole and C. Sanchez, *Adv. Mater*, 2011, **23**, 599.

70. E. Ruiz-Hernández, A. López-Noriega, D. Arcos, I. Izquierdo-Barba, O. Terasaki and M. Vallet-Regí, *Chem. Mater.*, 2007, **19**, 3455.

71. B. Julian-López, C. Boissiere, C. Chaneac, D. Grosso, S. Vasseur, S. Miraux, E. Duguet and C. Sanchez, *J. Mater. Chem.*, 2007, **17**, 1563.

72. A. López-Noriega, E. Ruiz-Hernández, S. M. Stevens, D. Arcos, M. W. Anderson, O. Terasaki and M. Vallet-Regí, *Chem. Mater.*, 2009, **21**, 18.

73. A. Baeza, M. Colilla and M. Vallet-Regí, *Expert Opin. Drug Delivery*, 2015, **12**, 219.

74. A. Baeza, D. Arcos and M. Vallet-Regí, *J. Phys.: Condens. Matter*, 2013, **25**, 484003.
75. N. Knezevic, E. Ruiz-Hernández, W. Hennink and M. Vallet-Regi, *RSC Adv.*, 2013, **3**, 9584.
76. M. Manzano and M. Vallet-Regí, *J. Mater. Chem.*, 2010, **20**, 5593.
77. M. Vallet-Regí, F. Balas, M. Colilla and M. Manzano, *Solid State Sci.*, 2007, **9**, 768.
78. E. Ruiz-Hernández, A. Baeza and M. Vallet-Regí, *ACS Nano*, 2011, **5**, 1259.
79. A. Baeza, E. Guisasola, E. Ruiz-Hernández and M. Vallet-Regí, *Chem. Mater.*, 2012, **24**, 517.
80. E. S. Lee, Z. Gao and Y. H. Bae, *J. Controlled Release*, 2008, **132**, 164.
81. Y. L. Zhao, Z. Li, S. Kabehie, Y. Y. Botros, J. F. Stoddart and J. I. Zink, *J. Am. Chem. Soc.*, 2010, **132**, 13016.
82. R. Liu, Y. Zhang, X. Zhao, A. Agarwal, L. J. Mueller and P. Feng, *J. Am. Chem. Soc.*, 2010, **132**, 1500.
83. R. Liu, P. Liao, J. Liu and P. Feng, *Langmuir*, 2011, **27**, 3095.
84. H. P. Rim, K. H. Min, H. J. Lee, S. Y. Jeong and S. C. Lee, *Angew. Chem., Int. Ed.*, 2011, **50**, 8853.
85. Q. Fu, R. Rama, T. L. Ward, Y. Lu and G. P. López, *Langmuir*, 2007, **23**, 170.
86. S. Laurent, S. Dutz, U. O. Häfeli and M. Mahmoudi, *Adv. Colloid Interface Sci.*, 2011, **166**, 8.
87. E. Ruiz-Hernández, A. López-Noriega, D. Arcos and M. Vallet-Regí, *Solid State Sci.*, 2008, **10**, 421.
88. M. Colilla and M. Vallet-Regí, *Responsive mesoporous silica nanoparticles for targeted drug delivery. Chemoresponsive Materials: Stimulation by Chemical and Biological Signals*, ed. H.-J. Schneider, Royal Society of Chemistry, Cambridge, UK, 2015, ch. 6, ISBN: 978-1-78262-062-4.
89. F. M. Martín-Saavedra, E. Ruíz-Hernández, A. Boré, D. Arcos, M. Vallet-Regí and N. Vilaboa, *Acta Biomater.*, 2010, **6**, 4522.
90. R. Jin, G. Wu, Z. Li, C. A. Mirkin and G. C. Schatz, *J. Am. Chem. Soc.*, 2003, **125**, 1643.
91. C. Y. Lai, B. G. Trewyn, D. M. Jeftinija, K. Jeftinija, S. Xu, S. Jeftinija and V. S.-Y. Lin, *J. Am. Chem. Soc.*, 2003, **125**, 4451.
92. S. Giri, B. G. Trewyn, M. P. Stellmaker and V. S.-Y. Lin, *Angew. Chem., Int. Ed.*, 2005, **44**, 5038.
93. Z. Luo, K. Cai, Y. Hu, L. Zhao, P. Liu, L. Duan and W. Yang, *Angew. Chem., Int. Ed.*, 2011, **50**, 640.
94. R. Liu, X. Zhao, T. Wu and P. Feng, *J. Am. Chem. Soc.*, 2008, **130**, 14418.
95. C. D. Austin, X. Wen, L. Gazzard, C. Nelson, R. H. Scheller and S. J. Scales, *Proc. Natl. Acad. Sci. U. S. A.*, 2005, **102**, 17987.
96. K. G. de Bruin, C. Fella, M. Ogris, E. Wagner, N. Ruthardt and C. Bräuchle, *J. Controlled Release*, 2008, **130**, 175.
97. A. M. Shauer, A. Schlossbauer, N. Ruthardt, V. Cauda, T. Bein and C. Bräuchle, *Nano Lett.*, 2010, **10**, 3684.
98. N. K. Mal, M. Fujiwara and Y. Tanaka, *Nature*, 2003, **421**, 350.

99. Y. Zhu and M. Fujiwara, *Angew. Chem., Int. Ed.*, 2007, **46**, 2241.

100. D. P. Ferris, Y.-L. Zhao, N. M. Khashab, H. A. Khatib, J. F. Stoddart and J. I. Zink, *J. Am. Chem. Soc.*, 2009, **131**, 1686.

101. J. L. Vivero-Escoto, I. I. Slowing, C. W. Wu and V. S.-Y. Lin, *J. Am. Chem. Soc.*, 2009, **131**, 3462.

102. Y.-T. Chang, P.-Y. Liao, H.-S. Sheu, Y.-J. Tseng, F.-Y. Cheng and C.-S. Yeh, *Adv. Mater.*, 2012, **24**, 3309.

103. R. de la Rica, D. Aili and M. M. Stevens, *Adv. Drug Delivery Rev.*, 2012, **64**, 967.

104. K. Patel, S. Angelos, W. R. Dichtel, A. Coskun, Y. W. Yang, J. I. Zink and J. F. Stoddart, *J. Am. Chem. Soc.*, 2008, **130**, 2382.

105. C. Park, H. Kim, S. Kim and C. Kim, *J. Am. Chem. Soc.*, 2009, **131**, 16614.

106. A. Schlossbauer, J. Kecht and T. Bein, *Angew. Chem., Int. Ed.*, 2009, **48**, 3092.

107. N. Singh, A. Karambelkar, L. Gu, K. Lin, J. S. Miller, C. S. Chen, M. L. Sailor and S. N. Bhatia, *J. Am. Chem. Soc.*, 2011, **133**, 19582.

108. E. Climent, A. Bernados, R. Martínez-Máñez, A. Maquieira, M. D. Marcos, N. Pastor-Navarro, R. Puchades, F. Sancenón, J. Soto and P. Amorós, *J. Am. Chem. Soc.*, 2009, **131**, 14075.

109. C. L. Zhu, C. H. Lu, X.-Y. Song, H.-H. Yang and X.-R. Wang, *J. Am. Chem. Soc.*, 2011, **133**, 1278.

110. D. E. Lee, H. Koo, I. C. Sun, J. H. Ryu, K. Kim and I. C. Kwon, *Chem. Soc. Rev.*, 2012, **41**, 2656.

111. J. M. Rosenholm, E. Peuhu, J. E. Eriksson, C. Sahlgren and M. Linden, *Nano Lett.*, 2009, **9**, 3308.

112. T. Wang, L. Zhang, Z. Su, C. Wang, Y. Liao and Q. Fu, *ACS Appl. Mater. Interfaces*, 2011, **3**, 2479.

113. C.-H. Lee, S.-H. Cheng, Y.-J. Wang, Y.-C. Chen, N.-T. Chen, J. Souris, C.-T. Chen, C.-Y. Mou, C.-S. Yang and L.-W. Lo, *Adv. Funct. Mater.*, 2009, **19**, 215.

114. S.-H. Cheng, C.-H. Lee, M.-C. Chen, J. S. Souris, F.-G. Tseng, C.-S. Yang, C.-Y. Mou, C.-T. Chen and L.-W. Lo, *J. Mater. Chem.*, 2010, **20**, 6149.

115. Q. He, J. Shi, X. Cui, C. Wei, L. Zhang, W. Wu, W. Bu, H. Chen and H. Wu, *Chem. Commun.*, 2011, **47**, 7947.

116. K. M. L. Taylor, J. S. Kim, W. J. Rieter, H. An, W. Lin and W. Lin, *J. Am. Chem. Soc.*, 2008, **130**, 2154.

117. J. L. Vivero-Escoto, K. M. L. Taylor-Pashow, R. C. Huxford, J. Della Rocca, C. Okoruwa, H. An, W. Lin and W. Lin, *Small*, 2011, **7**, 3519.

118. J. Kim, H. S. Kim, N. Lee, T. Kim, H. Kim, T. Yu, I. C. Song, W. K. Moon and T. Hyeon, *Angew. Chem., Int. Ed.*, 2008, **47**, 8438.

119. Z. Zhang, L. Wang, J. Wang, X. Jiang, X. Li, Z. Hu, Y. Ji, X. Wu and C. Chen, *Adv. Mater.*, 2012, **24**, 1418.

120. S.-R. Ji, C. Liu, B. Zhang, F. Yang, J. Xu, J. Long, C. Jin, D.-L. Fu, Q. X. Ni and X. J. Yu, *Biochim. Biophys. Acta, Rev. Cancer*, 2010, **1806**, 29.

121. Z. Liu, W. Cai, L. He, N. Nakayama, K. Chen, X. Sun, X. Chen and H. Dai, *Nat. Nanotechnol.*, 2007, **2**, 47.

122. K. Yang, J. M. Wan, S.-A. Zhang, Y. J. Zhang, S. T. Lee and Z. Liu, *ACS Nano*, 2011, **5**, 516.

123. H. Hong, Y. Zhang, J. W. Engle, T. R. Nayak, C. P. Theuer, R. J. Nickles, T. E. Barnhart and W. Cai, *Biomaterials*, 2012, **33**, 4147.

124. A. A. Bhirde, V. Patel, J. Gavard, G. Zhang, A. A. Sousa, A. Masedunskas, R. D. Leapman, R. Weigert, J. S. Gutkind and J. F. Rusling, *ACS Nano*, 2009, **3**, 307.

125. (a) M. Vila, M. C. Matesanz, M. J. Gonçalves, M. J. Feito, J. Linares, P. Marques, M. T. Portoles and M. Vallet-Regí, *Nanotechnology*, 2014, **25**, 035101; (b) C. Wang, J. Li, C. Amatore, Y. Chen, H. Jiang and X. M. Wang, *Angew. Chem., Int. Ed.*, 2011, **50**, 11644.

126. S.-H. Hu, Y.-W. Chen, W.-T. Hung, I. W. Chen and S.-Y. Chen, *Adv. Mater.*, 2012, **24**, 1748.

127. W. Chen, P. Yi, Y. Zhang, L. Zhang, Z. Deng and Z. Zhang, *ACS Appl. Mater. Interfaces*, 2011, **3**, 4085.

128. J. Kim, Y. Piao and T. Hyeon, *Chem. Soc. Rev.*, 2009, **38**, 372.

129. H. B. Na, I. S. Lee, H. Seo, Y. I. Park, J. H. Lee, S. W. Kim and T. Hyeon, *Chem. Commun.*, 2007, 5167.

130. A. Moore, Z. Medarova, A. Potthast and G. Dai, *Cancer Res.*, 2004, **64**, 1821.

131. H. Y. Lee, Z. Li, K. Chen, A. R. Hsu, C. Xu, J. Xie, S. Sun and X. Chen, *J. Nucl. Med.*, 2008, **49**, 1371.

132. J. Xie, K. Chen, J. Huang, S. Lee, J. Wang, J. Gao, X. Li and X. Chen, *Biomaterials*, 2010, **31**, 3016.

133. N. Kohler, C. Sun, A. Fichtenholtz, J. Gunn, C. Fang and M. Zhang, *Small*, 2006, **2**, 785.

134. M. K. Yu, Y. Y. Jeong, J. Park, S. Park, J. W. Kim, J. J. Min, K. Kim and S. Jon, *Angew. Chem., Int. Ed.*, 2008, **47**, 5362.

135. A. Gianella, P. A. Jarzyna, V. Mani, S. Ramachandran, C. Calcagno, J. Tang, B. Kann, W. J. Dijk, V. L. Thijssen, A. W. Griffioen, G. Storm, Z. A. Fayad and W. J. Mulder, *ACS Nano*, 2011, **5**, 4422.

136. Q. Quan, J. Xie, H. Gao, M. Yang, F. Zhang, G. Liu, X. Lin, A. Wang, H. S. Eden, S. Lee, G. Zhang and X. Chen, *Mol. Pharmaceutics*, 2011, **8**, 1669.

137. F. M. Kievit, F. Y. Wang, C. Fang, H. Mok, K. Wang, J. Silber, R. G. Ellenbogen and M. Zhang, *J. Controlled Release*, 2011, **152**, 76.

138. D. M. Huang, J. K. Hsiao, Y. C. Chen, L. Y. Chien, M. Yao, Y. K. Chen, B. S. Ko, S. C. Hsu, L. A. Tai, H. Y. Cheng, S. W. Wang, C. S. Yang and Y. C. Chen, *Biomaterials*, 2009, **30**, 3645.

139. B. D. Chithrani, A. A. Ghazani and W. C. W. Chan, *Nano Lett.*, 2006, **6**, 662.

140. Q. He, J. Shi, M. Zhu, Y. Chen and F. Chen, *Microporous Mesoporous Mater.*, 2010, **131**, 314.

141. S. P. Hudson, R. F. Padera, R. Langer and D. S. Kohane, *Biomaterials*, 2008, **29**, 4045.

142. J. Lu, M. Liong, J. I. Zink and F. Tamanoi, *Small*, 2010, **6**, 1794.

143. C. Slavador-Morales, E. Flahaut, E. Sim, J. Sloan, M. L. Green and R. B. Sim, *Mol. Immunol.*, 2006, **43**, 193.

144. L. Yan, Y. Wang, X. Xu, C. Zeng, J. Hou, M. Lin, J. Xu, F. Sun, X. Huang, L. Dai, F. Lu and Y. Liu, *Chem. Res. Toxicol.*, 2012, **25**, 1265.

145. A. K. Gupta, R. R. Naregalkar, V. D. Vaidya and M. Gupta, Recent advances on surface engineering of magnetic iron oxide nanoparticles and their biomedical applications, *Nanomedicine*, 2007, **2**, 23–39.
146. A. Nel, T. Xia, L. Mädler and N. Li, *Science*, 2006, **311**, 622.
147. V. I. Shubayeb, T. R. Pisanic II and S. Jin, *Adv. Drug Delivery Rev.*, 2009, **61**, 467.
148. T. D. Terley, *Biochem. Soc. Trans.*, 2007, **35**, 527.

CHAPTER 7

Magnetic Nanoceramics for Biomedical Applications

7.1   Introduction

The use of nanoparticles as new therapeutic and diagnosis tools represents an outstanding advance for biomedical sciences. Undoubtedly, a great contribution has been made by the development of iron oxide-based magnetic nanoparticles. The application of magnetic nanoparticles (MNPs) for diagnosis was introduced in the 1990s, as contrast agents for magnetic resonance imaging (MRI) of liver. Currently, there are several products on the market and undergoing clinical research that point toward better benefits for diagnosis (mainly in MRI) as well as for the treatment of tumors by hyperthermia, magnetofection and targeted drug delivery vehicles.[1]

The capability of MNPs for modifying the relaxation of the protons in living tissues, thus acting as contrast agents, is the most developed biomedical application of these nanodevices. MNPs in the form of superparamagnetic iron oxide nanoparticles (SPIONs) have found applications as contrast agents in the bowel (Lumiren® and Gastromark®), as well as for the imaging of liver and spleen (Endorem® and Feridex IV®). Several contrast agents based on ultrasmall superparamagnetic iron oxide nanoparticles (USPIONs) are currently under clinical research (for instance Combidex), due to their capability for reaching lymphatic nodules and contribution to the detection of metastases.

The basic structure of MNPs for biomedical applications involves magnetic nuclei of a few nanometers in size and a coating. The overall behavior, such as specific clinical applications, limitations, toxicity, *etc.* of the nanoparticles strongly depends upon their size nanoparticles, the chemical nature of the coating and the interaction between these two components. The size

RSC Nanoscience & Nanotechnology No. 39
Nanoceramics in Clinical Use: From Materials to Applications, 2nd Edition
By María Vallet-Regí and Daniel Arcos Navarrete
© M. Vallet-Regi and D. Arcos Navarrete 2016
Published by the Royal Society of Chemistry, www.rsc.org

determines the fate of the nanoparticles. Large MNPs (>200 nm) will be easily detected by the immune system and removed from the blood and delivered to the liver and the spleen.[2] Very small MNPs (<5.5 nm) can be excreted through the kidneys.[3] In general, nanoparticles with diameters ranging in size between 10 and 100 nm exhibit longer blood half-life, while being small enough to permeate capillaries.[4] The coating plays different significant roles that comprise colloidal stability, avoidance of opsonization, acting as a platform for the attachment of targeting moieties, *etc.* The biodistribution of the MNPs currently available is based on a passive mechanism, *i.e.* is dependent on the preferential uptake by the liver and spleen as a natural consequence of biological detoxifying mechanisms. This significantly limits the targeting of MNPs towards other tissues or organs when they are administered into the bloodstream. For this reason, one of the most active research lines in the field of MNPs addresses the specific functionalization of MNPs. The final aim is to shift from the passive biodistribution towards active vectorization to other targets, which is fundamental for the new MNP-based therapies and diagnosis of cancer.

Although passive biodistribution involves the accumulation of nanoparticles in the liver and spleen, MNPs can benefit from passive targeting through the *enhanced permeability and retention* (EPR) effect, due to the fenestrations in the vascular endothelium exhibited in by the blood vessels of many types of tumors. The combination of EPR effect with active targeting plays a lead role in the development of new diagnosis tools and therapies of cancer. To date, the enhancement in the degree of MNP localization by active targeting is moderate. There are only a few targeting moieties that have reached high enough localizations in tumors to be considered specific. Among them, luteinizing hormone-releasing hormone deserves to be highlighted as it produces 11-fold tumor localization over its non-targeted counterparts.[5] Understanding the targeting processes is not easy. In addition to the serious difficulties in discovering targets and developing specific vectors, different, less obvious factors, such as size, surface charge, colloidal stability, *etc.* determine the success or failure of a MNP-based system.

Iron oxide-based nanoparticles (both SPIONs and USPIONs) have been prepared by different approaches, such as controlled precipitation from soluble salts, aerosol assisted methods, *etc.* In the field of biomedical applications, most SPIONs and USPIONs are prepared by co-precipitation in the presence of the coating agent. In this strategy, the coating agent is mandatory for colloid stability. Primarily, it acts as capping agent in the nucleation and crystal growth processes, thus avoiding the excessive enlargement of the nanoparticles. Secondly, the coating exerts a colloid stabilization function based on interparticle repulsions of electrostatic and/or steric nature.

The decrease of magnetic properties with respect to their bulk counterpart is a consubstantial problem in the preparation of MNPs. This is due to the defects that the coating agent introduces into the crystalline structure, as well as due to the surface effects (non-collinear spins, spin canting, spin-glass-like behavior, *etc.*) that become greater as the volume/surface ratio decreases. In this sense, typical values of saturation magnetization for magnetite nanoparticles range between 30 and 50 emu g^{-1}, significantly lower than the 90 emu g^{-1} reported for magnetite in bulk.[6] Even so, the development of SPIONs and USPIONs has reached very important milestones in biomedicine. Some products have crossed the "valley of death" in the words of Declan Butler,[7] and are often used in clinical applications. However, developing the full potential of these systems involves more research to specifically target MNPs towards cell, organelles or even molecules and developing synthesis strategies to improve the magnetic properties of the MNPs without compromising the colloidal stability.

7.2 Structure and Magnetic Properties of Iron Oxide Nanoparticles

Iron oxide nanoparticles have been developed and used for biomedical purposes over the past decade. Interest in these kinds of nanoceramics keep on growing, mainly in the field of SPIONs such as magnetite Fe_3O_4 and maghemite γ-Fe_2O_3 with sizes between 5 and 20 nm. Magnetite is an iron oxide commonly found in nature. Magnetite shows inverse spinel structure with a *fcc* cubic close-packed oxygen array, and iron in both fourfold (tetrahedral) and sixfold (octahedral) coordination. Fe^{2+} cations occupy octahedral sites and half of the Fe^{3+} cations are placed in the tetrahedral sites, whereas the other half occupy octahedral ones. The unit cell has $Fd\bar{3}m$ space group with a lattice parameter $a = 8.3941$ Å. The chemical formula of the unit cell can be written as $(Fe^{III}_8)_A[Fe^{II}_8,Fe^{III}_8]_BO_{32}$, where A and B indicate tetrahedral and octahedral sites, respectively. The maghemite γ-Fe_2O_3 structure is similar to magnetite, *i.e.* an inverse spinel structure with *fcc* cubic close-packed oxygen array, and iron in both fourfold (tetrahedral) and six fold (octahedral) coordination. However, maghemite only has Fe^{3+} cations distributed in a cubic unit cell with $P2_13$ spatial group with $a = 8.33$ Å. Similarly to magnetite, in this unit cell the Fe^{3+} cations occupy 1/8 of tetrahedral sites, whereas the rest of Fe^{3+} cations occupy the octahedral, thus leaving vacancies in the octahedral sublattice. The chemical formula by unit cell can be written as $(Fe^{III}_8)_A[Fe^{III}_{40/3},\square_{8/3}]_BO_{32}$, where *A* and *B* mean tetrahedral and octahedral sites, respectively and $\square$ means vacancies.[8] Figure 7.1 shows the crystalline structure of Fe_3O_4 and γ-Fe_2O_3.

The classification of a material's magnetic properties can be made on the basis of the magnetic susceptibility (χ). χ is defined as the ratio of the induced magnetization (M) with respect to the applied magnetic field (H) as deduced from eqn (7.1)

$$M = \chi H \qquad (7.1)$$

In diamagnetic materials M is antiparallel to H and the result is that magnetic susceptibility χ shows very small and negative values in the range of -10^{-6} to -10^{-3} (χ is a dimensionless magnitude). These materials do not

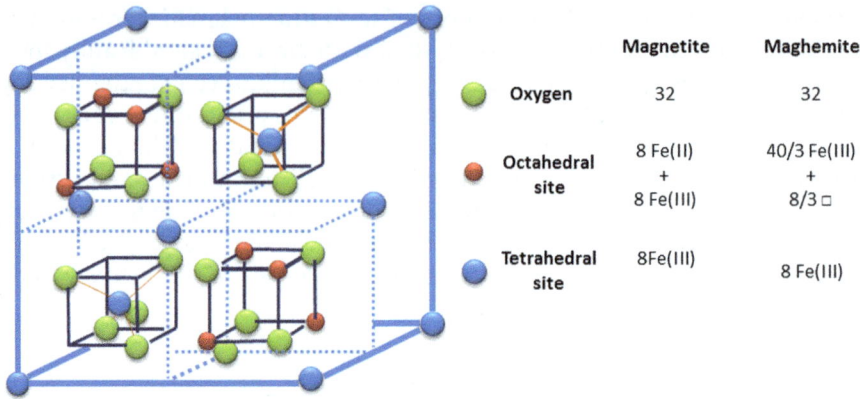

		Magnetite	Maghemite
●	Oxygen	32	32
●	Octahedral site	8 Fe(II) + 8 Fe(III)	40/3 Fe(III) + 8/3 □
●	Tetrahedral site	8Fe(III)	8 Fe(III)

Figure 7.1 Crystalline structure of magnetite and maghemite.

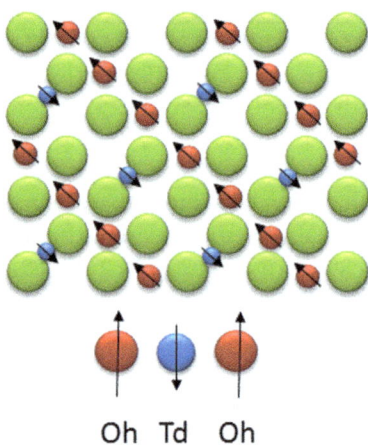

Oh   Td   Oh

Figure 7.2 Simple representation of the magnetic spins in a ferrimagnetic oxide.

retain the magnetism when H is removed. In contrast, paramagnetic materials are those whose magnetic moment M aligns parallel to the applied magnetic field H, thus exhibiting positive values of χ in the range of 10^{-6} to 10^{-1}. In addition to a parallel arrangement of M with respect to H, coupling interactions between the material electrons can occur. Under these circumstances the appearance of ordered magnetic states, denoted magnetic domains occurs resulting in ferri- or ferromagnetic materials. These materials can exhibit large spontaneous magnetization and their χ depends on their atomic structures, temperature and applied H.

The crystalline arrangements of magnetite and maghemite result in two magnetic sublattices, tetrahedral and octahedral, with different magnetic moments and antiferromagnetically ordered by superexchange interactions, thus leading to a magnetic moment different to zero, *i.e.* to *ferrimagnetic* ordering (Figure 7.2). Ferromagnetism and ferrimagnetism are cooperative

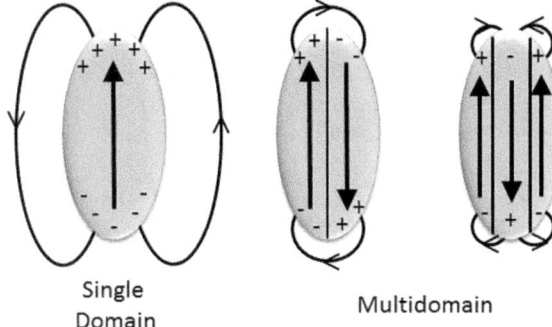

Single Domain

Multidomain

Figure 7.3 Scheme of the formation of domains and the effect on magneto-static energy. The magnetostatic energy can be approximately halved if the magnetization splits into two domains magnetized in opposite directions.

magnetic effects that cannot be shown by single atoms. However when the atoms get together in a solid, these behaviors occur.

Ferro- and ferrimagnetic solids adopt domain magnetic structures. It means that the solid is actually composed of small regions called magnetic domains, within each of which the local magnetization is saturated but with different magnetization directions with respect to each other. Domains are small (1–100's of micrometers), but much larger than atomic distances. The reason for the appearance of the magnetic domains is the magnetostatic energy of the material, which is the potential energy produced by the external magnetic field. As shown in Figure 7.3, the magnetostatic energy is minimized by the formation of magnetic domains, which close the magnetic flux, confining it within the material. Each domain behaves like a small permanent magnet. In a non-magnetized piece, domains are randomly oriented so that their effects are canceled and no measurable net magnetization is observed outside the material.

Domains, like grains of a polycrystalline material, are separated by a domain wall or edge, known as Bloch walls. The orientation of the elementary dipoles changes gradually, not suddenly, from a magnetic domain to the adjacent one. As a result, the width of the Bloch walls is quite high, on the order of 300 atoms. Because of the non-perfect spin alignment, the wall is always an area of greater internal energy than that inside the domain. However, what if the particle size were too small to form Bloch walls and multi-domain structure?

When the size of a ferro- or ferrimagnetic solid is smaller than the size of a particular magnetic domain, the material becomes *superparamagnetic*. Superparamagnetism is a form of magnetism which appears in small ferromagnetic or ferrimagnetic nanoparticles. In small enough nanoparticles, magnetization can randomly flip direction under the influence of temperature. The typical time between two flips is called the Néel relaxation time. In the absence of an external magnetic field, when the time used to measure the magnetization of the nanoparticles is much longer than the Néel relaxation time, their magnetization appears to be on average zero: they are said to be in the superparamagnetic state. In this state, an external magnetic field is able to magnetize the nanoparticles, similarly to a paramagnetic material. However, their magnetic susceptibility is much larger than that of paramagnetic material.

The decreasing of the particle size of Fe_3O_4 and γ-Fe_2O_3 to reach the nanometrical range involves magnetic changes in addition to the ferrimagnetic to superparamagnetic transition. For instance, saturation magnetization losses occur as the particle size decreases. The saturation magnetization is the maximum induced magnetic moment that can be obtained in a magnetic field (H_{sat}); beyond this field no further increase in magnetization occurs. This is due to the atomic disorder associated to the particle surface, which involves the phenomena of spin canting and magnetic dead layer.[9] The loss of M_S values also depends on the anisotropy constant K of the compound, related to the presence or otherwise of easy magnetization axes. Fe_3O_4 and γ-Fe_2O_3 are ferrimagnetic when the grain size is in the micrometer range and become superparamagnetic when the particle size is decreased below 20 nm. As mentioned earlier, the particle size reduction comprises an increment of the surface/bulk atomic ratio, which results in a decrease of particles' magnetization. This reduction is very significant when Fe_3O_4 and γ-Fe_2O_3 SPIONs are <10 nm, *i.e.* USPIONs. When SPIONs and USPIONs are intended for *in vivo* applications, the surface must be modified to provide specific functionalities aimed to resolve the diagnostic or therapeutic task. Functionalization of SPIONs and USPIONs with non-magnetic coatings results in an additional decrease of magnetic properties. For instance, coatings of poly(lactide co-glyclide) polymer over SPIONs can reduce the M_S values by up to 50%.[10] This result has been also reported for other coatings such as oleic acid, poly(vinyl pyrrolidone), *etc.* commonly used to modify iron oxide nanoparticles. In these cases the structural and magnetic characterizations reveal differences in the particle size. When particle sizes are determined by transmission electron microscopy or X-ray diffraction the particle size is larger than that derived from magnetic measurements, thus indicating the presence of a magnetically dead layer at the surfaces of functionalized SPIONs and USPIONs.

7.2.1 Magnetic Properties of Iron Oxide Nanoparticle Colloids

In order to form a colloidal suspension, the particle size must be much lower than 1 micrometer. The average size of magnetite and maghemite nanoparticles currently used in biomedical applications ranges between 4 and 18 nm, which is much lower than that of a magnetic domain. Therefore, colloidal magnetite and maghemite nanoparticles are totally magnetized as single-domain nanomagnets. Higher the magnetic moments result in better performance of the colloid.

In addition to the magnetization value, the single domain nanoparticles are characterized by another property: the anisotropy energy (E_a). The magnetic energy of a nanomagnet depends on the tilt of the magnetization vector direction with respect to the crystallographic directions. In this way, the directions that minimize the magnetic energy are named easy axes. Anisotropy energy and is defined by eqn (7.2)

$$E_a = K_a V \qquad (7.2)$$

where K_a is the anisotropy constant, which is closely related to the crystalline structure and V is the crystal volume. This model assumes that the anisotropy is uniaxial, which is not the real situation. However, for high-symmetry cubic systems, like ferrites with spinel structure, this approximation is assumable and so Fe_3O_4 and γ-Fe_2O_3 can be studied under this approximation.

The anisotropy energy also determines the Néel relaxation time (τ_N), which is another important parameter of the magnetic behavior in single domain nanoparticles. When these particles are found as dry powder, τ_N is characterized by the time of return constant to the equilibrium state after a perturbation. Under high anisotropy conditions, the crystal magnetization is locked in the easy magnetization axis, as in this situation the system presents lower magnetic energy. The Néel relaxation defines the fluctuations that arise from the jumps of the magnetic moment between different easy magnetization axes.

When the particle size of Fe_3O_4 and γ-Fe_2O_3 decreases, the energy barrier of the reversal magnetization also decreases. The consequence is that the thermal fluctuations lead to relaxation phenomena. In the case of the Néel relaxation, *i.e.* the magnetic moment fluctuations trough an anisotropy barrier, the relaxation time, τ_N, of the nanoparticles is determined by the ratio between the anisotropy energy (E_a) KV and the thermal energy kT, as follows:

$$\tau_N = \tau_0 \exp[KV/kT], \text{ being } \tau_0 \sim 10^{-9} \text{ s} \qquad (7.3)$$

K = anisotropy constant
V = particle volume
k = Boltzmann constant
T = temperature

For a characteristic time of measurement, τ_m, we can determine a critical particle volume (V_c) in such a way that the relaxation time for these particles $\tau_N(V_c)$ is equivalent to the measurement time, $\tau_N(V_c) = \tau_m$. Below this volume τ_m is longer than τ_N, thus revealing the *superparamagnetism* phenomenon. For magnetite SPIONs, the critical diameter is about 20 nm.

In the case of magnetic fluids, SPIONs are dispersed within a liquid medium forming a colloid. In this situation the return of magnetization towards the equilibrium state is determined by two processes. The first is the Néel relaxation (explained above) and the second is the Brownian relaxation,

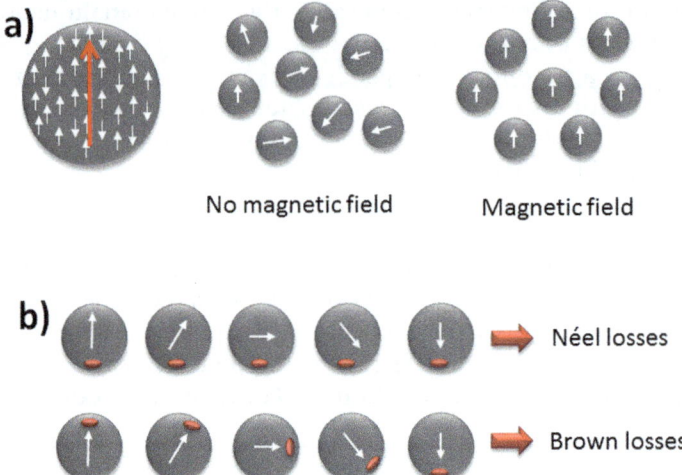

Figure 7.4   (a) Schematic representation of superparamagnetic particles in the absence and presence of a magnetic field; (b) rotation of the moment within the magnetic nanoparticles (MNPs), overcoming their anisotropy energy barrier that leads to Néel loss (top), and rotation of the MNPs themselves, which will create frictional losses with the environment and lead to Brown losses (bottom).

which is characterized by the rotation of the particles as a whole (Figure 7.4). The global magnetic relaxation rate is the addition of the Néel and Brownian relaxation,

$$\frac{1}{\tau} = \frac{1}{\tau N} + \frac{1}{\tau B} \tag{7.4}$$

where τ, τ_N and τ_B are the global, Néel and Brownian relaxation times, respectively. τ_N described by an exponential function as expressed in eqn (7.5), while τ_B is expressed as

$$\tau_B = 3V\eta/kT \tag{7.5}$$

η being the damping constant.

For large particles τ_B is shorter than τ_N because the Brownian component of the magnetic relaxation is proportional to the volume of the crystal, whereas the Néel relaxation is an exponential function of the volume. For this reason, the viscous rotation of the particle becomes the dominant process and determines the global magnetic relaxation, which is much faster than that for dry powder. In summary, the properties of a magnetic colloid are mainly determined by the particle diameter, the magnetization values and its τ_N, which is strongly dependent on the K_a.

In the superparamagnetic regime, the magnetization of small nanoparticles can randomly change due to temperature. In the absence of an external magnetic field, when the measurement time of the magnetization is longer than the relaxation time, the average magnetization observed appears to be zero. In this state an external magnetic field can magnetize the nanoparticles, as in the paramagnetic situation. However, the magnetic susceptibility is much higher than paramagnetic materials, obtaining magnetization values similar to ferro- or ferrimagnetic materials, but exhibiting coercive fields very close to zero. These features make SPIONs very interesting systems for biomedical applications, as they elicit a high magnetic response without leaving magnetic remanence when the magnetic field ceases.

7.3 Synthesis of Magnetic Iron Oxide Nanoparticles

7.3.1 Co-Precipitation Synthesis

In recent decades, several synthetic methods have been developed for the preparation of SPIONs. Among the different strategies, the co-precipitation of magnetite from soluble ferrous and ferric salts is the most widely used, perhaps because is one of the most efficient and easiest methods for this purpose. When preparing SPIONs for biomedical applications, we find two main challenges:

- defining the experimental conditions to obtain monodisperse and reproducible systems; and
- avoiding complex purification procedures.

The synthesis by co-precipitation fulfils the conditions to overcome these challenges and, consequently, both Fe_3O_4 and γ-Fe_2O_3 are commonly prepared by precipitation of ferrous and ferric salts previously dissolved in aqueous media. Thus, the chemical reaction for magnetite preparation can be written as:

$$Fe^{2+} + 2Fe^{3+} + 8OH^- \rightarrow Fe_3O_4 + 4H_2O$$

Therefore, for the completion of this reaction the pH must be between 8 and 14, while keeping a stoichiometric ratio $Fe^{3+}/Fe^{2+} = 2$ and a non-oxidant environment to avoid the $Fe^{2+} \rightarrow Fe^{3+}$ oxidation. In fact, magnetite is not a very stable phase in physiological conditions (humid and oxidative), and can be oxidized to γ-Fe_2O_3 and α-Fe_2O_3 as follows:

$$Fe_3O_4 + 2H^+ \rightarrow Fe_2O_3 + Fe^{2+} + H_2O$$

For this reason, many authors opt for the application of *maghemite* instead of magnetite, although maghemite exhibits lower magnetic properties. Maghemite already has all the iron cations as Fe(III) and cannot undergo further oxidation or transformation into the non-magnetic *hematite*, α-Fe_2O_3. As mentioned above, one of the advantages of the co-precipitation method is its efficiency. In fact, this method allows the preparation of a large amount of nanoparticles. However, the main drawback is the

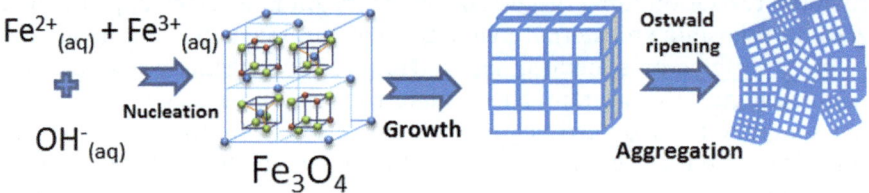

Figure 7.5   Stages of the co-precipitation method for the Fe_3O_4 preparation.

difficulty of controlling the particle size, because the nucleation and crystal growth processes are mainly ruled by kinetic factors, which are non-thermodynamic. If we want to obtain a single-mode particle distribution, the nucleation and growth processes must be separated in time. When both processes occur simultaneously (as commonly they do) the newly formed nuclei are used to enlarge the already formed larger particles, in a process known as Ostwald ripening. These kind of systems are polydisperse and are difficult to use in biomedicine.

One of the most common approaches to obtain monodispersed nanoparticles is to start from supersaturated solutions. Under these conditions, the nuclei form simultaneously and the subsequent growth results in monodispersed precipitated particles. The co-precipitation comprises nucleation, growth, maturation and aggregation processes which occur simultaneously (Figure 7.5). In fact, the co-precipitation process is not a trivial question. The co-precipitation reactions exhibit the following characteristics:

(1) The precipitated products generally are low-soluble species formed in supersaturated conditions;
(2) Supersaturated conditions rule the nucleation process, which is the key step to obtain narrow-size nanoparticles;
(3) Subsequent processes such as Ostwald ripening and aggregation dramatically affect the size, morphology and properties of the magnetic nanoparticles; and
(4) The supersaturation condition generally is a consequence of a chemical reaction. For instance, the obtaining of magnetite prior to formation of the corresponding hydroxides:

$$FeCl_2(aq) + 2FeCl_3(aq) + 8NH_4OH(aq) \rightarrow$$
$$Fe(OH)_2(s) + 2Fe(OH)_3(s) + 8NH_4Cl(aq) \rightarrow$$
$$Fe_3O_4(s) + 4H_2O(l) + 8NH_4Cl(aq)$$

In this case, the formation of non-soluble hydroxides (and subsequent dehydration to magnetite) is a consequence of the reaction of soluble iron chlorides with ammonia.

The fact of precipitating a product form a liquid solution does not guarantee the formation of monodispersed nanoparticles. The processes of nucleation and growth rule the size and shape of the particles. When the precipitation begins, numerous tiny crystallites initially form, but they trend to aggregate very fast, resulting in larger and more thermodynamically stable particles. As explained above, the key point is the supersaturation degree, S, given by eqn (7.6)

$$S = a_A \cdot a_B / K_{ps} \qquad (7.6)$$

where a_A and a_B are the activities of the reactants in solution and K_{ps} is the solubility product constant of the precipitated compound. In fact, the homogeneous nucleation rate, R_N, which is the amount of nuclei formed per unit of time and volume, is an exponential function of S.

$$R_N = A \exp(-16\pi\sigma_{SL}^3 v^2 / 3k^3 T^3 \ln^2 S) \qquad (7.7)$$

where σ_{SL} is the solid–liquid surface tension, v is the atomic volume of the solute and k is the Boltzmann constant. From eqn (7.7) it follows that R_N has negligible values until a certain critical saturation, $S*$ and temperature T are reached.

In another way, temperature determines the size of the nuclei. The higher the temperature the smaller the crystallites formed. When nucleation begins, there is an equilibrium critical radii $R*$, in such a way that

$$R* = \alpha/\Delta C \qquad (7.8)$$

where $\alpha = (2\sigma_{SL}/kT \ln S)v$ and $\Delta C = C - C_{equilibrium}$, that is the precipitation driving force.

Those nuclei with sizes that fulfil $R > R*$ keep on growing, whereas those with $R < R*$ are dissolved and cannot exist. Consequently, high values of ΔC and T mean that $R*$ acquires lower values and smaller particles can thus exist.

In addition to concentration and temperature, the particle size and shape can be controlled with certain success by adjusting the pH, the ionic strength, the counter-ions (perchlorate, chlorides, sulfates and nitrides) and the Fe^{II}/Fe^{III} ratio. In 1981 Massart reported the first controlled alkaline precipitation of SPIONs from $FeCl_2$ and $FeCl_3$.[11] In his original synthesis, Massart obtained magnetite particles of 8 nm in size, as determined by X-ray diffraction. Thereafter, other parameters, such as the influence of the base used (NH_3, CH_3NH_2, NaOH, *etc.*), the pH value, the presence of other cations and the Fe^{II}/Fe^{III} ratio was also studied. When all these parameters are adjusted, particles with sizes in the range 4–16 nm and having narrow diameter distributions can be prepared.

7.3.2 Microemulsion Synthesis

Water-in-oil microemulsion synthesis is a very interesting method for the preparation of SPIONs with a narrow size distribution and uniform physical properties, as this method allows the control of the size and the spherical shape of the nanoparticles. In 1943, Hoar and Schulman[12] discovered that certain proportions of water, oil, surfactant and an alcoholic or amine co-surfactant resulted in mixtures that were apparently homogeneous-like dissolutions. These mixtures were denoted microemulsions. In the study of Hoar and Schulman the oil phase was a long-chain hydrocarbide (non-polar) and the surfactant was hexadecyltrimethylammonium bromide, $C_{16}TAB$, which is a cationic surfactant with a polar head of ammonium and a non-polar chain of C_{16}. Currently, $C_{16}TAB$ is one the most common structure-directing agents used for the synthesis of mesoporous materials, as explained in Chapter 5. Although the mixtures are optically isotropic, microemulsions are not solutions as the surfactant molecules are not randomly distributed, despite its solubility in both water and oil phases. The surfactant, through ion-dipole interactions with the polar co-surfactant, forms spherical aggregates where the polar heads are oriented towards the center. The co-surfactant acts as an electronegative spacer that minimizes the repulsions between the positively charged polar heads of the surfactant (Figure 7.6).

The addition of water to the system provokes the expansion of the aggregates from the center; the water molecules are placed in the center of the sphere are also attracted by ion-dipole and dipole–dipole interactions. The arrangement of the surfactant molecules into these reverse micelles minimizes the interfacial tension in the aggregates, and consequently water-in-oil (*w/o*) microemulsions are thermodynamically stable. This stability distinguishes microemulsions from macroemulsions, which are thermodynamically unstable.

The synthesis of magnetic nanoparticles by microemulsions mainly consists of precipitation driven by supersaturation within a confined nanospace,

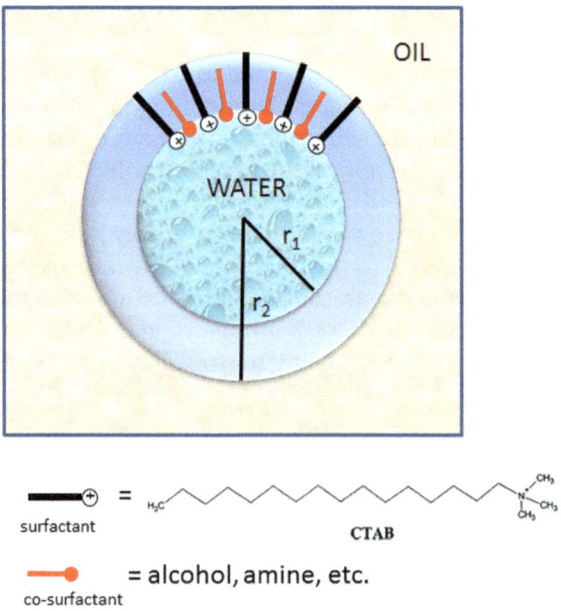

Figure 7.6　Model of a reverse micelle using CTAB as surfactant and stabilized with a co-surfactant.

i.e. the reverse micelle. The size of a reverse micelle can be easily controlled with the ratio

$$\omega_0 = [H_2O]/[S] \tag{7.9}$$

Experimentally, the micelle radius, R_M, varies linearly with the water content, especially when $\omega_0 \geq 1$. Although controlling the micelle size is relatively easy, the formation mechanisms of reversal micelles and their phase equilibria are inherently complexes, as they are systems of four or more components: water, oil, surfactant and co-surfactant(s). Microemulsions take place only in limited proportions, which are generally represented in Gibbs triangles.

7.3.2.1 Reaction Dynamic in Reverse Micelles

The small size of reverse micelles forces them into a continuous Brownian motion, even at room temperature. The collisions between micelles are very frequent and approximately one of each 1000 collisions leads to the formation of short-life dimers through the expulsion of surfactant molecules to the oil phase. During the ~100 ns of dimer life, two reverse micelles exchange the payloads of their aqueous nuclei before decalescence. Eventually, this process results in an equilibrium distribution of all the payloads.

Considering this model of interactions between the reversal micelles, their application as nanoreactors is understandable. Since the size of the aqueous nuclei can be easy tailored by means of ω_0 and the Brownian motion of micelles allows the distribution of the reactants, the chemical reactions can be performed in confined nanospaces and the reaction products are uniform in size and shape.

However, the complexity of these systems should not be underestimated, especially when ionic salts are dissolved in the aqueous nuclei of a reverse micelle. This is the case of magnetite and *maghemite* nanoparticles, which are commonly prepared from iron chlorides. In these cases, the solvated ions interact with the polar head of the surfactant and shrink the micelle radii R_M. This effect increases with the concentration and ionic charge. On the other hand, surfactants acts as capping agents that avoid the flocculation, but once the nanoparticles are formed the Ostwald ripening cannot be avoided. In other words, the micelles temporally confine the space but not throughout the precipitation process.

7.3.2.2 Synthesis of Iron Oxide Nanoparticles by Microemulsion Technique

When a microemulsion of reverse micelles containing iron salts, for instance $FeCl_2$ and/or $FeCl_3$, is added to another microemulsion containing a Brønsted base (NH_4OH, $NaOH$, *etc.*), the metallic cations are precipitated as hydroxides. After centrifugation and moderate heating, magnetite and maghemite nanoparticles can be obtained. This strategy also works for the preparation

Table 7.1   Methods for the preparation of iron oxide magnetic nanoparticles from microemulsions.[a]

Material	Raw materials	Surfactant	Precipitating agent	Reaction conditions	Nanoparticle size (nm)	Reference
Fe_3O_4	$FeCl_2$ $FeCl_3$	AOT	NH_4OH		2	13
Fe_3O_4	$FeSO_4$	AOT	NH_4OH		10	14
Fe_3O_4	$FeCl_3$ $Fe(NO_3)_3$	NaDBS	N_2H_4	90 °C	5	15
Fe_3O_4	$FeCl_2$ $FeCl_3$	AOT	NaOH	90 °C	70–120	16
Γ-Fe_2O_3	$FeSO_4$ TEA	CTAB	$NaNO_2$	45 °C	22–25	17

[a]AOT: surfactant aerosol OT; NaDBS: sodium dodecylbenzenesulfonate; TEA: tetra ethyl amine; CTAB: cetyltrimethylammonium bromide.

of other nanocrystalline mixed ferrites, such as $ZnFe_2O_4$, $CoFe_2O_4$, and $MnFe_2O_4$, *etc.* In the case of magnetite preparation, nanoparticles of as small as 2 nm can be prepared in a very narrow monodisperse distribution by changing the temperature.[13] In the same way, the size can be also adjusted with the cations and alkali concentrations. Table 7.1 shows some of the methods found in the literature for the preparation of iron oxide MNPs by microemulsion methods.

The microemulsion method allows the preparation of monodispersed coated SPIONs in a one-pot synthesis. For this purpose, organic bases are used, as these compounds act as precipitating agents and coatings. For instance, oleyamine has been successfully used for this goal.[18] The system works in a w/o microemulsion based on cyclohexane/Brij-97/H_2O, where Brij-97 is a non-ionic surfactant, so that it does not interfere in the iron oxide crystallization occurring in the aqueous nuclei. The incorporation of oleyamine in microemulsion B and its mixture with a microemulsion containing iron cations leads to the magnetite precipitation, coated with an oleyamine layer and a narrow size distribution of 3.5 ± 0.6 nm.

7.3.3   Solidification of Solid Solutions

These processes, also known as polyol methods are a useful chemical approximation to prepare nanoparticles with controlled size and well-defined shapes. This method, described in Chapter 2, has been also used for the synthesis of non-aggregated magnetite nanoparticles[19] using tetra-ethyleneglycol (TEG) as polyol solvent and iron(III) acetate as magnetite precursor. TEG acts as a reducing agent of Fe(III) to Fe(II), as well as a stabilizer of the particle size.

This methodology exhibits two advantages with respect to co-precipitation methods. The first is that the SPION surfaces are coated with hydrophilic polyol ligands and therefore are easily dispersed in aqueous media.

In addition, this process is carried out at relatively high temperatures, which favors nanoparticle crystallinity and thus leading to higher magnetic properties.

7.3.4 Aerosol-Assisted Methods

Aerosol-assisted methods are very attractive technologies because they involve continuous processes with a high production rate. In aerosol-assisted techniques, the precursors are iron salts solubilized in aqueous or organic solvents (commonly low molecular weight alcohols). The aerosol generation can be carried out by spraying (Venturi effect), piezoelectric-induced ultra-sounds, *etc.* The droplets of aerosol containing the precursors are trans-ported by an inert gas flow towards the reactor. This reactor is commonly a furnace where the solvent evaporation takes place and the precursors are pyrolysed. The resulting dried solid is collected by means of a filter, which can be physical (barrier) or electrostatic. This solid is composed of particles with a size strongly dependent on the size of the aerosol droplets generated at the beginning of the process. In this sense, maghemite and magnetite par-ticles have been obtained with particles sizes ranging in size between 5 and 60 nm, by using different iron salts as precursors.[20]

Aerosol laser pyrolysis produces non-aggregated nanoparticles of lower size and narrow distribution. The technique is based on the heating, by means of a laser, of a mixture in vapor phase of the iron precursors and a flowing gas. When the process conditions are properly adjusted, SPIONs with sizes between 2 and 7 nm with a very narrow distribution can be prepared.

7.4 Surface Functionalization of SPIONs

7.4.1 Stabilization of Iron Oxide Nanoparticles

The stabilization of SPIONs is a fundamental stage during the preparation of stable magnetic colloidal suspensions. The stabilization strategy must avoid the agglomeration of the SPIONs in the physiological environment and under the action of an external magnetic field. The stability of the mag-netic colloidal suspensions depends on the attractive–repulsive equilibrium between four different kinds of interactions:

(1) van der Waals forces: these interactions lead to strong and low-range attractive forces;
(2) Electrostatic repulsion forces, which can be shielded by the ions pres-ent in the physiological fluids;
(3) Magnetic dipolar forces: these forces induce anisotropic interactions that, globally considered in all directions, result in attractive forces; and
(4) Steric repulsion forces: these forces are absent in naked particles, but play a fundamental role in the strategies of SPION stabilization.

The steric and electrostatic repulsion forces are the main tools for the preparation of stable magnetic colloids. Controlling the intensity of both interactions is a key point to reach this aim. The steric forces are difficult to control. These interactions are widely described for polymers and mainly depend on density and molecular weight. The electrostatic repulsion can be followed by means of zeta potential measurements of the particles. This parameter is strongly dependent on the ionic strength and the pH of the solution. In the case of iron oxide nanoparticles, the Fe atoms at the surface act as electron acceptors, *i.e.* Lewis acids. These atoms coordinate to the oxygen atoms of the water molecules that are dissociated and hydroxylate the SPION surface. The Fe–OH groups are amphoteric and can react with acids and bases. Consequently, depending on the pH value of the solution the zeta potentials change from positive towards negative values when the solution is acid or basic, respectively. The isoelectric point of for Fe_3O_4 and γ-Fe_2O_3 is about pH 7. It means that in pH close to the physiological one the electric charge of the SPIONs is near zero. This fact means that at physiological conditions the interparticle repulsion is very low, the magnetic colloid is not stable and the particles flocculate. Therefore, developing strategies to increase the repulsive forces, both steric and electrostatic, is one of the main goals of the field of the development of SPIONs for biomedical applications.

7.4.1.1   Silica and Organosilica Coatings

Silica, SiO_2, and their organic–inorganic hybrid derivatives are among the most widely studied materials for stabilization coatings.[21,22] Silica provides chemical stability and avoids the aggregation of nanoparticles by means of two different mechanisms. The first is reducing the magnetic dipole attraction by shielding the particles against each other. The second mechanism is related to the isoelectric point of the SiO_2, which is reached at pH ~2. At physiological pH (7.2–7.4) SiO_2-coated iron oxide nanoparticles, Fe_3O_4@SiO_2, are negatively charged and the electrostatic repulsion prevents their aggregation.

However, among the SiO_2-based coatings that have reached the highest degree of development are their organic–inorganic hybrid derivatives. For instance, ferumoxil, a magnetic colloid clinically tested by oral administration, consists of nuclei of iron oxide coated by [3-(2-amino-ethylamino)propyl] trimethoxysilane with a total particle diameter ~300 nm. This product is used as a contrast agent and enhances the definition of the outlines of the uterus and lymphatic nodules. Other example are NanoTherm (MagForce, Berlin, Germany), developed for the treatment of tumors by hyperthermia. In this case, magnetite SPIONs are coated by aminopropylsilane (APTS). The amine groups provide positive charge at physiological pH, thus leading to electrostatic repulsion and colloidal stability. The coatings are performed by hydrolysis and condensation (*via* sol–gel) of the corresponding silane, for instance tetraethylorthosilane (TEOS) or an organosilane such as APTS in the presence of the SPIONs previously prepared. Alternatively, SiO_2 coatings have been prepared using silicic acid as starting material. In fact, some

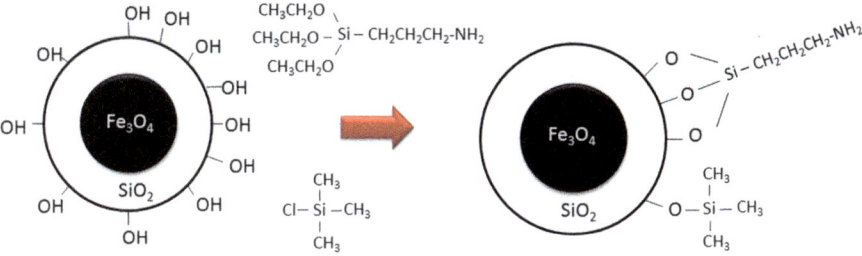

Figure 7.7 Silanization process of a SiO_2-coated magnetite nanoparticle.

studies demonstrate that this method is more efficient, as higher amounts of SPIONs are coated[23] and the size can be controlled between several tens to several hundreds of nanometers.

Finally, other synthetic routes such as microemulsions and aerosol pyrolysis have been used for this purpose. The microemulsion strategy has been used to prepare monodisperse silica-coated SPIONs and further entrapment of biological macromolecules in the nanoparticle pores.[24] The aerosol pyrolysis has been employed for the SiO_2-coating of SPIONs[25] as well as for the encapsulation of superparamagnetic nanoparticles into mesoporous silica. The latest systems have been demonstrated to behave as thermoseeds for hyperthermia treatment of tumors and matrices for drug delivery purposes.[26]

The presence of silica as a coating allows the incorporation of other chemical functional groups, through their bonding to the silanol (Si–OH) present on the surface. One of the most common strategies is the so-called silanization: SiO_2 is reacted with an organosilane, thus incorporating the desired functional group, as indicated in Figure 7.7.

7.4.1.2 Polymeric Coatings

Polymeric coatings have played a fundamental role in the transference "from bench to bedside" of iron oxide nanoparticles. Polymer-coated SPIONs can be prepared *via in situ* and post-synthesis routes. *In situ* coating involves the coating of the nanoparticles during the synthesis process. In these cases, the polymer also acts as a capping agent during the particle nucleation and growth, thus avoiding Ostwald ripening and aggregation. Post-synthesis routes consist of coating the SPIONs after synthesis. In the scientific literature there are a large number of polymers available for this task. In this section we focus on those that have found application in the clinical field or that have opened possibilities for the use of SPIONs in new biomedical applications.

7.4.1.2.1 Dextran. Dextran is a polysaccharide polymer composed of two units of α-D-glucopyranosyl, where the length and branching of the chains is variable (Figure 7.8). The biocompatibility of dextran used as a coating has been widely demonstrated and there are several medical

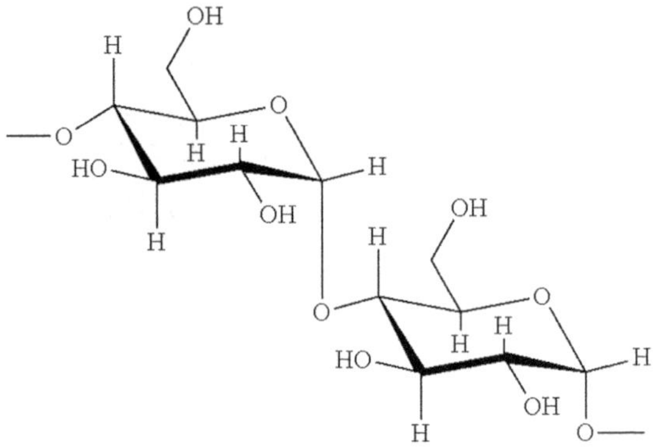

Figure 7.8   Dextran structure.

devices in the market based on iron oxide nanoparticles coated (both *in situ* and post-synthesis) with this compound. In 1982, the synthesis *in situ* of dextran-coated magnetite was reported for the first time.[27] The synthesis route consists of mixing a dextran solution with a second solution of $FeCl_2/FeCl_3$. Thereafter, NH_4OH is dropwise added under stirring and kept at 60–65 °C for 15 minutes. Afterwards, the large aggregates are separated with several cycles of centrifugation. One example of a medical device prepared by this route is ferumoxtran-10. Ferumoxtran-10 is a magnetic resonance contrast agent developed in the 1980s for magnetic resonance lymphography,[28] consisting of ultra-small super-paramagnetic iron oxide particles. Ferumoxtran-10 is marketed as Sinerem® in the Netherlands (Laboratoire Guerbet, Aulnay sous Bois, France), and as Combidex® in the USA (Advanced Magnetics, Cambridge, MA), and it has been successfully evaluated for improved detection of lymph node metastases in various clinical trials.[29]

Clinically approved dextran-coated nanoparticles for MRI diagnosis show a hydrodynamic diameter between 15 and 30 nm. They exhibit a long lifetime in blood, and can be captured by macrophages of deep tissues. Following injection, the nano-particles slowly escape from the vessels into the interstitial space, from which they are transported to lymph nodes by way of lymphatic vessels. Within the lymph nodes, the particles are internalized by macrophages, and these intra-cellular iron-containing particles cause changes in magnetic properties that can be detected by MRI.

Dextran is attached to the iron oxide nanoparticles through van der Waals forces, hydrogen bonds and electrostatic interactions. For this reason it is important that the size of dextran chain favors the polar interaction by hydrogen bond with the SPION surface. Although these interactions are significantly weaker than covalent bonds, the total bonding energy is high due to the large number of hydroxyl groups contained in the chains. However, dextran detaching can also occur. In order to avoid this situation, the most common method is adding cross-linkers during the coating process.

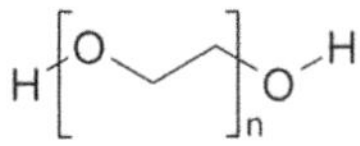

Figure 7.9 Polyethylene glycol.

7.4.1.2.2 Polyethylene Glycol. Polyethylene glycol (PEG) is a polymer that has greatly contributed to the clinical application of nanoparticles. PEG is biocompatible hydrophilic polymer, whose chemical structure presented in Figure 7.9.

Its main contribution to the field of nanoparticles in biomedicine relays in the increment of lifetime in blood when nanoparticles are coated with PEG.[30] Blood plasma has a high ionic force and large amount of serum proteins. The mission of these proteins is opsonizing any foreign particle and its removal from the blood system. When SPIONs are coated by the opsonins they are easily recognized and phagocytosed by the reticuloendothelial system (RES), which is a group of cells with the mission of capturing inert particles circulating within the organism. The RES is constituted of macrophages, Kupffer cells in the liver and reticular cells in the lungs, bone marrow, spleen and lymphatic nodes.

The survival of nanoparticles in blood, for at least a few hours, is mandatory for most of their therapeutic and diagnosis purposes.[31] PEG acts as a stabilizer in aqueous media and as opsonization inhibitor. In this way, PEG hinders recognition by macrophages and increases the lifetime in blood of the SPIONs and USPIONs. Feruglose (Clariscan) is an example of PEGilated magnetite used as a MRI contrast agent. It consists of USPION particles that are composed of single crystals (4–7 nm diameter) and stabilized with a carbohydrate PEG coating. The iron oxide particles have to be suspended in an isotonic glucose solution and the final diameter of an USPION particle is ~20 nm. Blood pool half-life is >2 hours in humans; the particles are taken up by the mononuclear phagocyte system and distributed mainly to the liver and spleen.

Several methods have been developed to attach PEG to iron oxide nanoparticles, including silane grafting,[32] polymerization at the surface[33] and modification through sol–gel approximation.[34] In order to control the polymer conformation and to provide stable covalent bonds between PEG and nanoparticles, bi-functionalized PEG-silanes have been developed.[35] These PEG-silanes are able to form self-assembling monolayers, thus enhancing the package density of PEG chains over the nanoparticles. In addition, PEG can be modified with amine or carboxylate groups, which expand toward the surrounding media and supply anchoring sites of other functional ligands such as drugs and targeting molecules.

7.4.2 Active Targeting Agents

Active targeting, also denoted specific targeting is perhaps the most promising strategy to take MNPs (and nanoparticles in general) towards diseased cells and tissues. Active targeting is based on the conjugation of targeting moieties that possess high affinity for biological entities, which are preferentially or specifically expressed in the diseased organs, tissues or cells. This affinity generally is a receptor–ligand- or antigen–antibody-like interaction and, in the case of active targeting towards tumors, it is commonly enhanced by the EPR effect. Consequently, the active targeting provides a very effective approach to improve the residence time of MNPs in tumors. Several aspects concerning active targeting are described in Chapter 6. In this section focus on those targeting ligands that have been conjugated to MNPs, including antibodies, peptides and other molecules that facilitate the accumulation of MNPs into tumoral tissues, as well as the endocytosis process within malignant cells.

7.4.2.1   *Monoclonal Antibodies*

Combination with monoclonal antibodies was the first approach to provide active targeting properties to MNPs.[36] This strategy is still currently used due to the high specificity for the accumulation of MNPs in diseased tissues. For instance, the mAb Herceptin (trastuzumab) is a specific antibody for the receptor HER2/neu and is used in combination with MNPs as a contrast agent in MRI diagnosis of cancer.[37] MNPs conjugated with Herceptin have shown a decrease in T2 of ~20% in tumors that express HER2/neu, for instance breast cancer.

Another example is found in the treatment of brain tumors. The early diagnosis of brain tumors is very complicated, mainly due to the difficulties of getting contrast agents through the blood–brain barrier (BBB). The BBB exhibits a tight cellular package that protects the central nervous system against pathogens. Unfortunately, other foreign bodies, such as MNP contrast agents are also blocked by this barrier. For this reason, MNP-based contrast agents have been conjugated with targeting molecules addressed to factors expressed in human multiform glioblastoma.[38] The antibody linked to the iron oxide MNPs was epidermal growth factor receptor (EGFR)vIIIAb, which selectively bonds to the epidermal growth factor deletion mutant EGFRvIII receptor expressed in malignant cells of human multiform glioblastoma.

7.4.2.2   *Peptides*

Peptide chemistry has also provided very interesting contributions to the field of active targeting. Among the numerous peptides available, perhaps the sequence Arg-Gly-Asp (RGD; see Figure 7.10) is one of the most widely studied as an active targeting agent. RGD peptide exhibits a high affinity with respect to $\alpha_v\beta_3$ integrin, which is an angiogenic marker, and therefore, RGD can be addressed to the newly formed blood vessels developed by tumors. The specific accumulation of iron oxide MNPs conjugated with the RGD sequence has been tested with certain success in breast tumors, malignant melanoma and squamous cell carcinomas.[39]

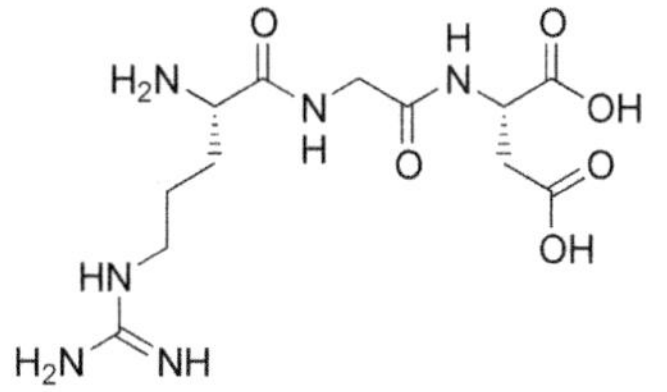

Figure 7.10 RGD (Arg-Gly-Asp) peptide structure.

F3 peptide specifically links to the nucleolin expressed in tumoral endothelium. This peptide has been also conjugated with iron oxide MNPs to accumulate them within brain tumors. F3-conjugated MNPs were used as theragnostic agents by the combination of photodynamic therapy and contrast agent for MRI diagnosis.[40] In this way the evaluation of the therapy could be followed by MRI in 9L gliomas in rats.

Chlorotoxin is a peptide originally isolated from the poison of *Leiurus quinquestriatus* scorpion. This peptide has shown specificity for the membrane-bound matrix metalloproteinase-2, which is a complex protein overexpressed in gliomas and other tumors.[41] Chlorotoxin conjugation with iron oxide MNPs also facilitates the accumulation of these nanovehicles into brain tumors.[42]

There are many other peptides of interest for active targeting purposes, such as CGKRK, which is a homing peptide that exhibits affinity for the tumor epidermal vasculature.[43] Other interesting peptides are polyarginine R11, which has demonstrated certain efficacy for accumulating MNPs in prostate cancer in mice;[44] amino-terminal fragment combined with MNPs in MRI contrast agents exhibits affinity for pancreatic tumors and allows them to be distinguished from chronic pancreatitis.[45] Finally, peptide P1C is able to target malignant cells in liver.[46] When P1C is conjugated with MNPs they accumulate in the periphery of the tumors, obtaining the best tumor-to-muscle contrast-to-noise ratio 12 hours after injection.

7.4.2.3 Other Molecules

A number of synthetic molecules have been designed as targeting moieties. They are addressed to link receptors sited at the cell membrane that are overexpressed in malignant cells. This is a direct consequence of the high demand for nutrients required by the ever-growing malignant cells.[47] Among the most important cases, the transferrin and folate receptors can be highlighted. Folic acid is an essential vitamin for DNA synthesis and it is one of the most studied molecules for active targeting[48] (Figure 7.11). Folic acid is directed towards the folate receptors CD71 in several malignant cells, commonly overexpressed in human oral carcinoma, metastatic breast, colorectal and other cancers, showing affinity to nanoparticles coated with folic acid.[49]

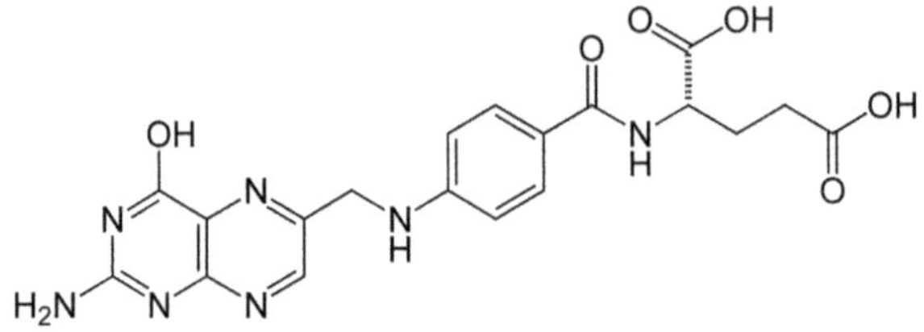

Figure 7.11   Folic acid structure.

7.5   Biomedical Applications of Magnetic Nanoparticles

7.5.1   Magnetic Nanoparticles in MRI

MRI is one of the most efficient non-invasive diagnostic techniques currently used in medicine. In most MRI applications, the protons of the tissues are used to create a signal, which is subsequently processed as an image of the body.[50] Firstly, a continuous and intense magnetic field B is applied to align the proton spins with the net moment precessing around B at a frequency ω_0, also denoted Larmor frequency. Commonly this field is about 1.5 T, although there are MRI devices designed to apply field up to 7 T. Thereafter, radiofrequency pulses are applied transversally to B_0 with the appropriated resonance frequency (Larmor frequency, ω_0) for the spins to get into resonance with the precession state (Figure 7.12). Finally, the radio-frequency magnetic fields are turned off and the spins get back to their equilibrium state through a process denoted relaxation. This relaxation is the combined result of two different mechanisms. The longitudinal relaxation T_1, or spin–lattice relaxation is due to the energy exchange between the spins and the surrounding environment. The transverse relaxation, denoted T_2 or spin–spin relaxation is due to the interactions of the spins among themselves, leading to the loss of phase coherence. Another important parameter is T_2* or apparent transverse relaxation time. T_2* is the combined relaxation time of the intrinsic relaxation time T_2 and the additional mechanism of phase loss due to magnetic field heterogeneities. Typically T_2* is shorter than T_2 and is often used instead of T_2 when SPIONs are used as contrast agents *in vivo*, as the magnetic field of the nanoparticles generates inhomogeneity in the protons field.

The radiofrequencies emitted during these relaxation processes are collected in a receiver coil. The contrasts between the different tissues are determined by the rate at which the excited protons return to their equilibrium state. This rate depends on the proton density as well as the physical and chemical features of the tissue, thus establishing the contrast differences.

Upon the accumulation of magnetic nanoparticles in the tissues, they provide a contrast enhancement (changes in the signal intensity) by means of shortening both the longitudinal and transverse relaxation of the surrounding

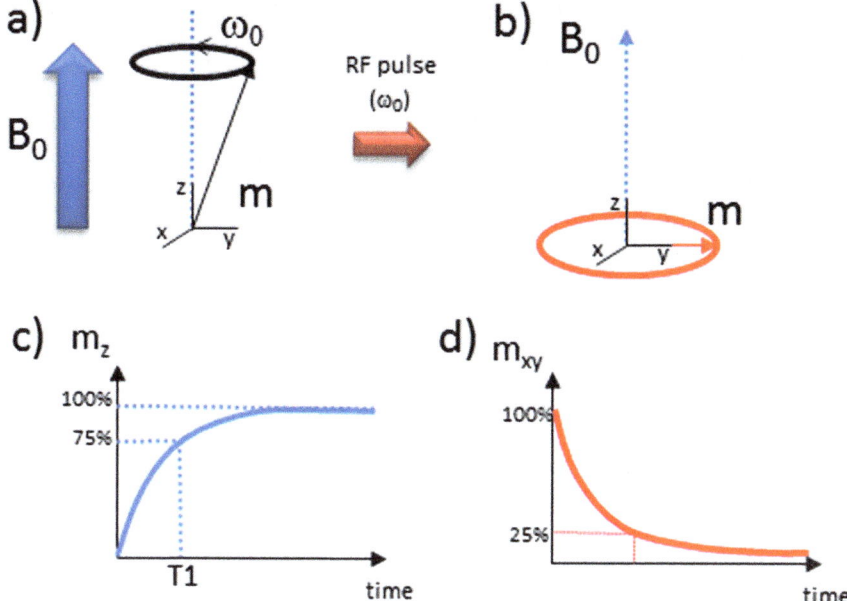

Figure 7.12 Fundamentals of magnetic resonance imaging: (a) an intense direct-current magnetic field B_0 is applied; the net moment precesses around B_0 at the characteristic Larmor frequency, ω_0; (b) a second external field is applied, perpendicular to B_0, oscillating at ω_0. This has the effect of resonantly exciting the moment precession into the plane perpendicular to B_0. In (c) and (d) the oscillating field is removed at time zero, and the (c) in-plane and (d) longitudinal moment amplitudes relax back to their initial values.

protons. However, the T_1 shortening processes require a very close interaction between the protons and the magnetic nanoparticle. Unfortunately, the needed coatings often hinder the MNP–proton proximity and this interaction is insufficient to lead a significant T_1 shortening. The effect of MNPs over T_2 reduction is due to a large difference in susceptibility between MNPs and the surrounding media, resulting in a microscopic magnetic field gradient. The diffusion of the protons through this gradient lead to the mismatch of their magnetic field (*i.e.* to the non-reversible loss of phase coherence), and thus the transverse relaxation times T_2 are shortened. Since the effect of MNPs is more intense over T_2 rather than T_1, MNPs are commonly used to provide negative contrast using pulse sequences weighted to T_2.

The efficiency of a contrast agent can be described by its relaxivity. Relaxivity is the proportionality constant between the measured relaxation rates, $R_1(1/T_1)$ and $R_2(1/T_2)$ over a range of concentrations for the contrast agent. The relaxivity of a sample does not depend only on the agent's properties, but also is very dependent upon variables such as the amplitude of the applied magnetic field, temperature of the environment, *etc.*

7.5.1.1 Magnetic Nanoparticles for the MRI Diagnosis of Cancer

MNPs have been widely studied as MRI contrast agents for the diagnosis of solid tumors. Currently, the clinical imaging of liver tumors and metastases through the uptake of MNPs by the RES has allowed the diagnosis of lesions as small as 2 mm.[51] In addition, the USPIONs have demonstrated efficiency for the identification using MRI of metastases in lymphatic nodules with diameters of 5–10 mm.[52] This non-invasive strategy has great significance for the determination of the degree of spreading in the lymphatic system, which is very important in determining disease in cases of colon, prostate or breast cancer.[53]

The class of the USPIONs includes several chemically and pharmacologically very distinct materials, which may or may not be interchangeable for a specific use. Some ultra-small (USPION) particles (median diameter <50 nm) are used as MRI contrast agents (Sinerem®, Combidex®, Clariscan™), e.g. to differentiate metastatic from inflammatory lymph nodes. USPIONs also show potential for providing important information about angiogenesis in cancer tumors, and could possibly complement MRI in helping physicians to identify dangerous arteriosclerosis plaques. Because of the disadvantageous large T_2*/T_1 ratio, USPION compounds are less suitable for arterial bolus contrast-enhanced magnetic resonance angiography than gadolinium complexes. The tiny ultra-small superparamagnetic iron oxides do not accumulate in the RES as fast as larger particles, which results in a long plasma half-life. USPION particles, with a small median diameter (<10 nm), accumulate in lymph nodes after an intravenous injection, e.g. by direct transcapillary passage through endothelial venules. Once within the nodal parenchyma, phagocytic cells of the mononuclear phagocyte system take up the particles. As a second pathway, USPIONs are subsequently taken up from then interstitium by lymphatic vessels and transported to regional lymph nodes. A lymph node with normal phagocytic function takes up a considerable amount and shows a reduction of the signal intensity caused by T_2 shortening effects and magnetic susceptibility. Caused by the small uptake of the USPIONs in metastatic lymph nodes, they appear with less signal reduction, and permit the differentiation of healthy lymph nodes from normal-sized, metastatic nodes. These super-paramagnetic iron oxide nanoparticles must be coated with a dense packing of dextrans to prolong their time in circulation. For intravenous administration on an outpatient basis, the solid product is reconstituted in normal saline and injected at a dose of 2.6 mg of iron per kilogram of body weight over a period of 15–30 minutes. This method of lymphography requires two MRI scans performed 24 hours apart. The first MRI scan is undertaken to evaluate the existence and location of the lymph nodes. Twenty-four hours after the injection of the contrast, the second MRI is performed to evaluate contrast enhancement of the identified lymph nodes.

Other clinical applications of USPIONs under study is their use to improve the determination of the outline of brain tumors. This is very useful to quantify the volume of tumoral tissue before resection. The contrast agents based on MNPs offer a more prolonged enhanced resolution of the tumor outlines compared with those contrast agents based on gadolinium chelates. The reason is that MNPs are internalized to a higher degree by the tumoral cells and their clearance takes a longer time.

7.5.1.2 Magnetic Nanoparticles for the MRI Diagnosis of Cardiovascular Diseases

MNPs have been proposed as MRI contrast agents for the diagnosis of myocardial lesions, atherosclerosis, risk evaluation of ischemia events and other cardiovascular diseases.[54,55] The strategy of functionalizing the MNPs by active targeting agents is playing a significant role in the milestones reached in this field. After identifying several peptides that link to atherosclerosis lesions, the vascular cell adhesion molecule (VCAM)-1 has been identified as a target of endothelial and macrophage cells responsible for atherosclerosis. The use of a VCAM-1 targeted peptide sequence has allowed the observation *in vivo* of the specific bond of MNPs with the subsequent enhancement of MRI contrasts in early vascular lesions.

7.5.2 Hyperthermia Treatment of Cancer

7.5.2.1 Cell and Molecular Basis of Hyperthermia

Cancer treatment by hyperthermia consists of the heating of cancerous tissues and organs to the range of temperatures between 41 °C and 47 °C. At this temperature interval, the selective and almost non-reversible destruction of malignant cells occurs. Depending on the temperature applied, hyperthermia can be considered as moderate or severe. Moderate hyperthermia takes place when the temperature of the treatment does not exceed 43 °C. In this case, effects on healthy cells are negligible or reversible after a few days or even hours. When the thermal treatment reaches the interval between 43 °C and 47 °C, severe hyperthermia is produced. In this scenario, the anti-tumoral treatment is more efficient but much less specific, thus damaging some healthy cells. Treatments comprising higher temperatures, commonly within the interval 47–56 °C, are denoted thermoablation and result in necrosis, coagulation or even carbonization of both healthy and malignant cells.

Hyperthermia improves the benefits of more conventional therapies such as chemo- and radiotherapy when applied together. Currently, clinical trials are mainly focused in the optimization and control of the thermal homogeneity at moderate temperatures (41–43 °C) over the target tissues. This task still requires great research and technical efforts.[56]

The cell and molecular basis of the hyperthermia selectivity over the cancerous cells were proposed by Overgaard in 1977.[57] Until then studies both *in vitro* and *in vivo* had shown that the hyperthermia treatment affected cancer cells at different levels: nucleic acids, cell mitosis, metabolism and lysosome activity. However, the degree to which each of these effects contributed to the final result of the treatment had not been determined.

7.5.2.1.1 Influence of Hyperthermia Treatment over RNA, DNA and Protein Transcription.

One of the first observed effects of the hyperthermia treatments was their action over the cell nuclear function, mainly the inhibition of RNA synthesis[58] and, to a lower degree, DNA and protein synthesis.[59] RNA synthesis inhibition also occurs in healthy cells. However, this effect is temporary and reversible, recovering the normal activity after 1 or 2 days when the thermal treatment does not exceed 43 °C. Higher temperatures in longer treatments can result in non-reversible damage at the nuclear level due to the denaturalization of chromosomal proteins.

7.5.2.1.2 Effects on the Cell Cycle.

The hyperthermia treatment blocks mitosis in both the healthy and malignant cells.[60] This effect is transitory and the cycle is re-established a few days later. In this sense, one of the most remarkable effects is that the malignant cells are more sensitive to chemo- and radiotherapy during the cell division process. Therefore, the external proliferative cells of the tumor are more easily destroyed by these therapies. In contrast, the hyperthermia treatment leads to a different effect. If the temperature reached is not enough to destroy the whole tumor, the cells most likely to be killed are the most internal and non-proliferative cells (steady stage). On the basis of these experiments, the idea that hyperthermia treatment selectivity could be due to factors related to the outer environment of the tumor cells was forged.

7.5.2.1.3 Effects on the Cell Metabolism.

One of the most remarkable effects of the hyperthermia is the strong decrease (or even inhibition) of the oxidative metabolism of tumor cells, whereas anaerobic glycolysis is only slightly diminished.[61] This metabolic disorder is not observed in healthy cells, thus providing a solid argument to explain the selectivity of hyperthermia for tumor cells, as the effects occurred at nuclear level cannot explain the selective condition and non-reversibility of this treatment.

7.5.2.1.4 Effects on Lysosomes.

Hyperthermia leads to an increase of the lysosomal activity in tumor cells.[62] This effect is also observed in healthy cells, but to a lower degree. However, the lysosomes become more labile under the heating in malignant cells. This allows certain selectivity in the cell destruction capability, which is more clearly shown with *in vivo* experiments rather than *in vitro*.

7.5.2.1.5 Influence of the Tumor Environment.

The higher sensibility of tumors to hyperthermia under *in vivo* conditions indicates that the selective destruction could be due to extracellular factors.[63] During the treatment by hyperthermia of a tumor, the cells sited at the inner non-proliferative part die first. In contrast, the peripheral cells seem to resist longer. This is explained in terms of the hypoxia and deficient feed of the inner tumor cells. The lack of oxygen and nutrients modifies the cell metabolism, changing towards a glycolytic process instead of an oxidative one, and accumulating lactic acid. The defective tumor vascularization facilitates the lactic acid accumulation, thus resulting in a significant pH decrease in the tumor region and an increase in the lysosomal enzymatic activity. When pH decreases to 5–5.5 the enzymatic activity of the lysosomes is at a maximum.

From these observations, the mechanism whereby hyperthermia results in a selective destruction of malignant cells can be explained. The different hypotheses can be grouped into two concepts:

(1) Nuclear function decrease: this mechanism is currently discarded, although it is important when hyperthermia is applied together with chemo- and/or radiotherapy. The non-selectivity and reversibility of the effects occurring in cell nuclei illustrate that the diminished synthesis of RNA, DNA and proteins are not the causes of a selective and non-reversible effect.

(2) Destruction of the primary cytoplasm. This concept explains the observed effects as consequence of the rupture of lysosomes, whose enzymatic payload destroys the cell cytoplasm. The selectivity of this mechanism for malignant cells is due to the difference in their environment with respect to healthy cells. The hyperthermia treatment modifies the metabolism, which shifts from oxidative towards glycolysis. The lack of oxygen, nutrients and impaired blood vessels lead to a pH decrease and consequently, to an increase of lysosome enzymatic activity. The rupture of the lysosomal membrane releases the enzymatic payload and leads to the destruction of the cytoplasm.

The heating of tumors temperatures causes the inactivation of the tumor cell in a dosage-dependent way. Although the dose–response curve is similar to the chemo- and radiotherapies, the specific intracellular target is still undetermined, in contrast to the well-known mechanism of DNA destruction driven by radiation. The hyperthermia affects most of the biomolecules, especially to the regulatory proteins of the cell growth and differentiation processes, as well as certain molecules of the receptors involved in the signal transduction pathways.[64]

Many effects are now known that significantly participate in thermal inactivation at the cellular level. Studies undertaken in the field of molecular biology demonstrate that, a few minutes after hyperthermia treatment, the expression of a special kind of proteins denoted *heat shock proteins* occurs. These proteins increase the survival of the malignant cells against subsequent treatments, leading to the *thermotolerance* effect.[65] The activity of certain regulatory proteins (kinases or cyclines) is also affected by hyperthermia, causing alterations in the cell cycle that can lead to cell death by apoptosis. In the same way it has been observed that after hyperthermia treatment, malignant cells are better recognized and destroyed by the immune system through the natural killer cells.[66]

7.5.2.2   Magnetic Nanoparticles for Hyperthermia Treatment of Cancer

Magnetic nanoparticles (MNPs) can exhibit heating power when they are subjected to alternating-current (AC) magnetic fields. These thermal effects are related to the losses occurring during the reversal magnetization processes of the MNP. During the interstitial hyperthermia treatment of cancer, it is very important to increase the temperature with the lowest possible amount of MNPs. Therefore, the MNP-specific loss power (SLP; measured as watts per gram of magnetic material) must be high enough. MNPs are excellent tools to be used as nano-thermoseeds for interstitial magnetic hyperthermia. By means of this therapy, magnetic material is implanted at the tumor site (commonly injected as a ferrofluid) to be subsequently heated by applying an AC magnetic field.

There exists a large number of magnetic materials that exhibit high SLP, which would ensure the local heating of the tumor volume. However, in the field of biomedical applications such as hyperthermia treatments, the number of candidates is significantly reduced because of biocompatibility requirements. In this sense, most of the research in this field is focused upon two iron oxides: magnetite Fe_3O_4 and maghemite γ-Fe_2O_3. Both of them have demonstrated excellent biocompatibility in humans. However, before validating a specific strategy for hyperthermia treatment, other questions must be resolved beyond the biocompatibility of the implanted material. For instance, in systems formed by magnetic nanoparticles the magnetic losses used to increase the temperature are due to different processes of reversal magnetization. These processes depend on the amplitude H and frequency f of the applied field. On the other side, the characteristics of the nanoparticle system also determine the magnitude and kind of losses. Among them, the most significant are the mean size, size distribution, particle shape and crystallinity.

SPIONs are the most attractive iron oxide systems for magnetic fluid hyperthermia. In SPIONs, the Néel relaxation is the primary mechanism of reversal magnetization that results in magnetic losses and thus in environment heating. Moreover, SPIONs guarantee high saturation magnetization values (MS), high magnetic susceptibility χ and no coercive force H_c. It means that we can obtain high magnetic response during the application of the external AC magnetic field, but without magnetic remanence (M_r) when the field is off.

In a suspension of superparamagnetic nanoparticles into a liquid with a determined viscosity η, the Brown relaxation mechanism takes place due to the reorientation of the particle as a whole which also contributes to the environment heating and possesses a relaxation time τ_B defined by eqn (7.10)

$$\tau_B = 4\pi\eta r_h^3/kT \tag{7.10}$$

being r_h the hydrodynamic radius.

When reviewing the data collected in the literature concerning different systems of magnetite and maghemite nanoparticles, a width scattering of

SLP results can be observed ranging between 10 W g^{-1} to almost 1 kW g^{-1}, as observed for magnetosomes of bacterial origin. In addition to the MNPs system features, this data scattering is also due to the different devices used for the application of the AC field for each experiment. In fact, many of the amplitudes and frequencies found in the literature are out of the range of applicability, from a clinical point of view. Therefore, establishing the limits for the AC magnetic field suitable for application in humans would be a good beginning.

7.5.2.3 AC Magnetic Field Devices: Limits for Hyperthermia Applications

The magnetic hyperthermia cancer treatment mediated by MNPs requires the implantation of nanoparticles and the subsequent subjection of the affected part of the body to a magnetic inductor. In principle, the SLP developed by the nanoparticles is an increasing function of the frequency f and amplitude H. However, increasing the SLP in the whole required volume of the tumor through the augmentation of H and f comprises very serious clinical and technical problems. Firstly, within an AC magnetic field, the induction of eddy currents in the body of the patient can lead to undesirable heating of healthy tissues, thus decreasing the selectivity of the hyperthermia treatment. This heating is strongly dependent on the diameter of the body part subjected to the AC field. According to the law of induction, the heating power is a function of the square of ($H \cdot f \cdot D$), D being the diameter of the loop subjected to induction.

Moreover, the degree of discomfort that each patient can withstand is also a limiting parameter. In 1988, Brezovich experimentally determined the limits for the product $H \cdot f$ using volunteers.[67] The experimental observations showed that for body regions with diameters ~30 cm and $H \cdot f$ product <4.85×10^8 A m^{-1} s^{-1} the volunteers felt heat but they could withstand the treatment for 1 hour or even longer without serious discomfort. For body regions of smaller diameter the $H \cdot f$ product can be higher. The first hyperthermia applicator equipment developed and marketed for human patients worked at a frequency of 100 kHz and amplitude up to 18 kA m^{-1}. More advanced devices comprise ferrite core-based applicators that overcome the technical limitations (size, current, voltage, *etc.*) associated with inductive coils. These other devices work at 100 kHz and amplitudes ranging between 0 and 15 kA m^{-1}. The last generation also comprises adjustable openings of 30–45 cm.[68] Currently, clinical trials in progress are undertaken with magnetite nanoparticles and tested in this hyperthermia applicator (Table 7.2).

7.5.3 Magnetofection

Gene transfection represents an alternative or complementary therapy in the treatment of cancer and other gene-based diseases.[69,70] This strategy consists of deliberately introducing nucleic acids into the nucleus of cells. In terms

Table 7.2   Clinical studies on hyperthermia treatment of cancer developed with magnetite nanoparticles (MagForce, Germany).

Indication	Number of patients
Glioblastoma multiforme ·	80
Prostate cancer	29
Esophageal cancer	10
Pancreatic cancer	7
Other indications	~20

of gene therapy, the incorporation of nucleic acids is mainly intended to replace deleterious mutant alleles with functional ones, or even to induce the malignant cell apoptosis. Moreover, the DNA reprogramming of bacteria and eukaryotic cells opens huge possibilities in the field of biotechnology.[71,72] A second approach consists of the transfection for RNA interference purposes. For this aim, small interfering (si)RNA sequences bind to targeted messenger RNA (m)RNA and initiate its degradation, which eventually results in gene silencing.[73] The advantage of this second approach is that targeting the gene delivery into the nucleus is not generally required. The RNA interference mechanism is a process of post-transcriptional gene silencing and the siRNA does not necessarily have to reach the cell nucleus.[74]

The introduction of a gene within both the cell nucleus and cytoplasm is not an easy task. The cell and the extracellular environment exhibit several mechanisms and barriers to hinder the DNA or siRNA incorporation (Figure 7.13). Traditionally, viral vectors have been used to facilitate this process, mainly due to the infection mechanisms of natural viruses to introduce themselves into cells.[75] However, concern about safety issues has impelled research toward a different approach. The use of non-viral vectors is one of the main strategies for gene delivery into the target cell nucleus. These nanosystems induce minimal host immune responses and they have been increasingly proposed as a safer alternative to viral vectors. In order to improve the efficiency of the gene transfer, a non-viral vector should fulfil the following requirements:

- capability to complex nucleic acids and protect them against serum nuclease enzymes at the extracellular compartment;
- to exhibit positive net electric surface charge at physiological pH to overcome the negative potential of the cell membrane, since otherwise the cell membrane impedes the incorporation of negatively charged phosphate-containing DNA;
- to possess a mechanism to protect DNA from the acid environment inside endosomes; and
- chemical stability to maintain integrity until the nucleus is reached.

Magnetofection is an excellent alternative to significantly reduce the transfection time from several hours to <60 minutes. The association of superparamagnetic nanoparticles with gene vectors enables the transfection into

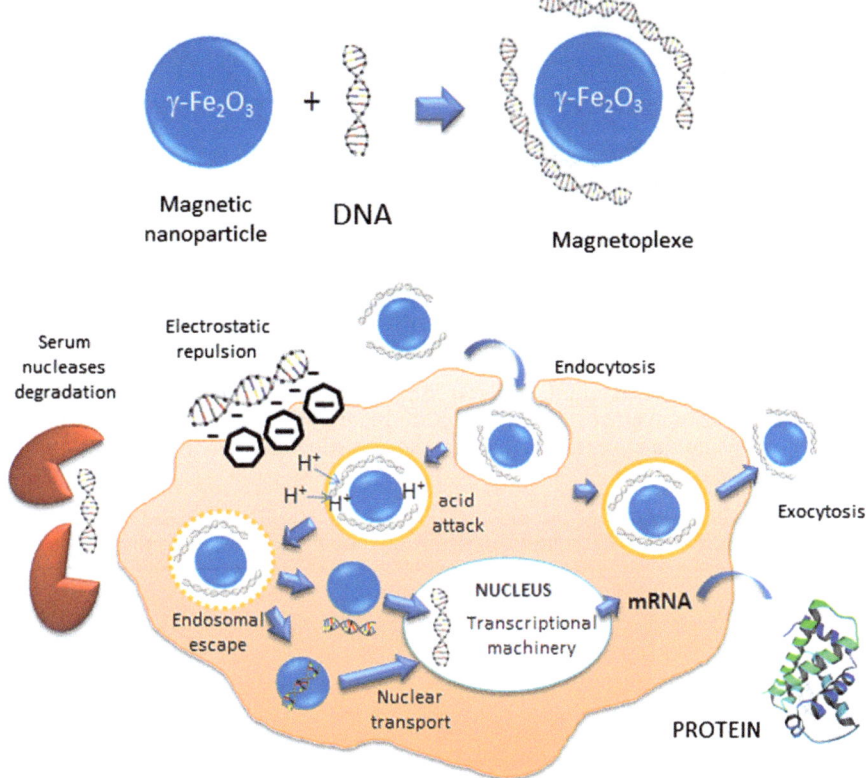

Figure 7.13 Cellular trafficking of magnetoplexes.

cells by the application of an external magnetic field, which targets and reduces the duration of the gene delivery, enhancing the efficiency of the DNA or siRNA vector.

The experimental procedures of this technology depend on whether the application is performed *in vitro* or *in vivo*. In the case of *in vitro* magnetofection, the DNA/magnetic nanoparticle complexes (magnetoplexes), normally in colloidal suspension, are introduced into the cell culture and then potent magnets, commonly made of FeNdB or SmCo alloys are placed below the culture plates, producing a magnetic field gradient. This gradient pulls the nanoparticles toward the bottom of the culture dishes, thus increasing the sedimentation rate of the magnetoplexes over the surface of the cells. The result is a significant enhancement of the internalization speed through endocytosis, as shown in Figure 7.14. Under *in vivo* conditions, the magnetic fields are focused over the target site not only to enhance transfection, but also to target the therapeutic gene to a specific organ or tissue within the body. Generally, the magnetic nanoparticles carrying in the therapeutic gene are injected and a direct-current magnetic field is externally applied to accumulate the magnetoplexes in the specific organ.

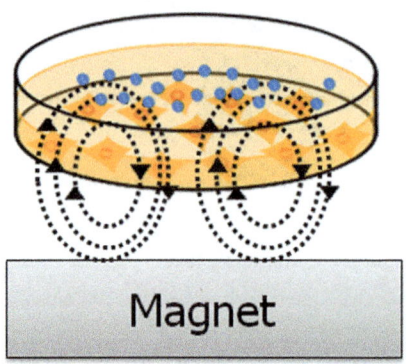
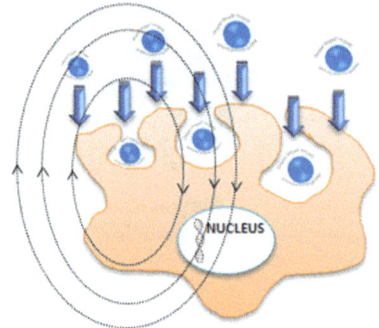

Figure 7.14   *In vitro* magnetofection technique. The magnetic gradient supplied by the magnet below the culture plate pulls magnetoplexes down, thus enhancing endocytosis and accelerating the DNA internalization.

Table 7.3   Some commercial products based on superparamagnetic iron oxide nanoparticles coated with polyethyleneimine (PEI).

Product	PEI molecular weight	Particle size	Company	Country
PolyMag	25 kDa	12 nm	Chemicell	Germany
transMAG^{PEI}	800 kDa	200 nm	Chemicell	Germany
Neuro-Mag		160 nm	Oz Biosciences	France

Magnetite and maghemite nanoparticles are often coated with polymeric materials to enhance their biocompatibility and colloidal stability, as well as to attain the capability for functionalization, as naked SPIONs do not effectively link DNA onto their surfaces. In this sense, the coating of SPIONs with polycationic polymers such as polyethyleneimine (PEI) makes magnetofection feasible.[76] In fact, there are several products marketed for *in vitro* gene magnetofection based on SPIONs coated with PEI of different molecular weights (see Table 7.3).

The size of the magnetoplexes influences the efficiency of the transfection, size it affects the rate of uptake and the toxicity. Endocytosis is more efficient with particles of ~50 nm and the particle movement within the cytoplasm is also favored by smaller sizes. However, in addition to the vector size, there are other parameters such as surface charge and the order in which the different elements of the magnetoplexes are assembled. It has been demonstrated that the combination of magnetic nanoparticles with PEI and DNA exhibits the best DNA transfection, when the different components are assembled by coupling magnetic nanoparticles with premixed PEI/DNA, although this strategy involved the largest magnotoplex among the different combinations studied.[77]

In addition to coating with PEI, other different synthetic approaches for the functionalization of iron oxide nanoparticles with dendritic macromolecules have been developed for a variety of applications. For intracellular uptake studies, iron oxide nanoparticles have been stabilized with carboxylated poly(amidoamine) (PAMAM) dendrimers through electro-static self-assembly onto the iron oxide surfaces.[78] A stepwise growth of dendritic PAMAM wedges on the surface of magnetite nanoparticles has been reported for gene delivery.[79] Another relevant approach is the modification of the iron oxide nanoparticles by combining the electrostatic layer by layer self-assembly technique with dendrimer chemistry. Positively charged iron oxide nanoparticles were modified with a bilayer composed of a negatively charged polyelectrolyte and a generation-5 PAMAM dendrimer.[80] In this sense, a more stable shell coating was obtained when iron oxide nanoparticles were assembled with multilayers of polyelectrolytes and an outer layer of PAMAM dendrimers to achieve a chemical crosslinking of the shells.[81]

Another approach for *in vitro* gene transfection is the preparation of a multifunctional nanosystem that consists of poly-(propyleneimine) (PPI) dendrimers covalently bonded to maghemite nanoparticles,[82] as shown in Figure 7.15. PPI dendrimers act as non-viral gene transfection agents and provide complexation sites for effective DNA binding in the proposed nanovector. It is noteworthy that the covalent bonding between the dendrimers and the magnetic nanoparticle ensures the vector integrity during the transfection process. Moreover, the presence of tertiary amines protects the genes from the acid attack in the endosomal compartment, since they have the capability of behaving as proton sponge. Three generations of PPI (GI, G2 and G3) were linked to the magnetite nanoparticles. The G3 magnotoplex demonstrated the best transfection efficiency due to its better behavior as a proton sponge, although this system was the largest one. This result indicates that the transfection efficiency depends on the capability of the dendrimer generation to protect and transport the new gene into the nucleus, and not only on the ease of the endocytosis process exhibited by the smallest nanovehicles.

7.5.4 Drug Delivery

Iron oxide MNPs have been widely considered as vehicles for drug delivery purposes. Certainly, the development of magnetic drug carriers has not reached the success achieved by polymeric or liposome carriers. However, their capability to act as multifunctional platforms impels an ongoing research effort for drug delivery purposes. In addition to the drug carrier function, MNPs can provide:

- magnetic targeting;
- drug–hyperthermia combined therapy; and
- drug-MRI theragnosis.

Magnetic targeting is based on localizing the drug carrier MNPs at the diseased organ or body region by means of external direct-current magnetic fields. Generally, the strategy consists of injecting into the blood the MNPs linked to the drug (commonly an antitumoral agent) and applying an intense

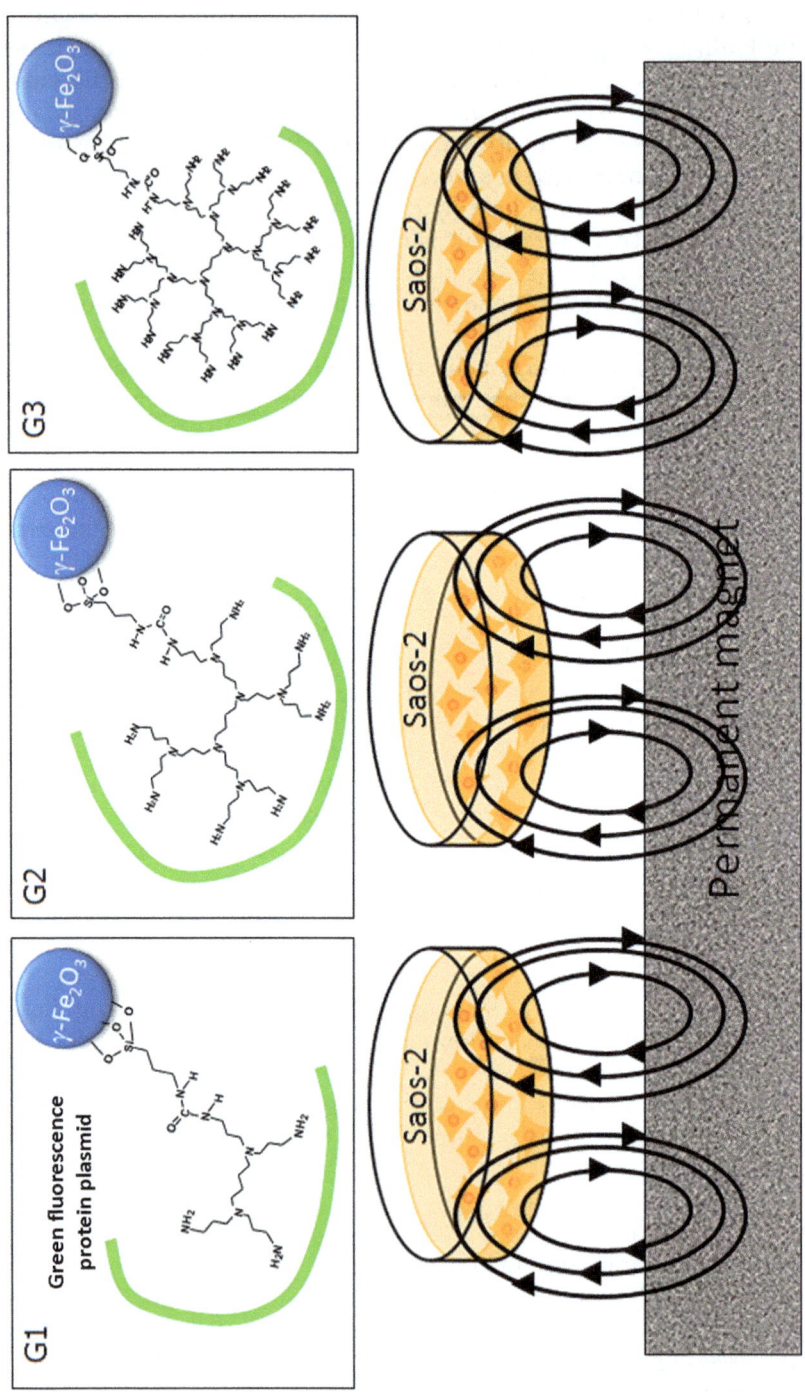

Figure 7.15 Dendritic functionalization of iron oxide nanoparticles through sol–gel chemistry with alkoxysilane poly(propyleneimine) (PPI) dendrimers. Magnetic nanoparticles functionalized with aminopropylsilane (APTS) (as a model), G1, G2 and G3 were prepared. Schematic structure of G3–MNPs material.

magnetic field in the tumor region, thus accumulating the MNPs by magnetic attraction. The final aim is to release the drug in a locally and selective way.

Although magnetic targeting seems to be a very straightforward method for local drug release, this strategy has serious drawbacks to overcome. For instance, the magnetic fields required to accumulate the MNPs are very high as they have to overcome the strength of the blood torrent. Calculations carried out by Grief *et al.*[83] indicate that magnetic targeting can be only achieved in surface regions of the body. Moreover, the risk of embolism must be also taken into account, due to the accumulation of MNPs within capillaries by the magnetic field. Clinical research performed with MNPs loaded with epirubicin and magnetically targeted toward solid tumors showed efficacy in 50% of the patients.[84] Several magnetically targeted systems have succeeded in removing tumors in animal *in vivo* studies. For instance, the incorporation of mitoxantrone in starch-coated USPIONs removed tumors of VX2-squamous cell carcinoma after 35 days of treatment in rabbits.[85]

The anchoring method to link the drug to MNPs is fundamental for the success of the systems. The MNP coating plays a lead role in this topic. The drug can be introduced embedded in a polymeric coating, in such a way that the release is ruled by diffusive process, thus strongly depending on the hydrophobic degree of the drug (generally high in the case of antitumoral drugs). This is the case of the system designed by Yang *et al.*,[86] who obtained poly(ethyl-2-cyanoacrilate) (PECA)-coated magnetite nanoparticles. These authors embedded two drugs: cisplatin and gemcitabine. Since the hydrophobicity of cisplatin is higher, this drug showed a sustained release, whereas gemcitabine was released in a faster way.

Zero early release is one of the most desired milestones when designing nanoparticles for anticancer drug release. Zero early release implies that the drug remains attached and inactive until it reaches the specific target. Once the vehicle arrives at the target, the drug is released, aided by a biochemical stimulus such as pH variations or redox enzymatic activity. In order to keep the cytotoxic drug anchored to the MNPs, strategies to covalently bond the drug to the MNP coating are used. In this sense methotrexate (MTX) has been covalently bonded to amine-functionalized MNPs.[87,88] For this aim, amide bonds are formed with the amine-functionalized MNP coating and the carboxylate groups of methotrexate, thus allowing the drug stability before the target is reached (breast and brain tumor cells in these cases). The cleavage of methotrexate is gathered by the pH decrease in the lysosome compartments within the malignant cells.

References

1. S. Laurent, D. Forge, M. Port, A. Roch, C. Robic, L. V. Elst and R. N. Muller, *Chem. Rev.*, 2008, **108**, 2064.
2. L. T. Chen and L. Weiss, *Blood*, 1973, **41**, 529.
3. H. S. Choi, W. Liu, P. Misra, E. Tanaka, J. P. Zimmer, B. Itty Ipe, M. G. Bawendi and J. V. Frangioni, *Nat. Biotechnol.*, 2007, **25**, 1165.

4. J. Gallo, N. J. Long and E. O. Aboagye, *Chem. Soc. Rev.*, 2013, **42**, 7816.
5. C. Leuschner, C. S. S. R. Kumar, W. Hansel, W. Soboyejo, J. K. Zhou and J. Hormes, *Breast Cancer Res. Treat.*, 2006, **99**, 163.
6. A. K. Gupta and M. Gupta, *Biomaterials*, 2005, **26**, 3995.
7. D. Butler, *Nature*, 2008, **453**, 840.
8. R. M. Cornell and U. Schwertmann, *The iron oxides: structure, properties, reactions, occurrences, and uses*, Wiley-VCH, 2003, p. 32.
9. A. H. Lu, E. L. Salabas and F. Schuth, *Angew. Chem., Int. Ed.*, 2007, **46**, 1222.
10. L. N. Okassa, H. Marchais, L. Douziech-Eyrolles, S. Cohen-Jonathan, M. Souce, P. Dubois and I. Chourpa, *Int. J. Pharm.*, 2005, **302**, 187.
11. R. Massart, *IEEE Trans. Magn.*, 1981, **17**, 1247.
12. T. P. Hoar and J. H. Schulman, *Nature*, 1943, **152**, 102.
13. H. S. Lee, W. C. Lee and T. Furubayashi, *J. Appl. Phys.*, 1999, **85**, 5231.
14. C. J. O'Connor, C. T. Seip, E. E. Carpenter, S. Li and V. T. John, *Nanostruct. Mater.*, 1999, **12**, 65.
15. Y. Lee, J. Lee, C. J. Bae, J.-G. Park, J.-J. Noh, J.-H. Park and T. Hyeon, *Adv. Funct. Mater.*, 2005, **15**, 503.
16. Y. Deng, L. Wang, W. Yang, S. Fu and A. Elaïssari, *J. Magn. Magn. Mater.*, 2003, **257**, 69.
17. V. Pillai, P. Kumar, M. J. Hou, P. Ayyub and D. O. Shah, *Adv. Colloid Interface Sci.*, 1995, **55**, 241.
18. J. Vidal-Vidal, J. Rivas and M. A. López-Quintela, *Colloids Surf., A*, 2006, **288**, 44.
19. W. Cai and J. Wan, *J. Colloid Interface Sci.*, 2007, **305**, 366.
20. T. Gonzalez-Carreno, M. P. Morales, M. Gracia and C. J. Serna, *Mater. Lett.*, 1993, **18**, 151.
21. M. D. Alcala and C. Real, *Solid State Ionics*, 2006, **177**, 955.
22. D. Ma, J. Guan, F. Normandin, S. Denommee, G. Enright, T. Veres and B. Simard, *Chem. Mater.*, 2006, **18**, 1920.
23. X. Liu, J. Xing, Y. Huang, G. Shan and H. Liu, *Colloids Surf., A*, 2004, **238**, 127.
24. H. H. Yan, S. Q. Zhang, X. L. Chen, Z. X. Zhuang, J. G. Xu and X. R. Wang, *Anal. Chem.*, 2004, **76**, 1316.
25. P. Tartaj, T. Gonzalez-Carreno and C. J. Serna, *Adv. Mater.*, 2001, **13**, 1620.
26. F. M. Martín-Saavedra, E. Ruíz-Hernández, A. Boré, D. Arcos, M. Vallet-Regí and N. Vilaboa, *Acta Biomater.*, 2010, **6**, 4522.
27. R. S. Molday and D. J. MacKenzie, *J. Immunol. Methods*, 1982, **52**, 353.
28. R. Weissleder, G. Elizondo, J. Wittenberg, C. A. Rabito, H. H. Bengele and L. Josephson, *Radiology*, 1990, **175**, 489.
29. Y. Anzai, C. W. Piccoli, E. K. Outwater, *et al.*, *Radiology*, 2003, **228**, 777.
30. R. Gref, Y. Minamitake, M. T. Peracchia, V. Trubetskoy, V. Torchilin and R. Langer, *Science*, 1994, **263**, 1600.
31. S. M. Moghimi, A. C. Hunter and J. C. Murray, *Pharm. Rev.*, 2001, **53**, 283.
32. M. D. Butterworth, L. Illum and S. S. Davis, *Colloids Surf., A*, 2001, **179**, 93.
33. C. Flesh, Y. Unterfinger, E. Bourgeat-Lami, E. Duguet, C. Delaite and P. Dumas, *Macromol. Rapid Commun.*, 2005, **26**, 1494.
34. Y. Lu, Y. D. Yin, B. T. Mayers and Y. N. Xia, *Nano Lett.*, 2002, **2**, 183.

35. N. Kholer, G. E. Fryxell and M. Q. Zhang, *J. Am. Chem. Soc.*, 2004, **126**, 7206.
36. S. Cerdan, H. R. Lotscher, B. Kunnecke and J. Seeling, *Magn. Reson. Med.*, 1989, **12**, 151.
37. D. Artemov, N. Mori, R. Ravi and Z. M. Bhujwalla, *Cancer Res.*, 2003, **63**, 2723.
38. C. G. Hadjipanayis, R. Machaidze, M. Kaluzova, L. Wang, A. J. Schuette, H. Chen, W. Wu and H. Mao, *Cancer Res.*, 2010, **70**, 6303.
39. C. F. Zhang, M. Jugold, E. C. Woenne, T. Lammers, B. Morgenstern, M. M. Mueller, H. Zentgraf, M. Bock, M. Eisenhut, W. Semmler and F. Kiessling, *Cancer Res.*, 2007, **67**, 1555.
40. G. R. Reddy, M. S. Bhojani, P. McConville, J. Moody, B. A. Moffat, D. E. Hall, G. Kim, Y. E. L. Koo, *et al.*, *Clin. Cancer Res.*, 2006, **12**, 6677.
41. J. Deshane, C. C. Garner and H. Sonthmer, *J. Biol. Chem.*, 2003, **278**, 4135.
42. O. Veiseh, C. Sun, J. Gunn, N. Kohler, P. Gabikian, D. Lee, N. Bhattarai, R. Ellenbogen, *et al.*, *Nano Lett.*, 2005, **5**, 1003.
43. L. Agemy, D. Friedmann-Morvinski, V. R. Kotamraju, L. Roth and K. N. Sugahara, *Proc. Natl. Acad. Sci. U. S. A.*, 2011, **108**, 17450.
44. A. S. Wadajkar, J. U. Menon, Y. S. Tsai, C. Gore, T. Dobin, L. Gandee, K. Kangasniemi, *et al.*, *Biomaterials*, 2013, **34**, 2618.
45. L. Yang, H. Mao, Z. H. Cao, Y. A. Wang, X. H. Peng, X. X. Wang, *et al.*, *Gastroenterology*, 2009, **136**, 1514.
46. G. Q. Wu, X. D. Wang, G. Deng, L. Y. Wu, S. H. Ju, G. H. Teng, Y. Y. Yao, *et al.*, *J. Magn. Reson. Imaging*, 2011, **34**, 395.
47. F. Danhir, O. Feron and V. Préat, *J. Controlled Release*, 2010, **148**, 135.
48. P. S. Low and A. C. Antony, *Adv. Drug Delivery Rev.*, 2004, **56**, 1055.
49. A. Schroeder, D. A. Heller, M. M. Winslow, J. E. Dahlman, G. W. Pratt, R. Langer, T. Jacks and D. G. Anderson, *Nat. Rev.*, 2012, **12**, 39.
50. H. Shokrollahi, A. Khorramdin and Gh. Isapour, *J. Magn. Magn. Mater.*, 2014, **369**, 176.
51. C. Corot, P. Robert, J. M. Idee and M. Port, *Adv. Drug Delivery Rev.*, 2006, **58**, 1471.
52. M. G. Harisinghani, J. Barentsz, P. F. Hanh, W. M. Deserno, S. Tabatabaei, C. H. van de Kaa, J. de la Rosette and R. Weissleder, *N. Engl. J. Med.*, 2003, **348**, 2491.
53. M. G. Harisinghani and R. Weissleder, *PLoS Med.*, 2004, **1**, e66.
54. D. E. Sosnovik, M. Nahrendorf and R. Weissleder, *Circulation*, 2007, **115**, 2076.
55. S. A. Wickline, A. M. Neubauer, P. M. Winter, S. D. Caruthers and G. M. Lanza, *J. Magn. Reson. Imaging*, 2007, **25**, 667.
56. J. Van der Zee, J. N. Peer-Valstar, P. J. M. Rietveld, L. de Graaf-Strukowska and G. C. van Ron, *Int. Radiat. Oncol. Biol. Phys.*, 1991, **20**, 1109.
57. J. Overgaard, *Cancer*, 1977, **39**, 2637.
58. J. A. Dickson and D. S. Muckle, *Cancer Res.*, 1974, **34**, 1263.
59. J. A. Dickson and D. M. Shah, *Eur. J. Cancer*, 1972, **8**, 561.
60. R. J. Martín and P. R. Schloerb, *Cancer Res.*, 1964, **24**, 1997.

61. B. Mondovi, A. F. Agro, G. Rotilio, R. Strm, G. Morica and A. R. Faneli, *Eur. J. Cancer*, 1969, **5**, 129.
62. J. Overgaard, *Cancer Res.*, 1976, **36**, 983.
63. K. Overgaard and J. Overgaard, *Acta Radiol. Therm. Phys. Biol.*, 1977, **16**, 1.
64. A. Jordan, R. Scholz, J. Schüller, P. Wust and R. Felix, *Int. J. Hyperthermia*, 1997, **13**, 83.
65. P. Burgman, A. Nussenzwigh and G. C. Li, in *Thermoradiotherapy and thermochemotherapy, vol. 1: Biology, Physiology, Physics*, ed. M. H. Seegenschiedt, P. Fessenden and C. C. Vernon, Springer, Berlin, 1995.
66. G. Multhoff, C. Botzler and M. Wiesnet, *Int. J. Cancer*, 1995, **61**, 272.
67. I. Brezovich, *Med. Phys. Monogr.*, 1988, **16**, 82.
68. A. Jordan, R. Scholz, K. Maie-Hauff, M. Johannsen, P. Wust, J. Nadobny, *et al.*, *J. Magn. Magn. Mater.*, 2001, **225**, 118.
69. A. L. Feldman and S. K. Libutti, *Cancer*, 2000, **89**, 1181.
70. K. Mancuso, W. W. Hauswirth, Q. Li, T. B. Connor, J. A. Kuchenbecker, M. C. Mauck, J. Neitz and M. Neitz, *Nature*, 2009, **461**, 784.
71. N. L. Rosi, D. A. Giljohann, C. S. Thaxton, A. K. R. Lytton-Jean, M. S. Han and C. A. Mirkin, *Science*, 2006, **312**, 1027.
72. M. Selbach, B. Schwanh€ausser, N. Thierfelder, Z. Fang, R. Khanin and N. Rajewsky, *Nature*, 2008, **455**, 58.
73. D. Moazed, *Nature*, 2009, **457**, 413.
74. D. Rischl and A. Zimer, *Nanomedicine*, 2009, **5**, 8.
75. C. E. Thomas, A. Ehrhardt and M. A. Kay, *Nat. Rev. Genet.*, 2003, **4**, 346.
76. R. Namgung, K. Singha, M. K. Yu, S. Jon, Y. S. Kim, Y. Ahn, I.-K. Park and W. J. Kim, *Biomaterials*, 2010, **31**, 4204.
77. M. Arsianti, M. Lim, C. P. Marquis and R. Amal, *Langmuir*, 2010, **26**, 7314.
78. E. Strable, J. W. M. Bulte, B. Moskowitz, K. Vivekanandan, M. Allen and T. Douglas, *Chem. Mater.*, 2001, **13**, 2201.
79. B. Pan, D. Cui, Y. Sheng, C. Ozkan, F. Gao, R. He, Q. Li, P. Xu and T. Huang, *Cancer Res.*, 2007, **67**, 8156.
80. S. H. Wang, X. Shi, M. Van Antwerp, Z. Cao, S. D. Swanson, X. Bi and J. R. Baker Jr, *Adv. Funct. Mater.*, 2007, **17**, 3043.
81. X. Shi, S. H. Wang, S. D. Swanson, S. Ge, Z. Cao, M. E. Van Antwerp, K. J. Landmark and J. R. Baker Jr, *Adv. Mater.*, 2008, **20**, 1671.
82. B. González, E. Ruiz-Hernández, M. J. Feito, C. López de Laorden, D. Arcos, *et al.*, *J. Mater. Chem.*, 2011, **21**, 4598.
83. A. D. Grief and G. Richardson, *J. Magn. Magn. Mater.*, 2005, **293**, 455.
84. A. S. Lubbe, C. Bergemann, H. Riess, F. Schriever, P. Reichardt, K. Possinger, *et al.*, *Cancer Res.*, 1996, **56**, 4686.
85. C. Alexiou, R. J. Schmid, R. Jurgons, M. Kremer, G. Wanner, C. Bergemann, *et al.*, *Cancer Res.*, 2000, **60**, 6641.
86. J. Yang, H. Lee, W. Hyung, S. B. Park and S. Haam, *J. Microencapsulation*, 2006, **23**, 203.
87. N. Kholer, C. Sun, J. Wang and M. Q. Zhang, *Langmuir*, 2005, **21**, 8858.
88. N. Kholer, C. Sun, A. Fichtenholtz, J. Gunn, C. Fang and M. Q. Zhang, *Small*, 2006, **2**, 785.

Subject Index

References to figures are given in *italic* type. References to tables are given in **bold** type.